BUILDING SCIENTIFIC MODELS

Building Scientific Models

GORDON J. LAING

Published in association with
the London School of Economics and
Political Science by Gower

Published by:
Gower Publishing Company Limited,
Gower House,
Croft Road,
Aldershot,
Hants GU11 3HR,
England

and

Gower Publishing Company,
Old Post Road,
Brookfield,
Vermont 05036,
USA

British Library Cataloguing in Publication Data

Laing, Gordon J.
Building scientific models.
1. Mathematical models
I. Title
511'.8 QA401

ISBN 0 566 00682 0

Printed in Great Britain by Blackmore Press, Longmead, Shaftesbury, Dorset

Contents

Foreword
by Lord Penney viii

Preface x

Acknowledgements xii

1 Classification of models and nomenclature 1

2 Some case studies 22

3 Some first steps in building models 38

4 Models using mathematical programming 51

5 Systematic trial and error models and heuristics 59

6 Simulation models and system dynamics 66

7 Models using game theory 78

8	The use of statistics	86
9	Decision theory and more about games	95
10	Repeated game play and deep strategy	106
11	Forecasting and time series models	122
12	Another method for building models	140
13	Modelling human behaviour	147
14	Cybernetic models	158
15	Subjective judgements	164
16	Software for computer modelling	177
17	Testing models and the use of data	182
18	Implementation, cost and performance	194

Appendices

1	Allocation using a 'transport' algorithm	203
2	Linear programming	212
3	Dynamic programming	221
4	Branch and bound models	246
5	Notes on an example of a simulation model for a computer	259
6	Some comments on simulation languages	266

7	The system dynamics approach	270
8	System dynamics using 'spreadsheet' software	276
9	A negotiation simulation game	287
10	Utility and risk aversion	291
11	A complex decision tree example	301
12	Strategy in decision analysis	311
13	Value judgements in decision theory	317
14	Forecasting models using the Box–Jenkins approach	322
15	The random division of linear quantities	348
16	The use of PERT and CPM	353
17	Subjective assessments in military battle models	365
18	Notes on statistical testing including t-tests, chi-squared and analysis of variance	369
19	Hypothesis test example	386
20	Fuzzy goal programming using LP techniques	389
References		399
Index		405

Foreword

LORD PENNEY

Some precocious young people who have barely completed their first decade of life tell their admiring grandmothers that this is the age of the computer. The grandfathers reflect on the children's words and probably wonder what they portend. Fresh young minds naturally register an image of the world that surrounds them. They cannot possibly be aware of the many extraordinary changes in human society since civilisation began ten thousand years ago.

High technology, and the future world society which futurologists predict, will have some of the elements perceived by the young. We are already taxed according to predictions made by government computers. If we join the Army, we draw out underclothes from stores stocked according to statistics coming from Army computers.

The computers which are going to influence and improve, but not control, the lives of people will either be data processors or will be calculating the behaviour of a model. The essential feature of a model is that it simplifies a system, a human situation, or a scientific problem to the relationships between the important variables and ignores the rest. The possible behaviour is then assessed by changing one or more of the variables and seeing what happens. The accuracy of the predictions depends on the skill of the maker of the model and the complexity of the problem he is trying to solve.

Architects and city planners make models to see how proposed new buildings harmonise with their surroundings. Engineers build models of aircraft and of bridges to study the behaviour in wind tunnels and, from the results, know how the aircraft or bridge will behave in certain circumstances. Traffic controllers make computer models of transport movements and seek to improve their operations at minimum cost. There are many more examples of the use and value of models. Models have become an essential part of the technology of maintaining our complicated society.

Those who make and use models, and those who wish to find out whether a particular model could help with their particular problem, will look for a book which explains the many different models now available. Dr Laing has a long and varied experience of making and using models and has admirably summarised his knowledge in this book. His approach is wide and involves several disciplines. Many readers will find in the book the method of solution of a problem which would be expensive or even impossible to solve in any other way.

Lord Penney OM, KBE
MA, PhD, DSc, FRS

Preface

This book aims to provide a general summary of methods available to model builders. It is intended to assist young scientists who wish to ensure that they have not overlooked useful techniques, and also to help older specialist experts who have decided that their approach to problem solving is becoming too narrow.

It is important, when presenting a proposed solution to a problem to be able to answer criticisms of the 'Why haven't you used this method?' variety. This can demand a very wide knowledge of model building methods. It is hoped that this text will help to overcome some of these difficulties, and that it will prove to be a useful reference for scientists, managers and operational research workers. A wide range of techniques and examples, together with some new ideas, are presented in the book. Additional examples are given in the appendices.

The breadth and scope of the available modelling techniques and the way in which different kinds of models are interrelated is outlined. A good understanding of this interrelationship should show the reader how several different types of model can be used simultaneously in order to provide the best validation at the least cost.

Model building, although at times highly mathematical, is essentially a very practical subject. The way in which models of the real world can be constructed must be learned both from formal teaching and from practical experience with its ensuing pattern of success and failure. As a part of this learning process the methods employed by the author to develop a number of both successful and unsuccessful models have been included. Learning what not to do in certain situations can be very important, especially if the wrong move seems to be the logical and obvious one at the time.

Several novel approaches are described, and the theoretical examples are very straightforward. The reader is asked to remember that the purpose of the examples is to demonstrate the use of the techniques in a simple manner, although the problems to which they are eventually applied in practice may be exceedingly complex.

Examples of the use of model building techniques for military studies have been included along with the commercial ones. The military aspects of model building require some attention. The USSR has published many text books on the use of quantitative methods and operational research for military purposes, whilst we in this country have published relatively few in this category.

Model builders tend to operate within only one discipline. Thus, for example, some aspects of model building discussed in this book may be new to those who work entirely in the operational research field. Knowledge of a wide variety of models provides a better understanding of the relationships between different kinds of models, and gives a better insight into wider aspects. A broader approach is often of great help in achieving a robust solution and cost-effective validation procedures.

The material is presented in a straightforward manner, and requires a knowledge of mathematics of little more than School Certificate level. The reader who carefully perseveres to the end of the book will acquire a good appreciation of some fairly advanced aspects of model building, and should be able to venture, unaccompanied, into new fields.

Gordon J. Laing

Acknowledgements

I am grateful to all who, at various times, have read all or part of this text. These include Professor A.S. Douglas of the London School of Economics, Professor K.C. Bowen of Royal Holloway College, Professor S. Vajda of Sussex University, Dr Ray Paul and Mr David Balmer (both of London School of Economics), Mr John Crookes of Lancaster University and Mr Trevor Lord, formerly of the Ministry of Defence. Any errors remaining are my own. I am also much indebted to Lord Penney for contributing a foreword and for his remarks.

I would also like to acknowledge the Ministry of Defence for permission to publish this book, and to thank Mr John Culshaw, the Director of DOAE, and Mr Douglas Andrews, the Deputy Director, for their encouragement.

Many scientists and agencies have been kind enough to provide material or photographs. Included amongst these are the National Aeronautical and Space Agency of America who provided various photographs; Dr Michaelis, the Editor of the *Journal of Interdisciplinary Science Reviews*, gave permission to reproduce tables dealing with the classification of models; the Trustees of Biometrika gave permission to reproduce the Box-Jenkins tables and graphs; the Society for General Systems Research in America gave permission to reproduce the metagame tables; and Investment Research Ltd at Cambridge provided share and commodity charts for publication.

However, there are very many others whom I should like to acknowledge but space does not permit. I have tried to indicate suitable references which can be studied should more detail be needed on any particular technique. Text books are very much a matter of individual preference and I apologise in advance if I have omitted a particular favourite from the list of references. Certainly the ones included have been well used by many, including the author.

1 Classification of models and nomenclature

There are many different dictionary definitions of the word 'model', but the scientific meaning usually concerns the representation of the real world by some mathematical expression or simplified physical form. Let us not spend our time here in philosophical arguments about what constitutes the difference between a model, a theory and a hypothesis. Others have done this with great competence and those interested should read Braithwaite.(1) We shall therefore only concern ourselves with trying to provide representations for a wide variety of events as an aid to problem solving. Anything which serves to this end, be it physical or purely mathematical, will be given the term 'model' within the confines of this text.

It is important to remember that models simplify the real world to an enormous extent so that the explanations of events which they provide are valid for only a limited set of circumstances. It can be argued that in order to provide all the answers which might be required in this universe, we need at least one other identical universe with which we can experiment. This might well be regarded as the viewpoint of the experimentalist rather than the mathematician, but it serves to emphasise the difficulty of decoupling our particular problem from all the complexities in which it is embedded.

The model builder, therefore, attempts to isolate his problem from all the other problems and effects which surround it. For a time, and sometimes for quite a long time, he may be successful. Eventually, however, it is likely that the magnitude of some variable, which was either insignificant originally or was represented by a constant in the calculation, will suddenly change dramatically. This may result in a disastrous failure. The variable which causes the problem may well have been excluded originally on the grounds of its stability and resistance to change with time. However, quite small changes in other variables, outside the model, may produce critical changes which suddenly require the inclusion of at least one more variable in the model if it is to perform effectively again.

Thus, the consideration of what variables shall be included in a model should never simply be left to the discretion of a computer and a statistical software package. These devices may be very valuable, or even indispensable, aids in the construction of a model, but they are not substitutes for common sense.

Occasionally, in teaching model building, quite a lot of emphasis is placed on the terminology. This can be rather unfortunate, since the use of intellectual jargon does not, in itself, solve problems. Nevertheless, a moderately detailed method of classification can be helpful in deciding which models to construct. Unfortunately, most of the terms currently employed are so broad as to be almost valueless in this respect. However, since people are not inclined to trust model builders who do not use the 'in' expressions, a list of some of the more common ones is included in the next section.

It is hoped that the reader will pass quickly through this section and, if he is not already aware of the terms, that he will not become too dazed in the process. Another method of classification will then be suggested and this may be rather more helpful in illustrating the interrelationships of the various types of model.

Existing terminology

Some of the most commonly used expressions applied to model building are the terms *deterministic* and *stochastic* or *probabilistic*. A *deterministic* model is one in which the variables are predetermined and do not permit any possible random change in circumstances, or 'state of the world'. A *stochastic*, or *probabilistic*, model is one in which the values of some of the variables appear in a random manner although certain of these values may occur more frequently than others.

The term *determino-stochastic* is sometimes used in cases such as the random flow of traffic through a fixed or predetermined network of roads. In the event that a model of this type is general, and can be applied to any network of roads, it might be more correctly described as a stochastic model for a deterministic environmental system. On the other hand, if the model describes a specific network, then the term determino-stochastic might be more appropriate since the model is concerned with the whole system which has both stochastic and determinate elements in it. Even so, it can also be argued that the term stochastic is adequate in that most stochastic models contain some determinate elements whilst, by definition, the reverse cannot be the case.

Other terms which are in fairly common use are *descriptive*, *analytic*, *explanatory* and *normative*. A *descriptive* model is one which describes the situation and may be of a predictive nature. For example, it may not be necessary to know exactly how a system functions, but merely to know the outcome of a particular situation which may arise. Such a model is said to be descriptive in that it may only rather vaguely describe the interactions. An *analytic* model refers to a method whereby the complete behaviour of a system can be described in detail and also usually in precise mathematical terms. The effect of changes on the model are completely understood so that the entire situation can be understood and explained if not completely quantified. The expression *normative* is often applied to models which are concerned with establishing the important factors in the system and those matters which should attract our attention; e.g. the model may indicate that we should be more concerned with market share than with profit.

The term *iconic* model is also sometimes used to describe the general outline of the system. The meaning is much more broad than that of a descriptive model and iconic models are, by nature, static pictures of a situation. Examples would be a routine fixed procedure chart, or the organisation chart for a company. The expression 'conceptual framework', or *conceptual* model, is often used by operational research workers to convey a rather similar broad pattern of events and functions. The term 'interaction diagram' or 'interaction model' is also sometimes used for the same purpose.

Other nomenclature sometimes used are the terms *homogeneous* models and *heterogeneous* models. A *homogeneous* model can be expressed in terms of a single variable objective whereas a *heterogeneous* model has several variables and usually more than one objective. An example of a heterogeneous model might be a company planning to maximise the sales of two products, or to maximise the market share and profit for a single product.

Now let us consider the variables themselves, which are the metaphorical building blocks of our model. *Variables* may be grouped into *controllable* and *uncontrollable* varieties. As the name implies, controllable variables are ones which we can control, such as the number of employees in a factory, or the quantity of nuts and bolts we decide to purchase. Uncontrollable variables are ones over which we can exert no real control and are highly reactive or are changing with time.

In building up a model by writing down the interaction of the variables using a conceptual diagram to illustrate the effects, it is often useful to place a *C* for controllable variable, or a *U* for uncontrollable variable, as a reference against the variables as they are listed or in the corner of the boxes of the conceptual diagram.

The *objective function* of a model is a mathematical expression stating the variables it is desired to optimise. Suppose that in an inventory analysis it is the intention to minimise the total annual cost K. The equation which relates K to those factors which affect it, such as storage, demand rate, etc. is known as the objective function. Quite frequently the objective function is the equation which relates the variable in the ultimate box in the chain of the conceptual diagram to the other variables. More generally it can be written:

$$\text{Objective function } F = (C\ U)$$

Variables are also often described as *endogenous* or *exogenous* in character. An *endogenous* variable is one which forms part of the expected pattern of nature as described by the model whereas an *exogenous* variable is one which in a sense acts from without, and usually comes as a surprise to the experimenter. At school it was usually an exogenous variable which frustrated the efforts of the chemistry master when his experiment failed to work satisfactorily, probably in the form of a blotting paper ink pellet which landed unnoticed in a beaker and introduced an unexpected impurity therein. In later life, these variables appear to become much more subtle and illusive, but are unfortunately still with us. There is some variation in the use of these terms and occasionally exogenous variables may simply mean the input variables.

The term *random variable* has already been explained but, additionally, the expressions *discrete random variable* and *continuous random variable* are also often used. An example of a *discrete random variable* would be the number of peas found in a pod, i.e. the event occurs in discrete quantities. A *continuous random variable* may take any value in the same way that a thermometer will continuously measure the temperature of the environment in which it is placed.

In some ways, the use of the word random may not be a good choice as a means of indicating a variable which changes in a probabilistic manner. A thermometer placed in a living room can hardly be said to vary in a random manner in the same way as a random number chosen from a random number table where every number has an equal chance of appearing. Clearly, in this example, the temperatures around the normal condition are much more likely to occur than any others. For this reason the term *stochastic variable* is sometimes used in preference to the expression random variable in order to describe a variable whose value changes in a biased probabilistic manner.

It is worthwhile examining in a little more detail the way in which other attempts have been made to classify models, particularly in the mathematical or operational–mathematical areas, since some of them are particularly orientated to the methodology of problem solving. Ackoff and Sasieni (2) have suggested a rather general method of classification according to the nature of the problem. Listing five groups they say that the first consists of problems in which the *pattern is obvious* enough to be understood after inspection. The simple inventory model described in a later section would come into this category.

In the second group, the *pattern is also clear but it is by no means obvious how to represent it symbolically.* A third group covers problems whose *structure is not apparent but can be deduced from the data.* The fourth group deals with *problems where experimental data are required* because the structure is not apparent, whilst the last group is concerned with situations where *no data are available and no experiments are possible.*

It is clear that by categorising the problems in terms of knowledge about them, Ackoff and Sasieni believe that a useful step will have been taken towards their solution and this seems to be true in most cases. Rivett in conjunction with Ackoff (3) has listed seven main types of models for operational research but he seems inclined to divide these models into those which are *reactive* or *non-reactive.* Basically, this means those with *feedback* from *the input* and *those without.* For example, the time taken to purchase and replenish a stock of goods, which is technically known as the lead time, may change slightly, or even considerably, with the size of the demand. This approach is obviously most important for *adaptive* models which are required to learn from experience and to make any necessary corrections.

Adaptive models are certainly a most important class, and they are often closely linked to *behavioural* models which involve a human response to a situation or stimulus. The development and use of adaptive models has increased with computer usage, since computers enable data to be accessed quickly and this permits the modelled system to be updated speedily.

Behavioural models are also gaining in importance as our understanding of psychology rapidly increases. Models for personality, motivation, cooperation, etc. have been developed and are playing an increasing part in management science. Interesting models based on analogies have been deployed. For example, Kurt Lewin has developed a model in which human reaction and behaviour can be viewed in an analogous manner to the behaviour of particles in electrical field theory.

Another group of models which is often classed separately are *economic* models. These are in turn subdivided into *macro-economic* models and *micro-economic* models. The object of *macro-economic* models is to paint a broad brush picture of the whole economy whilst *micro-economic* models are only concerned with describing in detail smaller parts or sections of the whole.

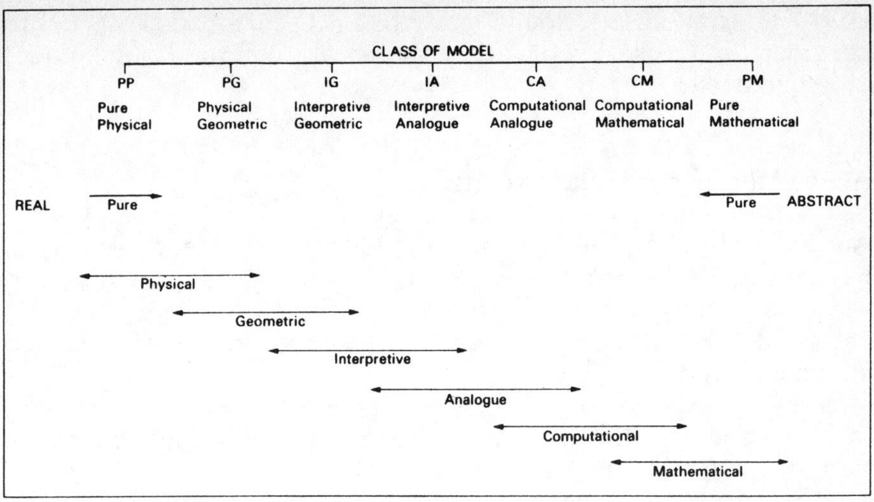

Figure 1.1 An illustration of the way in which models form a spectrum from physical reality to purely abstract

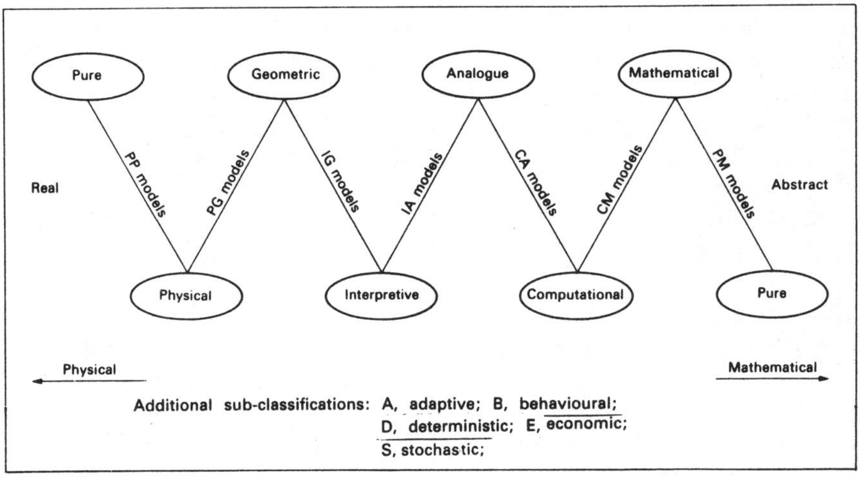

Figure 1.2 Another way of showing the main model classifications

Having dealt with some of the varied terminology often used in connection with modelling it may be helpful to consider a more generally applicable method of model classification.

A new method of classification

An understanding of how models differ from each other is important in learning how to use them and in formulating the best approach to problems. The method of classification described here has been developed to formalise such an understanding. (4)

First, a spectrum of modelling techniques which range, at one end of the spectrum, from a full-scale reproduction of physical processes to purely mathematical formulae at the other. In between these two extremes the model can exist with various degrees of reality.

Figure 1.1 shows a way of considering the various possible classes. Some of the divisions may seem a little arbitrary and are a matter of choice, but the intention is to illustrate the changeover from one form of modelling to another and how each class frequently interlinks with an adjacent class.

A single model may incorporate the features of several separate classes and hence may be described as a 'complex' model. Whether or not a complex model should be employed depends on the nature of the problem and this type of modelling should not be sought after merely because it indicates an advanced 'state of the art'. In some cases, complex modelling is desirable, but the most simple direct method to obtain an answer should be developed wherever possible.

Sometimes two or more different types of model may be used separately to solve the same problem. This is a valuable technique since it can enable the scaling laws and assumptions for one model to be tested, or at any rate partially tested, by means of the other. For obvious reasons this is sometimes described as 'collaborative' or 'corroborative' modelling. Moreover, as mathematical models are normally much cheaper than physical models, costs are often greatly reduced by this approach, and one attempts to use only sufficient models towards the physical end of the spectrum to provide adequate validation.

Figure 1.2 shows another way in which the classes can be visualised and remembered. Indeed the dubious mnemonic '*P*urple *P*etunias *G*o *I*nto *a* *C*olourful *M*ixed *P*resentation' may be helpful in remembering the words in the circles in this diagram. Figure 1.2 also contains some additional classification descriptions often applied to models.

Tables 1.1 and 1.2 will give a rough idea of the way in which the different classes of model are used. It will be noted that around the analogue node in Figure 1.2 the models cease to have physical shape and thereafter are in paper or abstract form. A more detailed explanation will be given in the subsequent sections in the precise nature of the different classes and how they can be used in a complementary manner on some occasions.

Table 1.1
Principal Classifications

Type of Model	Comment	Example
PP Pure Physical	Real full-scale models	'Reconstruction of the crime' using real people and objects
PG Physical Geometric	Geometrically scaled model retaining the general physical characteristics of the full-scale version	A model of a building designed to study strength of the structure, etc.
IG Interpretive Geometric	Model geometrically scaled but with different physical characteristics to the original	A scaled model human dummy, the design of which entails the use of mathematics
IA Interpretive Analogue	Model is unlikely to physically resemble the original but is likely to give an analogous effect	A shape designed to indicate an effect, such as ball bearings of different weights and sizes designed to simulate the movement of a human body under shock waves
CA Computational Analogue	These models are concerned with mathematical simulation and use very simple mathematics	The probability of ships waiting for unloading, as in a computer simulation model
CM Computational Mathematical	Mathematical estimation based on observation and mathematics which are not especially complicated	The determination of the coefficients of an equation for a market share model by analysis of observations
PM Pure Mathematical	Mathematical models using pure and 'applied' mathematics	The Newtonian laws of physics

Table 1.2
Sub-Classification

Special Category		Comment	Example
A	Adaptive	Adapts to latest data	Forecasting
B	Behavioural	Explains behaviour patterns	Interactions of groups under stress
D	Deterministic	Chance does not affect model	Newtonian Laws of Motion
E	Economic	Indicates influence of economic factors	Prediction of effects of change in bank rate
S	Stochastic	Model makes allowance for chance events	Probability of critical build-up in reactors

The meaning of the various classes will now be briefly discussed, with some case studies being given in subsequent sections.

Pure physical or PP models

A full-scale representation, or replication, of a physical situation, or object, is a pure physical, or PP, model. A theatre play, for example, provides a pure physical model of a situation. It may be imagined that this type of model is of little interest to scientists, but this is not true. Consider, for instance, the value of this type of model in crime reconstruction. It is true, in practice, that it may be less used than some fiction writers suggest, since its purpose in fiction often seems to be to introduce a second macabre event. One of the practical difficulties of using this model in real life is the reintroduction of all the correct elements to provide a good reconstruction. However, it does normally allow the space–time aspect to be examined in detail. For example, it can show the time it takes to complete particular activities, and the methods most likely to be employed. Army manoeuvres and some types of full-scale field trials are good practical examples of the use of this type of model.

PP modelling is frequently used in assessing the cause of accidents. It is often found that injured persons have great difficulty in remembering precisely what happened immediately before the accident took place. In these cases it is sometimes possible to obtain a more complete picture of events by having a 'stand-in' re-enact the activity under relatively safe conditions and, if possible, under the supervision of the injured person. These methods are not always successful, but when they are, their findings may be of great value.

Full-scale models are also sometimes used to provide an indication of spaciousness and distances. In designing consoles, control panels and seating, men cannot be reduced in size so the panels and seating are 'mocked up' (imitations are made on the full scale) with the knobs and instrument dials in the correct positions. Various sizes of people are taken in turn and seated at the controls to check whether the switches are readily accessible. This kind of man–machine engineering work forms part of a branch of science known as ergonomics.

A particularly interesting example of PP modelling was the full-scale all-metal mock-up of the lunar excursion module. The module served as a 'moon base' for the first astronauts on the moon before they eventually took off in order to join the orbiting command module and return to Earth. The astronauts were able to use the 20 foot high model shown in Figure 1.3, to practise embarking and disembarking with full equipment, and to develop efficient procedures.

Generally, therefore, PP modelling is used in order to obtain the opinions of man about some aspects of the situation, and since man cannot be scaled down in size, full-scale models are necessary. Children can be used to some extent for decisions on very costly constructions, but usually it is not too difficult to produce reasonably cheap mock-up systems. Thus, in summarising these remarks, it can be stated that pure physical modelling is usually concerned with space and/or time assessments in relation to man.

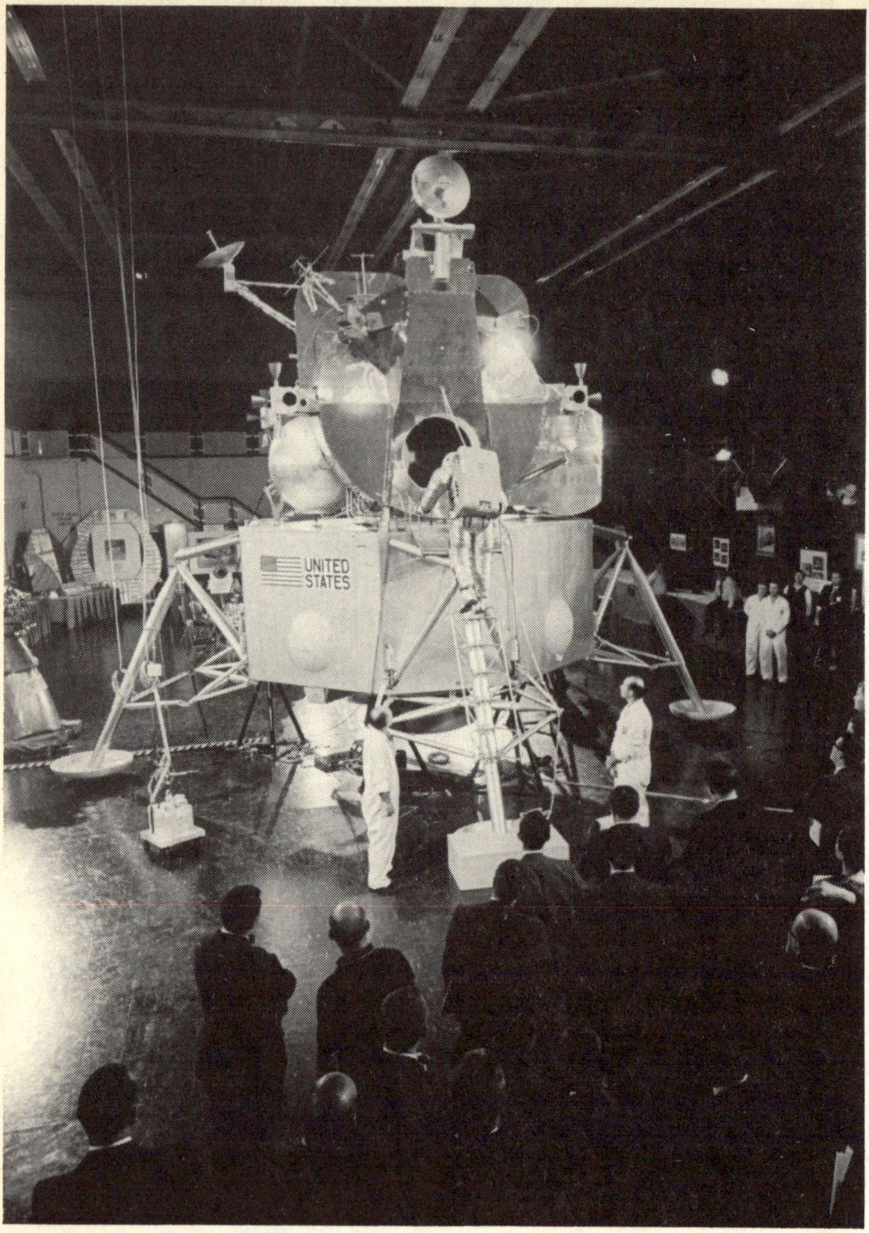

Figure 1.3 A mock-up of the lunar excursion module used by American astronauts to practise embarking and disembarking procedures

Physical geometric or PG models

In physical geometric modelling a carefully reduced or enlarged replica of an object is constructed and the properties of the model are the same as the original whatever scale is used.

Consider, for example, a simple girder or beam subjected to a uniform pressure all over its uppermost surface. Now suppose the pressure is gradually increased until the beam fails. If a model girder or beam is constructed from exactly the same material to a scale only one half the size of the original, and if the pressure is gradually increased until this beam also collapses, it will be found that the pressure required to cause failure is similar to that for the first beam. Theoretically for static loads the pressures should be identical, but in practice the effect of skin stresses, surface flaws etc., tend to introduce deviations when the size is greatly reduced. The more complicated the model the more pronounced this effect generally becomes. It is for this reason that tests involving models of rivetted joints, or plates with a multiplicity of holes or stress concentration zones, are rarely meaningful when the model is reduced below about one-quarter full size.

A way in which this problem can be overcome, at any rate to a limited extent, is by determining what is known as the scaling effect. This is done by constructing the model to various scales in order to measure precisely what changes take place. Thus, a series of different sizes of models may be constructed to say 1/50th full size, 1/10th full size and 1/5th full size. The performance of each size of model can then be studied and any departure from the predicted scaling observed at each stage. This often enables the appropriate scaling corrections to be made in order to determine the effects which occur on the full scale. It is in fact a fairly common procedure for estimating scaling errors.

Probably this was one of the original methods of scientific modelling for the models would probably be constructed from the same materials as the full-scale object. It is also a fairly obvious procedure when constructing working models with moving parts. Certain corrections may be necessary in attempting to assess power capability, and dynamic aspects with this type of model. These will be considered later when dealing with other methods.

PG models are also used in wind tunnel experiments. In this case it is not necessary to scale all the physical properties, but only the drag and lift forces etc.

Geometrically scaled models are, of course, used by planners, architects, builders and others to provide three-dimensional visualisation of projects, space in rooms, between and around buildings etc. In this case the scaling of the physical properties of the material are unimportant since only the concepts of space and finished appearance are under consideration.

This type of model is therefore primarily used for space visualision, assessing the strength of structures, or determining functional operation and performance of dynamic machinery.

Interpretive geometric or IG models

Interpretive geometric models have the appearance of being geometrically similar to the original object, but are constructed from different materials in order to produce the desired scaling effect. In this way the effects produced replicate in some manner those which would be observed on the full scale.

It is one of the difficulties of using models for studying physical effects that not all the variables necessarily change in a similar manner as the model is reduced in size. This may mean that in certain respects the model must be changed in order to maintain other desired characteristics. For example, gravity cannot easily be scaled. It is difficult to carry out experiments in centrifuges, and it is normally not possible for us to work in laboratories in space. The changes in gravity as a result of the heights or depths man can achieve on this planet are usually too small, or too inconvenient, to establish scaling factors. Thus, although it is generally possible to scale the forces due to pressures in a geometric manner, the gravitational forces are less easy to control. This will be discussed in more detail at a later stage.

Consider also the effect of air flowing over a model. Normal reflection effects may be reasonably reproduced on a model scale, but flow round the model will not scale in the same way. In other words the Reynolds number, which is the coefficient for air of fluid friction, or viscosity as it is called, does not scale in the same way as the normal geometric characteristics of the model. Some changes can be produced by roughening the surface of the model, but this effect is not always satisfactory in that it is often required to reduce the viscosity. Variations in the air density or gas used in the experiment can produce the necessary corrections, but this is an alteration to the environment rather than to the model. It will readily be appreciated from the foregoing that the surface finish of models used in wind tunnel experiments requires some care.

Hence, although IG models conform to a geometric scale, they also incorporate features which compensate for some variable, or variables, which do not scale in a geometric manner. In certain cases physical geometric (or PG) models may be used in an interpretive manner by carrying out a little arithmetic in the data analysis, or by changing the environment to an 'interpretive' condition. In the latter case, however, the model used is better described as a PG–IA models where the second pair of letters describes the environment.

Interpretive analogue or IA models

The ability to construct an analogue model implies some knowledge of the mathematical equivalence of the system. An interpretive analogue model behaves in an analogous fashion to the full-scale object under examination, but does not necessarily have the same geometric shape, and may not be constructed from the same materials as the original.

War games are a good example of 1A modelling. With this kind of model real people fight imaginary battles. The opposing field commanders sit in separate rooms with maps of the terrain. They have knowledge of the disposition of their own forces, but only a limited knowledge of the disposition of those of the enemy. An umpire observes the moves made by the various forces and decides the outcome of each small skirmish, or major battle, on a basis of the quality of arms and disposition of forces in each situation. Business games, which can be used for training executives, are also examples of this kind of modelling.

Interpretive analogue models can vary considerably in their character and appearance. They are obviously much cheaper than making exact copies of a real situation, even supposing this were possible. Let us suppose that it is required to study the way in which masses of logs will flow down a river. It may not be necessary in the model to use logs, or objects shaped like logs, at all. Provided the river has reasonable depth, a great deal of information can be obtained by simply throwing a few hundred garden canes into the water and noting their speed and travel, probable accumulation zones etc. Naturally, such a simple model may not yield complete information with regard to log jams etc., and a better model would be obtained by using rough planks of similar width and length to the logs. However, for many purposes this more expensive method may not be necessary.

Interpretive analogue models can also be used to explain phenomena which cannot be readily observed directly. An example of this would be the way in which a stream of moving cars may be used to visualise the difference between the particle flow in a sound wave and a shock wave. The cars will be able to stop, or adjust their speed, if the car in front brakes suddenly, so long as their speed is not too great. This corresponds to the particle flow in a sound wave. Above a certain speed, however, the cars will pile into each other, and this is very similar to the conditions in a shock wave where the particles are unable to adjust themselves to an impending disturbance.

It will be appreciated that although IA models may be useful in a descriptive sense, as outlined above, their major scientific value lies in deducing the possible mechanism of systems which are only partially understood. Once an understanding of the way in which forces act on a body has been obtained, an analogue model can be constructed which will receive the same forces, and respond in the same way, as the original object. If the IA model is now placed in a new environment it will be possible to predict the behaviour of the original body by observing the response of the model.

Interpretive modelling has to be carried out very carefully in view of the mathematical extrapolations necessary to deduce the appropriate system. It is, perhaps, best employed in repetitive situations where a few experiments can also be carried out with the real object in order to check the accuracy of the method. Once a satisfactory scaling method has been achieved, this type of model can be extremely useful and effective, in that a great deal of information may be accumulated at relatively low cost.

Computational analogue or CA models

It is rather less easy to define the more mathematical models in a simple distinctive manner than those at the physical end of the spectrum of models. The distinctions between the various types are probably most easily illustrated by means of some examples.

Computational analogue models are often used where an analogy cannot be obtained by allowing reactions in the physical environment but where, from a knowledge of the behaviour of part of the system and a knowledge of the environment, it is possible, by repetitious computation, to build up a better understanding of the system as a whole. A typical example of this is to be found in modelling the queuing at the check-outs in a supermarket. The arrival times of the customers at the check-out, and the probable time the customers spend in receiving service from the cashier may be known to take place in certain random ways. The times for these actions may have been obtained experimentally by observing customers in the same, or similar supermarkets, and noting the arrival and service times. If the arrival times are written on slips of paper and placed in a hat, and the service times are placed in another hat, by drawing out the pieces showing probable arrival and service times it is possible to obtain an imitation of the real situation and also a variety of other conditions, such as improved service or increased trade.

Sometimes a rather long queue may form, but at other times the arrival times may be favourable and the queue will be relatively short. This method of replicating the possible situations is known as a simulation process, since the various conditions are built by allowing the process to happen 'on paper'. It will be seen that this method is therefore a special form of analogy in that, by a computational process, a variety of probable situations have been estimated entirely by testing the possibilities of the system over a period of time. This process of simulation is usually carried out with the aid of a computer and for obvious reasons falls into the CA category of models.

Computational analogue models are useful for studying dangerous situations which cannot be allowed to happen. It is possible, using this method, to study in complete safety the way in which a nuclear power generator may become critical. The probability of collision of atomic particles can be computed, and given a certain number of collisions, the probability of further collisions can be studied and the whole range of possibilities considered.

This type of model, although still concerned with modelling physical situations, no longer employs a physical model for the task. The analogy is achieved either purely 'on paper' or inside a computer. CA models are sufficiently far across the 'spectrum' of model classifications given in Figure 1.1 to have moved entirely away from a physical system. It is, nevertheless, the most primitive of the non-physical models and merely replicates the exact physical actions on paper, or inside a computer. These models often use statistics to obtain the appropriate random samples, or to determine the length or number of simulation experiments necessary to deduce the probable conditions and variations, but essentially the process is simple and automatic. It is possible to construct large simulations of very complex processes, but some care must be taken to avoid excessive complication as, even with modern electronic computers, which operate at enormous speed, the running time may become too long.

Simulation models can grow to such a size that their validity becomes very difficult to assess. Errors which result in a million lorries being parked in six loading bays at the end of the day are easy enough to detect, but the real problems lie in the detection of errors which are not immediately obvious. For these reasons it is often advisable to build up simulation programs in gradual stages, starting with the most simple case, which it may be possible to check using mathematics. Complexity should then be built into the model a small piece at a time so that the effect of each change can be carefully studied to ensure that consistency and the correct magnitude of the results are obtained. Care should be taken to build complex simulations with a number of modular programs. They should be constructed so that modification is relatively easy and elaborate 'flagging' for this purpose can be a great help. Programs should be well documented and attention given to files and file handling procedures. This activity may seem unimportant at the time but is essential if the model is to provide continued in-service satisfaction.

Computational mathematical or CM models

Quite frequently we have some understanding of the system under examination and can make mathematical predictions about part of it by assuming the relative importance of certain features or 'variables', as they are usually called. For example, in deciding whether to sell a new product in a chain of supermarkets, there are many factors to be determined before stocking the product. If it is decided to stock the new product, how many people will buy it and what will be the effect of advertising in the local papers? Secondly, what effect will a special featured display stand have? Will promotion such as free gifts help? What will happen if it is decided to cease to stock other rather similar lines? Let us represent the number of people who purchase the new product out of curiosity without any advertising as X_1, the number who purchase after advertising as X_2, the expenditure on advertising as C_1, etc. and suppose that a suitable model can be gradually constructed. If the results are promising, then it is likely that by making simple tests in perhaps only one of our chain of supermarkets, it will be possible to find the values and more precise relationships of the variables X_1, X_2, X_3, etc.

A model constructed in this way often requires large numbers of calculations in order to deduce the value of the variables or coefficients in the model. The derivation of formulae by means of statistical techniques such as analysis of variance, multiple regression, or multivariate analysis, are good examples of CM modelling. The importance and interdependence of all the variables in these models is determined by a process of computation.

Dynamic programming provides another computational method for arriving at an optimal solution using the minimum amount of calculation. Even so, the amount of calculations necessary may still be extremely large and an electronic computer is likely to be necessary. In the case of certain very complicated problems it is possible that even the use of a large electronic computer will not provide a solution in a reasonable time.

The basic idea in dynamic programing is to divide the problem into a series of stages. First, the best solution is found between one stage and the next. Progress is then made to the next stage and the various possibilities for this stage are tried in conjunction with the optimal solution for the previous stage and the best value determined. This process, when repeated throughout all the different stages, leads to an optimal solution to the problem.

Hence, although some facts are known about the system being studied in the case of CM models, a fair amount of computation and statistics may be necessary to develop the model to the state where it can be used for prediction purposes.

Pure mathematical or PM models

A pure mathematical model can be described as a complete representation of a system by one, or many, equations and/or constraints which satisfy the known data and the past and present behaviour of the system. In this context it should be noted that by 'pure mathematical model' I am not trying to distinguish between 'pure' and 'applied' mathematics and mean simply that no iterative computational activities or physical features form the basis of the model.

It seems probable that we will never be able to build a really comprehensive purely mathematical model since, as knowledge grows, new facts emerge which negate the original premises on which the model was founded. Some good examples of this kind of mathematical modelling are Newton's laws of motion which were subsequently superseded by Einstein's model. Now even Einstein's model seems unsatisfactory.

There has been much discussion between such eminent philosopher–scientists as Sir Karl Popper, Thomas Kuhn, I. Lakatos and others [5-7] on the adherence to, and the abandonment of, scientific theories and paradigms. It is not advisable to enter into these arguments here, but it should be noted that, although models may be abandoned and new ones accepted, this is sometimes because the original concept lacked a simple additional complementary idea which would modify its results and hence its predictive ability. Thus although today we believe

Newton's models to be incorrect, it is possible that science, or what we call the scientific method, has not been in operation long enough to preclude a return to the concept of a more mechanistic universe, and a modified form of Newton's model.

Pure mathematical models are the result of the mathematical development of basic argument, or concepts or paradigms which are at present accepted as true and fundamental. Having said this, one can at once say that pure mathematics means 'everything' and 'nothing' since the 'everything' depends on the paradigm.

It is probably also worth noting that, although pure mathematics may be used in the development of the model, in one sense all models are 'applied'.

It is initially the aim of almost every scientist to develop PM models since they should contain the minimum of uncertain data. Many such models may start off with only very basic assumptions, but subsequently make a great many unsatisfactory simplifications or further assumptions during the development of the model. The final result is a PM model, but in such a degraded form that it fails to perform satisfactorily.

The segregation of scientific disciplines increases the problems since the pure mathematician, who is most able to create such a model, is often amongst the least competent to make the best simplifying assumptions. Some mathematicians, such as Sir Geoffrey Taylor and Lord Penney, have shown the ability to visualise the importance and realism of their assumptions and in consequence have been able to make great advances in scientific theory.

2 Some case studies

The following three short case studies, which have been drawn from different parts of the modelling spectrum, have been included in order to try to widen the horizons of the model builder and to show how a narrow approach can prevent real understanding in certain situations. The studies may also help to show how it is possible to be wrongly encouraged into using methods which will not provide a solution to the problem. The need for the model builder to provide some help to the sponsoring authority to ensure the right problem is being solved is also indicated.

Case study no. 1

A great deal of time could be spent talking about the different kinds of model and very little idea might emerge about how to construct an effective model which will answer all the relevant questions.

The following case study is included in order to show how effort can be wasted and to demonstrate the way in which a solution to the problem was eventually obtained.

In 1948 a certain government department had the task of designing some new buildings in Central London. The Second World War had recently ended and in view of the uneasy situation in Europe it was considered desirable to make the buildings reasonably robust with regard to nuclear blast effects. Accordingly, information was requested from the appropriate defence department with regard to the precise impulsive load buildings of a certain specified size and shape would sustain in the event of a nuclear attack.

In some respects impulse alone seemed a rather strange parameter to request, and with hindsight its value as a sole design parameter might well have been questioned. However, the defence group concerned set about the task of obtaining this information by using models. There are various ways in which the impulse measurements could have been carried out and it is not easy to state categorically that one method is better than another. In this case it was decided to use a modified form of ballistic pendulum, possibly because of previous experience in using such an instrument for air blast impulse measurements. A model of the building was constructed and was attached to an arm so that the whole was free to swing like a pendulum, the height of swing being a measure of the impulse. This device is rather crudely illustrated in Figure 2.1.

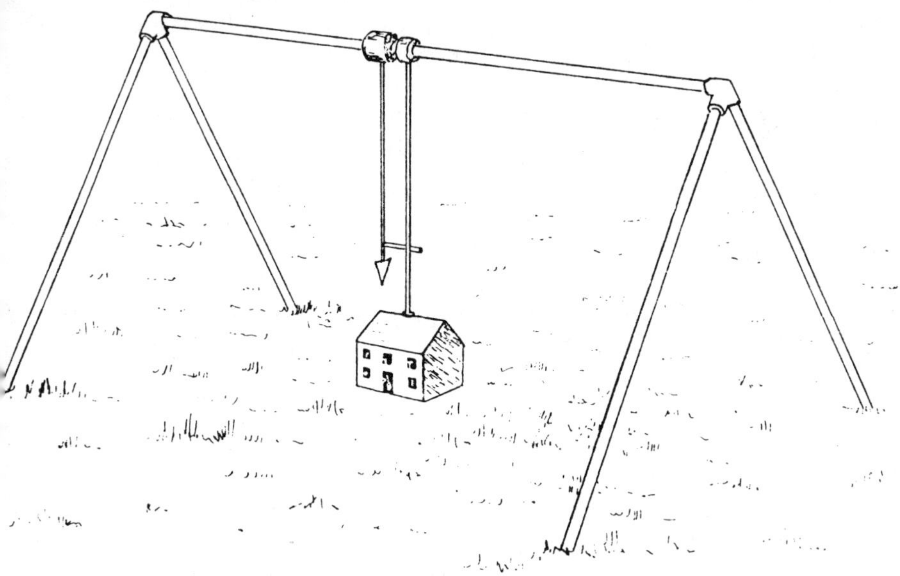

Figure 2.1 Modified ballistic pendulum for blast impulse measurements on buildings

Shock waves from explosives can be modelled satisfactorily by using miniature charges and the general nature of the scaling laws for shock waves are at once obvious from an examination of Figure 2.2.

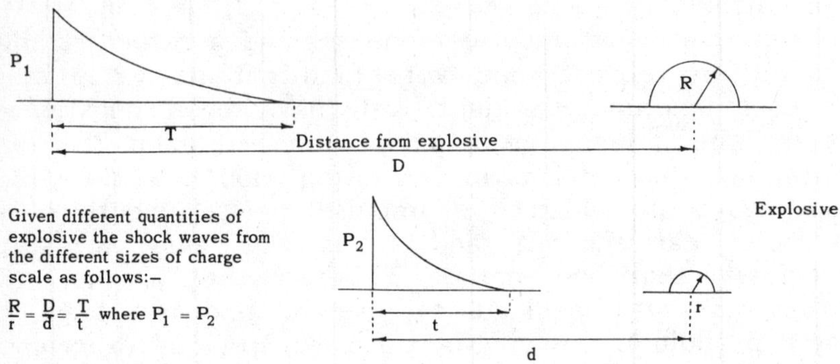

Figure 2.2 Scaling laws for explosives and shock waves

Reducing the quantity of explosive has an effect on the distance at which the same pressure occurs, and on the duration of shock wave, rather like looking at the system through the wrong end of a telescope. In other words length and time are scaled linearly. However, the pressure remains the same.

It was therefore possible to arrange for a shock wave from a small high explosive charge to have a reasonably similar shape in relation to the model building structure as a nuclear blast wave would have to a large office building. The impulsive force imparted by the blast wave to the model on the ballistic pendulum was deduced by observing its height of swing. Some slight errors were introduced by the assumption that the height of swing was directly related to the impulse, but these were fairly small in this case.

The impulsive measurements were then duly forwarded to the construction experts who decided that these data were very interesting, but they also decided that it was difficult to relate the values to damage effects. After some discussion they decided to ask for pressure-time data so that they could have a better understanding of the way in which the forces built up on the structure.

As a result of this new request for additional information several new model buildings were constructed.

These consisted of wooden blocks in which a number of holes had been bored to simulate the air flow through a building with broken windows. The models were mounted solidly on the ground and piezo-electric pressure gauges were fitted into the front, top, side and back of the models in order to obtain the pressure-time history. It was necessary to use some mathematics in order to determine the size of holes required to give the correct air flow through the building, and although the main outline was geometrically scaled, the holes in the block gave an analogous flow. Consequently, this new model could really be described as being of the IA type.

The designers again duly received these data with the appropriate 'thank you', but once again did not appear to know how to apply them. They requested information with regard to the flow time into rooms so that they could decide whether the front panels would collapse before the pressure had a chance to get inside and provide some support. They also asked for miniature rooms to be constructed in the model and for tiny pressure gauges to be placed within the rooms. In addition they also wanted metal front wall panels with strain gauges fitted so that they could deduce the pressure acting on the front panels.

It was rapidly becoming clear that unless another approach to the problem could be found the investigating group might have a job for life. Even so, metal panels were constructed and these were coated with brittle lacquer and stressed in order to determine the stress concentrations. In this technique, known as the 'brittle lacquer technique', a resinous material is coated onto the model which is then stressed. The resin cracks under stress and areas of heavy stressing are shown up by a high concentration of cracking in the resin. With a certain amount of eye-strain, and some difficulty, a number of tiny electrical resistance wire strain gauges 17x10 thousandths of an inch in size (smaller than a pin-head) were constructed and attached to the rear of the metal panel.

Now the data began to give a feeling of reality, although the metal panels 'rang' like a bell, unlike concrete or brick and are much less elastic. It seemed that still more information would be needed. What questions might the designers ask next? A careful appraisal of the situation was necessary and it seemed advisable to restate the requirements more exactly. What was needed was not impulse, it was not the pressure-time load on the model, it was not even the way in which the walls oscillated under an impulsive load, but was quite simply whether or not they survived the blast.

The obvious logical procedure was to consider whether or not it would be possible to construct a simple model from the correct materials, and to decide what effect would be obtained on this PG model structure if it were exposed to a blast wave. It was time to do some quick and simple calculations. It is surprising how often these methods succeed where complex thought fails.

It is rather difficult to convey problems in retrospect, and geometric scaling now sounds quite a simple process. At the time, however, this method was uncertain, and calculations were made for beams and panels etc., in order to determine whether the tensile and compressive strengths should be the same on the model scale as it was on the full scale. It was found that from static loading aspects, at any rate, this method of scaling was satisfactory and, provided the flaws and skin stress effects were not too pronounced, that a geometric model should fail at the same pressure on the surface (e.g. dynes per square centimetre) as the full-scale version of which it was the counterpart.

Here then was a fortunate situation, for the blast wave and the explosive also scaled geometrically with the exception of the pressure dimension, which remained the same. Many problems still existed. The full-scale wall panels were made from brick or concrete, and most of the concrete experts seemed to be of the opinion that the strength of concrete could not be controlled sufficiently accurately for model construction.

The concrete experts commented usefully regarding the difficulties of controlling the quality in view of the importance of blending times, 'curing' times and temperatures. Indeed, the problems seemed formidable. Despite the difficulties, there remained the certainty that if they could be solved a simple and direct answer would be possible. It was decided to carry out some experiments with plaster of Paris as this was a fairly pleasant material to use, although it would undoubtedly have some of the complications that might be experienced with concrete. A number of test specimens were made using very carefully controlled quantities, mixing and pouring times. The specimens were allowed to set, or 'cure', in temperature-controlled rooms for a specified number of days, and were then tested. The results were astonishingly good and all were within a few per cent of the mean. A number of model plaster of Paris wall panels with windows were then constructed and these were tested in order to ensure that their strength was consistent.

This success brought about a resurgence of interest by the concrete experts who ably completed the task by constructing models in concrete to carefully controlled standards. The stringent conditions of manufacture they enforced ensured that the constructed model panels failed at the same pressure as the full-scale panels. The problem was now virtually solved, and by using small high explosive charges against the models in the field, it was possible to tell those responsible for designing the new building exactly how much of it would survive and would continue to provide shelter after a nuclear strike.

Since that time the civil engineering group (now part of AWRE Foulness) have produced the most elaborate concrete structures containing wires, designed to simulate reinforcing materials. A wide variety of scales have been employed and the method has had many applications. These have ranged, for example, from the design of shelters to nuclear reactor safety studies.

This example of modelling has been described in detail not only because it is an interesting study, but also because it illustrates the importance of selecting the most direct method. It is true that the final model ignored certain aspects of scaling, but they were generally found to be relatively insignificant for this kind of model.

Case study no. 2

Let us now consider another example which shows how a considerable amount of information was accumulated directly and within a relatively short time. In this case the questions were clear-cut and the model builder was given immediate freedom to choose the actions.

In 1955 only a very limited amount of information was available about the effects of very large nuclear weapons on military equipment such as tanks, guns or aircraft. A rather serious question arose with regard to just how effective such weapons would be against normal conventional military equipment.

Lack of knowledge of the response behaviour precluded the exclusive use of mathematics since this would require verification, had any suitable theory been available. The chance of obtaining suitable data seemed remote, since the deployment of a megaton weapon against a squadron of aircraft or tanks was ruled out in view of the enormous cost.

The question was asked 'could the data be obtained using a series of experiments with models?' The problem did not seem immediately soluble since robust equipment, such as tanks and lorries, might be affected more by being overturned, or thrown some distance by the blast, than by the simple panel failure effect mentioned in the first case study.

It was not easy to see exactly what kind of modelling could be employed. However, from a simple analysis it seemed likely that if straightforward PG scaling were to be used, the disturbing forces would scale as a square law with the size of the model, whilst the restoring forces would scale as a cube law. This meant that the models would be thrown a relatively large distance by the blast, thereby probably causing excessive damage and making the results almost impossible to interpret. At first sight, therefore, it seemed important to have the acceleration correct. If the correct acceleration forces could be induced in the model these could be measured by means of accelerometers. This would then make it possible to study the effects on the full scale, since similar forces could be obtained by dropping the equipment from an appropriate height.

As already explained, when a scaled model such as a tank is constructed, the disturbing and restoring forces scale differently. The blast forces striking it vary roughly as the presented area, whereas the restoring forces preventing motion depend upon the weight–which depends on the volume. Hence one effect obeyed a square law of the linear dimensions of the model and the other a cube law.

After some thought it seemed that one way to overcome this difficulty was to increase the weight of the model proportionately as the size was reduced. The only reasonably priced high density material available was lead, and it was found that by reducing the scale to one-seventh full size it was possible to produce a model tank whose weight per unit volume was seven times that of the real object. Scaling even further was virtually impossible, as at this level the model was practically solid lead, and although only about two feet long, it weighed as much as a small motor car. Small instruments were fitted in order to record the acceleration in three directions and determine the forces acting on the model.

Figure 2.3 A model centurion tank approximately 2 ft long weighing almost as much as a small car. This complex model employed IG scaling for the body and PG scaling for the tracks and side plates.

However, although the acceleration and distance the model moved would be valuable measurements, it was important to extract the maximum information from the trial. Decisions were therefore taken with regard to the probable weakest links in each model and these parts were strength scaled in the normal PG manner. For example, in addition to the density scaling in the model tanks, the tracks were strength scaled, and the suspension frequency scaled so that the appropriate amount of destructive energy would be available to the tracks (see Figures 2.3 and 2.4).

Figure 2.4 Models used to study human reaction to blast waves. Those left of the picture used IG scaling whilst the model one on the right is an example of IA scaling.

Eighty-one models of tanks, guns, miniature dummies, V-bombers etc. were constructed using this density scaling principle. They were all fitted with three axis accelerometers and the weakest parts were made of frangible materials. The model aircraft were mainly solid material but they had a number of frangible panels in the fuselage, whilst the wings, ailerons and elevators, etc. were scaled to have the correct natural frequency and also to fail at the correct strength.

However, although such an array of models was an impressive sight there was little point in using them unless some corroborative evidence could be obtained. It was therefore decided to design a separate experiment in order to test whether the modelling laws proposed for the experiment were likely to be correct. To this end a number of special, or 'idealised', targets were constructed. These consisted of cubes of various sizes and density fitted with accelerometers. The smallest cube had six–inch sides and the largest had sides of about three feet in length. The cubes were placed at various distances from the explosion in such a way that, if the scaling were correct at each distance, they should all be thrown exactly the same distance by the explosion despite their variation in size and density.

The array of models and cubes was tested against a low-yield nuclear weapon with dramatic effect. The scene after the explosion was similar to that which might have been expected when looking down on a battlefield stricken by a megaton weapon. However, even more exciting and important was the fact that the cubes had rolled back in line as the theory had predicted. The model tanks, lorries, equipment, men, field defences and aircraft were in varying states of disarray depending on their distance from the explosion. A very clear picture was immediately obtained of the way in which a large nuclear weapon might affect military strategy. These data could not have been obtained so cheaply or easily in any other way.

The modelling used in these tests involved a mixture of IG and PG techniques and required a fairly good understanding of the system. Note how, in this case, confirmatory evidence for the modelling theory was obtained by the use of a third type of model (the idealised cubes, which were in fact IA models). It provides a useful example of how this type of physical modelling should be undertaken.

Case study no. 3

Let us now look briefly at the construction of a purely mathematical model, choosing in this case a fairly simple example which is easy to follow and yet which has proved useful in practice.

All mathematical models, however sophisticated, are based on hypothetico-deductive reasoning. The more realistic the original premise on which the model is based, the stronger the model, providing also, of course, that no foolish assumptions or simplifications are made in the calculations themselves. In many day-to-day problems, which are basically of a mechanical or applied physical character, Newtonian physics often provides a good starting point.

As an example, let us take a case study and consider the existing state of knowledge with regard to the mechanism of damage from bomb fragments some five years after the Second World War had ended. The problem had been extremely important during the war, and a great deal of experimental data existed as a result of very many series of differing kinds of field trials. Unfortunately, however, the results of the various workers appeared to be completely inconsistent. A worker who might be concerned with measuring fragment penetration into strawboard (packs of cardboard bound together) would claim that the depth of fragment penetration depended linearly on the impact velocity and on the cube root of the mass of the fragment, whilst another worker studying fragment penetration into a wood would announce with some degree of truculence that penetration varied linearly with fragment mass, and also depended upon the square of the fragment velocity. Other workers (such as Sir Solly Zuckerman's wound ballistics team) had still different formulae which did, in fact, involve velocities to the four thirds power. Meetings of the experts seemed to increase, rather than decrease, the confusion since each worker had strong views about the accuracy of his results.

A careful examination of the results indicated that much of the data had, in fact, been obtained by highly competent workers, and that the scatter within any particular experiment was usually very small. It therefore seemed that experimental errors alone were unlikely to account for the wide variations suggested in the empirical relationships which had been deduced from the results. The next thought, which was not quite so obvious, was could it be that all the separate workers were right, or nearly right, and if so, how?

At that time several quite sophisticated theories for the mechanism of fragment penetration were being advanced, but as they satisfied the criteria set by only one part of the data or another, they were not thought to be of especial value. Under these circumstances it seemed inadvisable to delve too deeply into the existing theories or try to extend them, and it appeared likely that a completely fresh examination of the problem might be worthwhile.

It was thought that it might be useful to try to express fragment penetration in terms of simple Newtonian physics. In the first place it is necessary to consider logically what happens during the process. One moment the fragment is flying through the air at high velocity, and the next moment it is stationary in a sheet of armour or a piece of timber. One second the fragment has a great deal of kinetic energy and the next second it has none. How is this energy absorbed? Clearly the answer lies in the fact that the fragment performs work on the material as it impacts.

First, let us examine the case where the fragment is moving so slowly that it is almost static. Now ask the question 'what causes a material to fail when subjected to a large static force?' If it is a thin plate, then it may be that the shearing stress is an important parameter, but if the impact material is thick, the penetration may depend on the nature of the principal stresses, on the compressive stress, the shear stress or even the tensile stress. The shearing strength of the material may be a function of the tensile strength or the compressive strength, and in unconfined modes a material subjected to a tension or compression may fail in shear.

At low speeds of fragment impact it therefore seemed reasonable to take the tensile, compressive or shear strength of the material as a penetration parameter since this most closely corresponded to static conditions. The question arose what other parameters were likely to prove important? Presumably the presented area of the fragment to the impact material might be important at low velocities, and the work done on the target material under these conditions could probably be directly related to the penetration.

If, on the left-hand side of the equation, we write the kinetic energy $\frac{1}{2}\, m.v^2$ (where m is the mass and v is the velocity of the fragment) and on the right-hand side of the equation we tentatively write down the parameters absorbing this energy we find that they can be arranged to express the work done on the material $f.A.P.$ (where f is the ultimate stress at material failure, A is the presented area and P is the penetration). The equation is in dimensional balance (i.e. energy units on both sides) and we have:

$$\frac{1}{2}\, m.v^2 = f.A.P.$$

So under near static conditions

$$P = \frac{C.m.v^2}{f.A}$$

where C is a constant.

Hence, at low velocities the penetration is a function of the mass of the fragment and its velocity. This model seemed reasonable for near to static conditions, but for high velocities, when high rates of strain are involved and a great deal of energy has to be transmitted through material very quickly, the same parameters are unlikely to be important. It was noted, for example, that stress engineers tend to use strain energy in calculations when high rates of strain are involved. It therefore appeared to be worthwhile considering what the formula would look like if a term for the strain energy, S, was included in the right-hand side of the original equation in place of the stress. This, then, might result in an effective formula for penetration effects at high velocities.

Thus, for dimensional balance, the equation could be:

$$\tfrac{1}{2}\, m.v^2 = S.A.P$$

This was not too convincing for several reasons. In the first place it implied that the depth of penetration increased with velocity until at certain velocities the depth become phenomenal. It could be argued from experience that this does not happen in air or in water and that a wedge of particles forms at supersonic velocities in air. Here then was a clue. Perhaps at certain velocities in solids this 'wedge', or something similar, would build up. If this happened, the shape of the fragment, rather than simply its area alone, might be important. This shape factor might well be dimensionless.

Thus if this is accepted, in order to produce a reverse effect on the velocity so that a ludicrous penetration is not obtained, the A on the right-hand side of the foregoing equation can be converted to P^2. Thus

$$\tfrac{1}{2}\, m.v^2 = S.P^2.P = S.P^3$$

so that

$$P = \frac{K.m^{\frac{1}{3}}.v^{\frac{2}{3}}}{S^{\frac{1}{3}}}$$

where K is a constant which includes the effect of the shape of the fragment.

This formula superficially appears to be realistic, but we cannot be certain of its validity without either a much more detailed mathematical analysis of particle movement, or other corroborative evidence. In this particular example the experimental data were available and fitted the hypothesis very well indeed over an extremely wide range of materials. It was therefore concluded that there was a changeover from one mechanism to another at a certain velocity.

The next question was at what velocity this changeover in the mechanism occurred. It seemed fairly clear that the changeover velocity must be related to the speed at which major disturbances could be propagated through the material. Within the elastic range of the material we know that $E = \rho . v^2$, where E is Young's Modulus for the material, ρ is the density and v is the velocity at which sound waves can be propagated. It may be noted that this formula is rather like unit kinetic energy, and Cottrell (8) has since explained it by relating the velocity of sound transmission to the vibration of the outer shell of electrons in the atoms, since these are responsible for propagating small disturbances. However, not a lot of energy can be transmitted by this mechanism, and with larger energies slip occurs on cleavage planes. Since this stressing to failure is a kind of energy transmission, it seemed reasonable to use the formula $E = \rho v^2$ to determine the velocity of disruption on the material provided that we used the secant modulus to failure E_{sec} to calculate the rupture velocity, since the part of the material which fails must follow the full stress-strain curve to failure.

Thus

$$v = \sqrt{\frac{E_{sec}}{\rho}}$$

where E_{sec} is the secant modulus (i.e. modulus to failure) and ρ is the density of the target material.

It was found from subsequent experiments that this dual mechanism explained the situation very well and that the velocity at which the changeover from one mechanism to the other occurred closely corresponded to the velocity as calculated from the secant modulus of the material. It was, however, also found that in practice it was desirable to use the dynamic secant modulus (i.e. to determine the modulus at a high rate of strain). Additional support for this experimental hypothesis was provided by firing airgun pellets at photo-elastic materials, which included various plastics, and gelatine of various strengths. By using this technique, and examining the material through crossed polaroids, it was possible to study the propagation of the stress waves and this gave supporting confirmatory information regarding the dual nature of the process.

Evidence which supports a new theory or hypothesis may often be found, almost by accident, at a later date. In this particular instance a metallurgical research group, which was engaged in studying the properties of armour made from a mixture of plates of different materials, found that the best resistance to attack for a twin plate armour of duralumin and steel was obtained when the duralumin plate was on the outside. Apparently this had caused some surprise within the research group as they thought the steel should be on the outside of the armour to obtain the best resistance to attack. However, the true state of affairs could readily be predicted from the foregoing theory, since the strain energy of duralumin is quite large and this is what is initially required to absorb the fragment energy. Later on, as the projectile slows down and enters the second plate, the maximum tensile, or shear strength, properties of the material become important and these are better in the case of steel. This kind of confirmatory data is important and the investigator should be alert to 'good' or 'bad' evidence which must be examined in an impartial manner.

Perhaps it is worthwhile studying the methodology used in this example.

1 Is there a simple known expression for the states before and/or after the event, or for the so-called boundary conditions?
2 Write down as far as possible the mathematical expression for the before and after states.
3 Is it possible to deduce further data about conditions at other times from a knowledge of these states?
4 Can the relative importance of the probable parameters be estimated from other sources and similar kinds of problem?
5 Can experimental data be obtained?

The object of this simple example is to show how effective formulae can be built up using simple but logical argument. In practice this is the way we have to undertake many problems, although the mathematics involved may be much more elaborate. It may be satisfying to develop formulae which are mathematically complicated and superficially much more rigorous, but it should be appreciated that the chances of constructing such formulae are less likely the longer the problem has been considered by mathematicians. Moreover, complicated, and apparently rigorous, formulae contain underlying assumptions. For example, even the relatively simple formulae developed by Bernoulli for the deflection of beams contain assumptions such as that the material is isotropic and that sections plane and perpendicular before bending remain plane and perpendicular after bending. These much used formulae for beams are generally very satisfactory and give reliable predictions of the effect of loading as long as the strain is proportional to the stress, but under conditions of plastic flow they are unsatisfactory and other methods must be employed. In this case models which are specifically designed for plastic flow conditions must be developed, or the strain estimation carried out by the method of 'finite elements'.

3 Some first steps in building models

The reader may say 'I see how the models in the case studies were developed' but may then ask 'how do I go about the task of building a model myself?' The answer to this is, of course, that there is no sure-fire method. Indeed, if there were such a method, it is doubtful whether by now there would be any problems left to solve. It is suggested, however, that the reader may find the following approach useful for many types of modelling.

Understanding the problem

The first and most important step the analyst must take is to make absolutely certain that he fully understands the nature of the problem. Even problems can, in themselves, be a problem. There are basically two types. These are the *cognisant* problem and the *non-cognisant* problem. In other words, there are those problems which we know about, and those about which we know very little. In the case of the cognisant problem there is an awareness of its existence and the area where the trouble lies. In the non-cognisant problem there is a lack of certainty exactly where the trouble lies, although by definition there is usually an awareness that a problem does exist in some form.

For example, the existence of certain problems may be self-evident, or easily deduced, and analysts may be asked to suggest remedial action. If this happens, it will be found that quite often there has been a correct assessment of the broad area of the problem and hence there is a reasonable knowledge of the nature of the difficulties. Occasionally the analyst may find that the situation has not really been diagnosed correctly, and the causal factors (i.e. the major input variables) are outside the area originally considered to be important by management, or the study sponsor. In such cases management is not really cognisant of the nature of the problem.

Quite frequently there are problems whose existence is completely unexpected, but which can flare up at any time. This type of problem can be described as *latent non-cognisant* and is very difficult to detect.

In order to make absolutely certain that he fully understands the true nature of the problem the analyst must ask questions until he is in danger of being thought a moron. After this it is usually advisable to go away and think about things quietly for a period. As a consequence of this further study the analyst may feel inclined to ask still further questions. He should be seeking to examine the situation in every possible way and to be continually considering completely new, more realistic, approaches. This should not mean creating models of ever-increasing complexity but rather a movement towards simplicity as a consequence of greater understanding of the nature of the complications.

Conceptual diagrams

After the initial discussions have been completed it is usually worth producing a conceptual flow diagram. This, perhaps less fancifully, might be called an action diagram. The object is to display the problem diagrammatically and to indicate which variables react with others. A very simple conceptual diagram is shown in Figure 3.1. In the case of large complex problems such a diagram may contain several hundred variables. It is unlikely that the diagram will, in itself, solve anything, but provided it is adequately discussed with management, it will go some way to ensuring that the problem has been fully understood. Other diagrams such as those used by

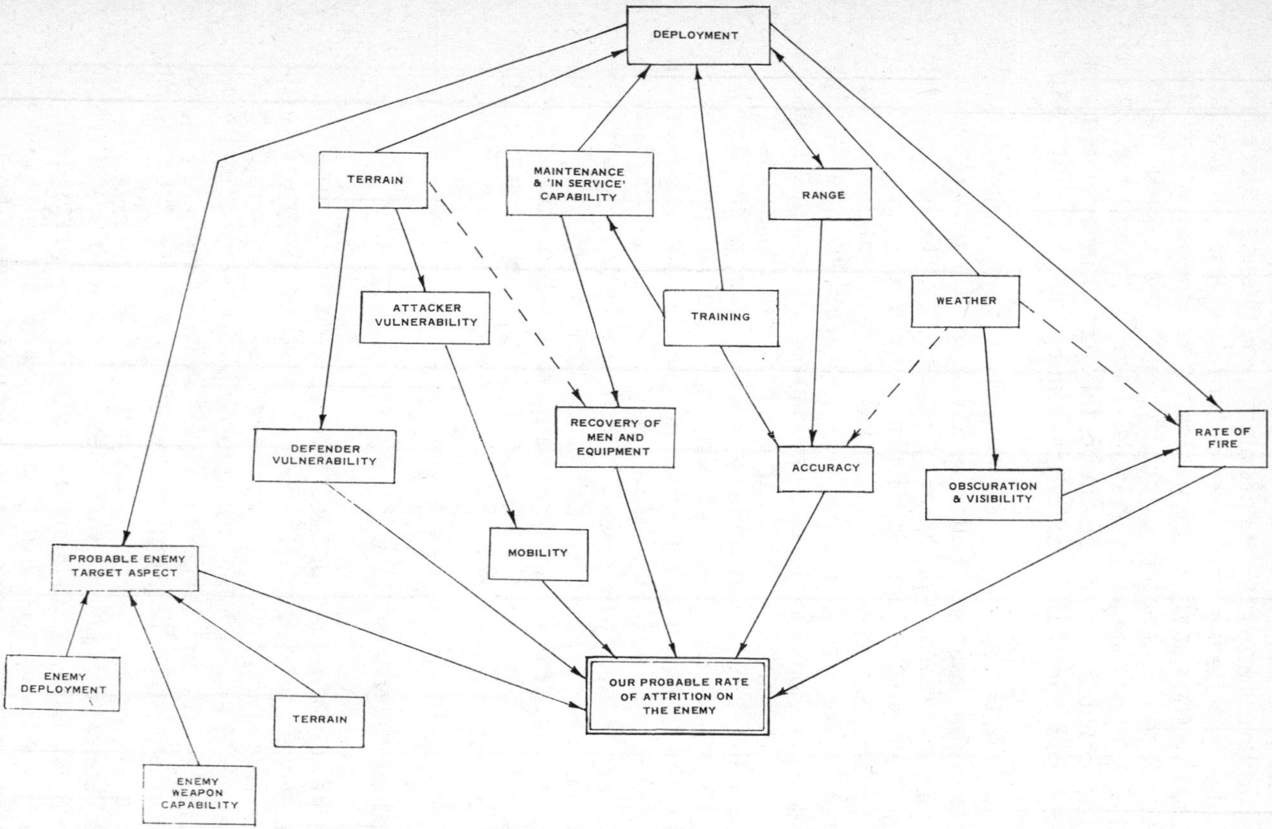

Figure 3.1 A conceptual flow diagram showing the factors involved in the estimation of lethality effects in a battle model

Bowen (9) which in some cases are the 'dual' of the normal type of conceptual diagram, can occasionally help to show the problem in its entirety when the number of variables is relatively small. Structural mapping of the type used by Rivett (10) can also be useful. The methods of Eden (11) may be useful but they require careful application and to some extent need the attention of a specialised expert.

A useful next step is to list the variables in order of their believed importance and to include only the ones of primary importance in the model. Even so, it is necessary not to completely forget those variables which are deemed to be of secondary importance but which have not been included at this stage.

One useful feature of the conceptual flow diagram is that, when completed, it should become more feasible to postulate what will happen to the other variables if any one of them is, say, halved or doubled in value. It is at this stage that the first indications may be obtained that the problem is of the non-cognisant variety. This may be discovered by the fact that a variable, whose importance had not been seriously considered by management, has a surprisingly large effect.

It may be advisable to try to carry out a simple experimental input - output analysis by changing the input values of some of the variables and measuring the output, or effect, on the others. In many cases this may be quite impossible in view of the expense, or hazard, it involves. Such situations can sometimes be examined by undertaking studies on a model scale, or carrying out limited full-scale pilot studies, although even these can prove to be prohibitively expensive.

Selecting the method

Clearly, it is important from cost considerations that in most cases a model should be chosen which is as far towards the mathematical end of the different categories of models as is deemed expedient. The approach is to select the kind of model most likely to produce a satisfactory explanation at least cost. If a mathematical model is used, it is almost certain that some physical experiments will need to be conducted in order to establish a satisfactory 'proof'.

It may be imagined that this stage of the programme is closely related to the branch of statistics entitled 'design of experiments' which involves Latin Squares, Analysis of Variance, Chi and t-tests, etc. In fact, whilst statistics may be of value in the final evaluation of the model, it is usually of limited value at the design of experiment stage in view of the complexity of the corroboration procedures when several kinds of modelling are employed. For example, if physical scaling is used, as in the high-density geometric models described in the second case study, then interpretive geometric models may also be used to establish the validity, or otherwise, of the mathematics used for the scaling laws employed in the construction of high-density geometric models.

The decision as to how many models shall be constructed for theory validation is complex and many-dimensioned. Perhaps the simplest way to look at this is to understand that we are really trying to validate the model for the range of values over which we wish to use it. Thus, although it is always nice to know at what stage our model will break down, in view of time and cost considerations we are normally limited to tests in the working range of the model. Even within the working range of the model itself our tests may have to be limited, and the greatest validation effort may have to be placed on a relatively short, but very active, part of the working range, with 'spot' checks to validate points on a few other dimensions at the less used parts of the working range.

It should not be supposed that the design of these tests needs to be any less careful than if one had been able to follow the straightforward statistical design of experiments procedures. Indeed, it needs a great deal more thought and a good understanding of the limitations and probable errors of experimental procedures.

Perhaps it may seem a little strange that I am talking about how to validate a model before we have built one. However, in most cases the consequences which are involved are such that an adequate validation schedule is necessary before a model can be used in practice. As a consequence, the cost and adequacy of various types of validation procedure must be considered in conjunction with the cost of contructing any of the different kinds of model which may be used.

As mentioned earlier, it is usually desirable, from cost considerations, that the method used should be as far towards the mathematical end of the model category scale as possible. However, we may only have a vague understanding of the mathematical nature of the problem and may decide that it is necessary to use a physical model. In this case it is likely that it will have to be scaled down to a small size. We must know exactly how this reduction in size will affect the parameters we wish to measure. Perhaps the crudest approach would be to construct a range of models of different scale sizes in order to estimate the scaling law and to establish whether the effects are linear. Dimensional analysis, which is discussed further at a later stage, may be helpful in assessing scaling effects.

Self argument and analysis

It is important, when the kind of model has been approximately decided, to try to argue out precisely what is happening in the scaling process. For example, in Case Study no. 2, the written or mental argument might go as follows: 'Alright, so I scale the model down in size, and I want my model to be thrown the scaled distance by the blast wave as a full-scale equipment would be thrown by a much larger weapon. Oh, there is quite a problem here! Surely the model will be thrown much further, relatively speaking, because the disturbing forces scale as a square law (since they depend primarily on the presented area), whilst the restoring forces (or some of them) seem likely to depend on the weight of the model and this will vary as a cube law.' At this stage a number of things will click into place. One may think 'so that is why the models of ocean liners sailed by enthusiasts on the pond in the local park tend to cut through the water with all the appearance of speed boats'. In other words, explanations may begin to emerge for a number of different events which one may have thought a little odd at the time, but which originally did not appear to merit any particular attention individually.

We may, of course, be interested in quite a number of kinds of effect. Probably in the type of damage model just mentioned we would like the accelerations and impact decelerations to scale satisfactorily. Thus in looking at the air pressure and gravity restoring forces alone we have probably barely started to consider the total problem. Surface friction must be considered from the scaling aspect, as must overturning moments, material behaviour on impact, and similar actions. If the model is thrown through the air we must decide whether retardation due to the air is important. Also is the model of a sufficient size that Reynolds Number and viscosity effects are not important, even for the initial blast flow over the model?

It was necessary, in the example outlined in Case Study no. 2, in Chapter 2, to consider all these effects and to check that all the main aspects scaled in the same way, or that any errors were likely to be secondary. This involved a large number of individual studies.

Subsequent to the actual experiments involving the deployment of the models against a low yield nuclear weapon, a symposium was held to discuss the results. A senior mathematician was asked to present a paper at this symposium 'proving' or 'disproving' my findings. He did, in fact, present a paper showing great mathematical prowess, but which in essence consisted of a dimensional analysis rather similar to one I had already included in an earlier report on the subject. His paper did not include the individual analyses of scaling effects such as overturning, friction, impact, etc., which I had carried out separately to try to ensure that all the important parameters scaled in the same way since dimensional analysis often tends to provide 'indications' rather than 'proofs'. The mathematician ended his lecture, having 'proved' my findings, with the good humoured comment that whereas he had obtained his result by mathematics, I must have obtained my result by witchcraft. My only reply to such a comment must be that, if I did so, it was mighty hard work casting the spells!

Data care and simplicity

The remarks in the foregoing section are included for a very specific reason. There is, indeed, a considerable danger in simply using rather fancy mathematics as an argument to justify observations. Dimensional analysis is a loose sort of tool enabling a variety of answers to be obtained, only one of which may be correct. Wherever it is possible, each feature of scaling should be examined individually. There is, of course, nothing wrong with mathematics in itself and in situations of great difficulty it can sometimes be the only way to progress. Unfortunately, when we apply mathematics to the problems of the world, there are many dangers in the assumptions and approximations we are likely to make *en route* to an answer. The writer is aware of at least one world-famous mathematician who, when presented with some experimental data, dashed away to his office and emerged some hours later waving a piece of paper and shouting excitedly 'I've proved it'. A number of rather red-faced technicians then had the unfortunate task of explaining that the unusual 'phenomenon' they had observed was only due to an error in their instruments.

The lesson in all this is that, given an observed phenomenon, mathematicians, being possessed with great ingenuity, are certain ultimately to find a 'proof' or explanation for the phenomenon irrespective of whether the phenomenon exists or not. Wherever possible, we should try to ensure that our mathematical theories are based on a thorough knowledge of physical effects. Part of the brilliance of really great contemporary mathematicians, such as Lord Penney and Sir Geoffrey Taylor, is their ability to know which physical effects are likely to be of primary importance, and to base their mathematics on these rather than on secondary effects or unrealistic assumptions.

Thus far in this section a fair amount of time has been spent in considering physical models, so let us now examine some rather more mathematical approaches.

A saying which is worth bearing in mind at this stage, and which might have been expected to be attributed to an American agent for the fashion model 'Twiggy', had she had one, is that 'the bestest models do not necessarily have the mostest'. A vast number of variables and a large amount of mathematics do not necessarily indicate a superior mathematical model, although it may demonstrate considerable mathematical ability. Indeed, many of the best models seem to have a kind of natural simplicity and only when the basic paradigms become strained does there seem to be a need for complication and elaboration. However, it must be stated that there are a great many problems for which very complicated formulae give highly satisfactory answers. Perhaps the essential feature about even these complicated models is that they are built up from very simple building blocks using very precise and pure mathematics to develop the theory with only the minimum assumptions.

Complexity and vertical thinking

As mentioned earlier, one of the most difficult aspects of mathematical modelling is deciding exactly what is permissible. An assumption which may not seem to have serious effects when made for one theory may prove disastrous under the extended relationships of another. Yet failure to make sufficient simplifying assumptions can make the mathematics so complicated that it either completely precludes a solution, or alternatively leads to a very cumbersome model. It seems that many of the great mathematicians have had the ability to visualise the physical meaning of the problem, as well as to execute the mathematics. The lack of this ability in lesser mortals can be a serious problem and lead to quite competent mathematicians being subjected to ridicule by the more practically minded. It is therefore an ability which should be sought at all costs.

There are various schools of thought about how one should go about constructing a mathematical model. One school suggests that it is advisable to begin by reading all the relevant attempts to solve the problem before starting the work. I have never greatly favoured this approach, but to some extent such things are a matter of opinion. It is, perhaps, worth remembering that if the previous methods had been suitable it is unlikely that there would be a requirement to study the problem further. Secondly, the one certain asset a new research worker has with regard to a problem, namely a mind completely uncluttered by knowledge of the methods used in previous attempts to achieve a solution, is lost for ever. Even without this knowledge of previous studies he may well be constrained by ideas and training to an extent he may well not fully appreciate. This is, of course, one of the reasons why ideas from people trained in other disciplines can sometimes be so helpful. De Bono's comments on 'vertical thinking' (12) are very much to the point.

Serious difficulties can arise when one or more of the initial ideas or assumptions made by previous workers is wrong, and yet the existing theory appears to have provided a satisfactory solution to the problem for a long time. It requires considerable vision and courage to disregard such a theory, especially if the assumptions have reasonable plausibility. Yet at certain stages this may have to be done if progress is to be made.

A technique which has been mentioned previously, and which can be useful for constructing both mathematical and physical models, is known as dimensional analysis. This is briefly described in the next section.

Dimensional analysis

Dimensional analysis is one of the most used, and also misused, tools in model construction. It tends to appeal to the mathematical and non-mathematical alike. Used carefully it can be a valuable adjunct to the techniques available to the model builder. [13-15]

Basically the method of dimensions is rather like looking at a jigsaw puzzle and trying to fit the pieces together. Unfortunately, the very power of this technique of splitting the problem down into small pieces can also prove to be a weakness in that there may be many similar pieces which can be assembled in a variety of ways, and consequently give quite a large number of different 'pictures' of the problem - of which only one will be the correct solution.

The measurement of any quantity in physics can be reduced to the measurement of mass, time and length, which are denoted respectively by M, T and L. The dimensions of a physical quantity are the exponents of M, L and T which occur in the formula expressing that quantity. Hence, area would have the 'dimensions' L^2, and volume the 'dimensions' L^3. Similarly velocity, which is a measure of length travelled in a certain time, will have the dimensions L/T or LT^{-1}. Acceleration can therefore be written as LT^{-2}. The dimensions of other physical parameters can be derived fairly readily and for example the force F is derived as follows:

Using Newton's second law:

$$\text{force} = \text{mass} \times \text{acceleration}$$

so that we can write

$$F = MLT^{-2}.$$

Thus pressure could be expressed as FL^{-2} or as:

$$MLT^{-2}/L^2 = ML^{-1}T^{-2}.$$

Work done is equal to the force multiplied by the distance over which it acts, so:

$$\text{force x distance} = MLT^{-2}L$$

$$= ML^2T^{-2}$$

The principle of dimensional homogeneity provides an important analytical technique, and basically this simply means that the relationships between quantities having the same physical character must be dimensionally homogeneous. The reader will recall how the dimensional balance was effectively preserved in the example where the equations relating fragment mass and velocity to depth of penetration were discussed; although at that time we were considering 'kinetic energy' on one side of the equation and 'work done' on the other.

Now let us consider another simple example using the homogeneity principle. It can be argued that flow through a capilliary tube Q is of the form:

$$Q \propto \frac{\rho R^n}{z\,\ell}$$

where R is the radius of the tube, ρ is the pressure difference across a length ℓ, the viscosity is denoted by z (and has the dimensions $FL^{-2}T$) and n is unknown.

The flow Q when expressed in volume per unit time will have the dimensions L^3T^{-1}. Hence the proposed formula can be rewritten in dimensional quantities as:

$$(L^3T^{-1}) = \frac{(FL^{-2}L^n)}{(FL^{-2}TL)}$$

$$= (L^{n-1}\ T^{-1})$$

However from the principle of dimensional homogeneity we know that the indices of the fundamental quantities must be the same on both sides of the equation.

Equating the indices of the dimension L gives:

$$3 = n - 1$$

$$\text{i.e.} \quad n = 4$$

Hence this indicates that R should be raised to the fourth power in the original equation.

Now suppose that we did not even know the general form of the original equation. In this case we would write:

$$Q = \rho^a R^b z^c \ell^d$$

where a, b, c and d are constants and may be negative.

Again using the dimensional quantities gives:

$$(L^3T^{-1}) = (FL^{-2})^a (L)^b (FL^{-2}T)^c (L)^d$$

$$= (F^{a+c}.L^{-2a+b-2c+d}.T^c).$$

Equating the indices of like terms:

$$-2a + b - 2c + d = 3$$

$$a + c = 0$$

$$c = -1$$

The powers of a and z are therefore explicitly determined as 1 and -1 respectively, but the individual powers of R and L are not determined. Clearly

$$R^b \ell^d = L^3$$

and

$$b + d = 3$$

so we can write

$$R^{3-d} \ell^d.$$

Experimentation or further mathematical argument is therefore necessary to determine d.

It also often happens that some of the important parameters in a model are dimensionless and are more usually termed 'coefficients'. Examples are the Reynolds Number (which is the ratio of the inertial forces to viscous forces), the Mach Number, pressure coefficients, drag coefficients, the Prandtl Number and so on. These dimensionless coefficients may be capable of physical interpretation and may enable formulations of solutions to new problems by analogy or by reference to similar problems. It is therefore possible to talk of 'kinematic' or 'thermal' similarity and to gain insight into one problem from the solution of another.

The method of dimensions is therefore clearly of great value in developing scale models of physical situations. In this way it is possible to use physical models to test mathematical formulae derived by the dimensional analysis and theoretical argument. Its application in statistics is considered by Finney [16].

4 Models using mathematical programming

Mathematical programming has become increasingly important as a consequence of the rapid development of computers. It is a different activity to what is normally described as computer programming, and is concerned with the optimisation of a mathematical function. Computers are generally, but not necessarily, used as an aid with this technique in view of the large number of calculations which are likely to be needed.

Matrices and mathematical programming

Another branch of mathematics which has increased in usage in recent years is matrix algebra. One way in which this can be explained in terms of a physical system is by considering a model of an electrical system *M* with inputs x_1 and x_2 and related outputs y_1 and y_2 (Figure 4.1).

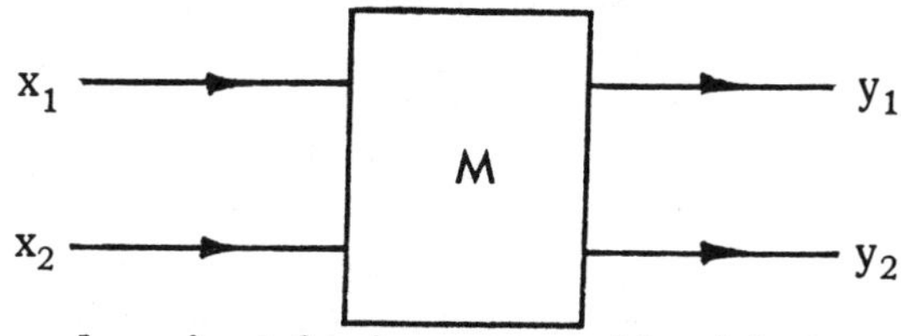

Figure 4.1 An electrical system *M* with inputs x_1 and x_2 and related outputs y_1 and y_2

Thus we could say

$$y_1 = a_1 x_1 + b_1 x_2$$

and

$$y_2 = a_2 x_1 + b_2 x_2$$

where a_1 a_2 b_1 b_2 are coefficients.

Such a system can be represented in matrix form as:

$$\begin{bmatrix} y_1 \\ y_2 \end{bmatrix} = \begin{bmatrix} a_1 & b_1 \\ a_2 & b_2 \end{bmatrix} \begin{bmatrix} x_1 \\ x_2 \end{bmatrix}$$

Some matrices can be large and complicated, but since the algebra for handling them is well developed, the method has great value where a number of set relationships of parameters is known to exist and to provide a description of the system as a whole. There are many interesting cases of the use of matrices in statistical mathematics. For example, the coefficients a_1 a_2 b_1 b_2 etc. often represent the probabilities of the occurrence of certain events. Matrices are often used in business for the construction of mathematical models for guidance in making decisions (see, for example, Appendix 1).

One mathematical model requiring considerable computation in some cases is known as *linear programming*, or more briefly as LP (see, for example, Vajda (17)). In many respects a linear program can be described as a pure mathematical model (in that within the broad description explained earlier both 'pure' and 'applied' are contained in this category), but since it usually has to be solved by a rather lengthy computational process, it is perhaps reasonable to consider it as being very close to the CM class of models.

It is necessary to understand the kinds of problem that can be solved by linear programs, since there is a belief amongst some that LP is a golden elixir for all troubles. Such is not the case, although in the past a great many problems have been squeezed into a state so that linear programming can ultimately be applied. Basically, linear programs are used to optimise *continuous* flow problems at some instant in time. They are frequently used for scheduling, although they are far from ideal in this role and are often used largely as a consequence of a lack of other suitable methods, or knowledge of other feasible methods. The main difficulty is that fluctuations in the flow problem require a large variety of cases to be investigated, some of which may not be immediately obvious. For example, it may be expedient to delay starting a particular job and leave some men and machines unused for a few days if it is likely that another highly valuable job will arrive which must be completed immediately in view of a penalty clause in the contract.

A number of quite small linear programs are soluble by graphical methods and so it is worth briefly describing this technique as an aid to understanding the types of problem for which the technique is most useful. Linear programming is really a method of blending certain items in an optimum manner, and so a straightforward blending problem has been considered in the example given in Appendix 2. Linear programming was originally employed by economists to find the cheapest satisfactory blend of food diet for certain underdeveloped countries or other areas where famine might exist.

An important use for LP has been found in oil refinery operations where decisions have to be made about the amount of different types of crude oil to be distilled, and the actions to be taken with regard to the operation of the reforming, cracking, lubricating oil, processing, gasoline and fuel blending plants. It will be evident that this is essentially a 'flow' system. The problem is to decide the optimum 'blends' to give the maximum profit, and in so doing to determine the best outputs for the various plants which are, of course, subject to capacity restrictions. There is little point in merely working through this type of problem in detail, as it is simply a development of the 'blending' problem shown in Appendix 2 to a larger scale, with profit as the objective function to be maximised. A small refinery might require some 50 to 60 variables and 25 to 30 constraints or equations. A problem of this size required a running time of about 15 minutes in 1968 on an ICL 1905 computer with a 32K core using a FORTRAN program. Today, using a modern computer, it can be completed in seconds.

Linear programming can also be useful in chemical production plants, where problems are rather similar. A very simple example of this type of formulation is contained in Appendix 2. Linear programming therefore provides an optimal solution for flow-type problems, but frequently one is concerned with batch production, or using machines sequentially for different jobs. In this case the deployment of effort may be arranged optimally at any particular instant by using linear programming, but the sequencing of tasks can create a huge number of combinatorial possibilities requiring investigation, depending on which items are processed first. For example, as explained previously, it is quite possible that a better sequencing arrangement may be achieved by not allowing all the resources to be used immediately, and by holding back a number of these items so that they are allocated later in order to give a better use of resources over the whole period of time considered.

Special computational algorithms such as the ICL software packages NIMS and the UKAEA package WASP were amongst the early programs developed for sequencing types of problem. Packages for studying sequencing problems often do not incorporate LP or even a straightforward combinatorial examination of the items, as this would take far too long. In practice they often employ simple semi-heuristic methods such as taking the biggest items or the smallest items first and then trying minor adjustments. Occasionally they use 'hill climbing techniques', and these will be described later. It is clearly often necessary to use some form of combinatorial analysis with these methods even though the full analysis may not be carried out, and this can lead to very long duration computer runs if the number of resources is large. A method known as 'branch and bound', which is described later, can also be useful.

Formulating a linear program

It is worthwhile providing some indication of how problems can be formulated as linear programs. This may seem rather unnecessary unless one has actually faced this task, but in practice problems soluble by linear programming may come in many shapes and guises. Indeed, the problem to be solved is rarely identical to a simple model formulation. Moreover, as all linear programs seem easy after formulation, it is difficult to appreciate the initial difficulties unless one has actually tried to apply the method in some slightly novel fashion.

The first task is to decide which are the unknown quantities and to make sure that these are the same kind of thing. For example, the quantities in question may be the *quantities* of *different types of food* in the well known nutritional model developed by the economists, or the *quantities* of *various substances* made in the different plants in an oil refinery. A good initial step is therefore to write down the problem as a kind of network showing *the way in which the flow occurs* and labelling the unknown quantities of these flows as x_1 x_2 x_3 etc. The cost or profit to be minimised or maximised depends on a certain number of these flows and so the coefficients of the objective functions indicate the cost or profitability of the various flows.

The problem may not obviously fall into flows, and some flow problems cannot be solved by LP if they involve time advancement or other dimensions. Not all the problems which are soluble by LP can immediately be visualised in the correct form. The following checklist may be useful in developing the model:

1 Note the unknown variables.
2 If the problem is complex, begin by taking a much simplified example.
3 Consider how the problem can be represented in terms of a flow which needs to be quantified.
4 Try to lay out the problem in a diagram or matrix.
5 Build any necessary further complications into the model.
6 Assess the computation time.
7 Attempt some form of practical tests.

Advanced mathematical programming

There are many specialised, or advanced, methods of mathematical programming which cannot be dealt with in this kind of general text on modelling. However, it is important to be aware of their existence and the kinds of problem for which they may be suitable. Almost all large computer installations have means of obtaining these specialist software packages even if they do not already have them on file. The real difficulty lies in formulating the problem correctly in an appropriate form. In doing this care must be taken to ensure the original problem is not too extensively 'manipulated' in order to get it into the appropriate condition.

Integer linear programming packages are available where some of the variables must be in integer units. For example, we can have half a gallon of oil, but a solution which deploys half a man here and half a man somewhere else at the same time is meaningless, unless the effect can be neglected because the total number of men deployed in relation to the number of tasks to which they are to be allocated is very large. Where integer solutions are required it is necessary to use special packages which explore the solution in integer steps around the corners of the feasible region. This is not particularly easy when the problem has many dimensions and a badly written integer linear programming package can take a long time to run. (18,19)

Zero-one integer linear programming is used where each integer value can be either 1 or 0. This form of programming is useful in certain decision situations where the decision involves taking a certain action, or not taking that action. In such cases half an action is not possible, and the decision is based on a straightforward yes-no situation. The value of this approach lies in the analysis of complex situations where many decisions have to be made simultaneously. However, one of the principal difficulties of the method is often in the assessment of the value of the payoffs, or losses, entailed by the various actions, since these are frequently partially subjective, or are not easy to quantify precisely. In other words, many 'right' decisions may have been made for the 'wrong' reasons and vice versa. (20,21)

Zero-one integer linear programming with interference. This is a special form of zero-one linear programming where, if certain decisions are made, some activities become impossible. This form was developed (22) to deal with interference on firing ranges where the decision to deploy for a trial on one particular range could affect a trial on a nearby or adjacent range. In the relatively compressed areas allowed for field trials in the UK this effect can be important, especially since some establishments have as many as 40 to 50 principal weapon-firing locations within their perimeter.

Various types of algorithms can be used to simplify the computation time for complex linear programs. The model builder does not normally need to know a great deal about these, but he should be aware of their existence so that he can assess the feasibility of the computational approach. The *decomposition algorithm* of Dantzig and Wolfe (23) involves splitting the problem into separate blocks, considering only certain constrained equations, finding a solution for these and the limits of the feasible zone, introducing constraints, which can modify the 'solution', reformulating the problem in these terms and then finding the final solution. The *principle of intermediate substitute algorithms* can be used to simplify the problem and to create an intermediate problem which is often soluble by other methods.

Fractional or hyperbolic programming is used when the objective function it is desired to maximise, or minimise, takes the form of one set of 'x values' divided by another set of 'x values'. *Parametric programming* can be used as an aid to the solution of fractional problems. A new parameter λ is introduced and this is made equal to the value of the fractional objective function. The numerator is equated to u and the denominator to v and the basic variables are expressed in terms of u and v. The original constraints are then transformed from the basic variables into terms of u and v, and since maximising u/v is the same as minimising v/u (if u is not zero), we can make the slope from the origin of the v/u curve to the transformed constraint zone as small as possible. A variety of other methods of dealing with fractional programming may be used. Charnes and Cooper (24) and Isbell and Marlow (25) have also developed satisfactory approaches.

Non-linear programming can be carried out using linear approximations known as *separable programming* or by using *quadratic programming*. The so-called *geometric programming* can be used to solve certain special classes of non-linear problems. The approach with this method is to find the solution to an associated dual problem which has the advantage that it is simpler to resolve than the principal. The solution to the primal is later deduced from the solution to the dual. (26)

Stochastic programming is sometimes used when the parameters of the problem are described as random variables rather than deterministic quantities. For example, in the 'nutrition type' of linear program we may not know precisely how much the nutrients will cost, or exactly how much nutrient each foodstuff contains. Three different methods are described by Tintner, [27] Dantzig [28] and Charnes and Cooper. [29] Each method has advantages in certain circumstances. The basic idea of all stochastic programming models is to convert the probabilistic nature of the problem into a deterministic form. *Chance-constrained programming* is the approach adopted by Charnes and Cooper to solve this problem. The process leads to non-linear constraints and ultimately necessitates the use of separable programming in order to complete the process.

5 Systematic trial and error models and heuristics

Many models, which rely heavily on executing a great many calculations quickly, have now been developed as a consequence of the enormous power of computers in comparison to all other previous methods of computation. Dynamic programming and branch and bound models are examples of this type of model which fall clearly into the computational mathematical (CM) group.

Dynamic programming

Dynamic programming, or DP, relies on routine, algorithmic calculation. When undertaken by hand, dynamic programming can be very tedious, even with quite small problems, but using a computer takes much of the strain from the method. Nevertheless some problems require so many calculations that even on a fast computer they can take many hours to run. The reader may feel that this method is not very sophisticated in that a very large number of calculations have to be performed. This is, to some extent, true. The calculations are, however, carried out in a systematic manner so that only the minimum number are made and only a fraction of the total possibilities are calculated.

Quite a few rather fancy mathematical symbols are used to express a dynamic program, but the idea is basically very simple and the principles of the deterministic dynamic programming can be grasped quite quickly. Imagine four states or stages A, B, C and D. At each stage it is possible to make certain choices, i.e. in state A we can start from substate, or condition 1, and progress to stage B and substate 1 or 2. Similarly, we can be at substates 1 or 2 at stage C, but at D in this particular example only one state is possible.

Now suppose that we attach a cost of going from one stage to the next for each route and the cost is different depending on which substate is chosen. Suppose that we are concerned with progressing from A to D through stages B and C. As the problem is quite small it is possible to draw a diagram with arrows pointing towards the next substate with the cost shown against the arrow as illustrated in Figure 5.1.

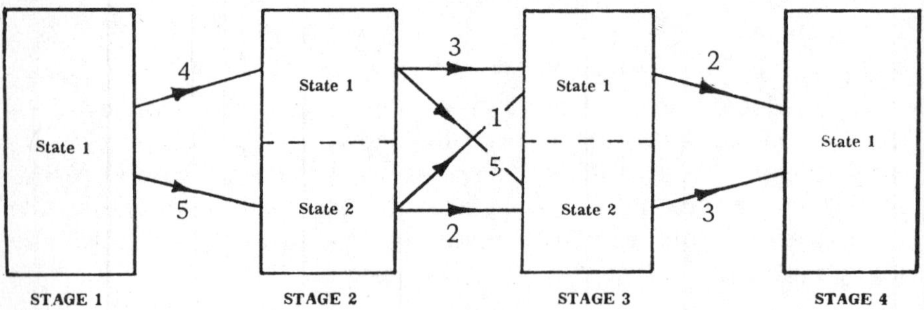

Figure 5.1 All the possible choices available in this simple problem

The problem is so small that the solution is immediately obvious, i.e. the cheapest, or optimal, route is A to B2, to C1, and then to D. The cost is 8 units. Let us consider the general method of solving this kind of problem. It will be seen that there is only one route to B1 and the cost is 4 units.

Thus, least cost of going from A to B1 is 4, and least cost of going from A to B2 is 5.

Least cost of A to C1 is the minimum of either

$$\begin{Bmatrix} 4 + (\text{B1 to C1}) = 7 \\ 5 + (\text{B2 to C1}) = 6^* \end{Bmatrix}$$

i.e. A to B2 to C1

and least cost of A to C2 is the minimum of either

$$\begin{Bmatrix} 4 + (B1 \text{ to } C2) = 9 \\ 5 + (B2 \text{ to } C2) = 7^* \end{Bmatrix}$$

where the asterisk denotes a minimum value.

Least cost of A to D is the minimum of either

$$\begin{Bmatrix} 2 + (\text{Min cost of A to } C1) \\ 3 + (\text{Min cost of A to } C2) \end{Bmatrix}$$

$$= \text{minimum of } \begin{Bmatrix} 2 + 6 = 8^* \\ 3 + 7 = 10 \end{Bmatrix}$$

i.e. A to B2 to C1 to D

= 8.

It will be seen that in the latter stages of the calculation we are no longer concerned with all the values in the earlier stages. For example, in moving to stage D from stage C, we are only concerned with the least cost route to C1 and the least cost route to C2. All other possible routes to C1 and C2 can be dropped from the calculation. Thus the possible route which exists through B1 is not considered at all after reaching the next subsequent stage, since it is clearly not on the least cost route.

This continued selection of minimum saves a great deal of calculation in a very large problem, and is known as the *principle of optimality.*

Another way of looking at DP is to visualise the problem as one of optimal routing. For example, if the shortest distance from London to Manchester lies through Derby 123 miles distant from London, it will not be possible to find a shorter distance to Manchester by going through any other town such as Nottingham which is also 123 miles from London, i.e. the shortest route from London to Manchester contains the shortest route from London to Derby. Basically, therefore, dynamic programming involves the consideration of various possible solutions to a problem in a scientific manner.

It is immaterial in dynamic programming whether one starts from one end or the other, as long as the data are available. For example, if in production scheduling it is definitely known that different quantities of items must be delivered each month and in certain months only a limited number can be produced; also that there is a cost of setting up each production run, and a variable cost of holding the items, how does one decide the best way to schedule the production runs? In this case it is obviously sensible to work backwards from the final delivery date since it is known that a definite number of items must be delivered at that time. The very last month's delivery could probably be manufactured in one go, so in this case there would be no holding costs from the previous months. Moving backward one month from the last, we can determine the profit resulting from manufacturing and selling the items that same month, and compare it with the profit of manufacturing more than the requirement for that month and holding the surplus to the next month. It is thus possible to continue working backwards to the present, continually selecting the maximum profit policy. Some further examples of dynamic programs are given in Appendix 3.

What useful hints can be given with regard to the formulation of DP models? In the first place it is advisable to try to decide the advantages or disadvantages of the 'backward' or 'forward' formulation, i.e. starting from some time in the future and working backwards to the present position, or starting at the present position and working forward into the future. In some production or inventory problems, where it is necessary to finish up with certain items of stock, then the backward formulation is probably a fairly obvious method. Some problems are less easy to decide but common sense is usually a good guide.

It is then necessary to state the known features of the problem and to start to write down the mathematical model. Not everyone is the kind of mathematical genius who can write down the answer to a complex DP problem in a few seconds, but it should be possible to achieve a satisfactory formulation some time in a period between ten minutes and a couple of hours. It is important to make a start on the problem and to write a few things down rather than try to formulate the entire problem mentally. Errors or omissions can easily be rectified later and the model gradually increased in its complexity.

Once a possible formulation has been achieved, check through it carefully to ensure that the principle of optimality has been followed. Try to make some estimate of the size and number of calculations necessary. If the problem has to run 'by hand' on a desk calculator, or desk computer, you may soon learn one of the sad facts of life about the large size of many apparently simple DP formulations, and even on a large digital computer it is usually necessary to check that the problem is soluble within the allotted time schedule.

Branch and bound methods

Another type of programming, known as branch and bound programming, also falls within the CM (computational-mathematical) group of models. This approach involves finding a solution and exploring alternative solutions in such a way that only the solutions most likely to give better answers are considered. Thus initially a possible solution is obtained, even though this is unlikely to be the best one. A series of different alternatives is then examined, each alternative only being continued in the investigation for as long as it is feasible for it to yield a solution better than the best result so far obtained. In any practical problem the number of possibilities are finite and the method enables some possibilities to be ruled out at quite an early stage in the proceedings. This can appreciably reduce the computational task.

An example of a straightforward branch and bound program is given in Appendix 4 together with an example of a zero-one branch and bound program. Careful examination of these examples will ensure that the concept of branching and bounding is well understood. At each iteration new branches are added to the tree in a sensible way so that no promising solutions are bypassed and each new solution yields the next largest upper bound for the value of the objective function among the unexplored branches.

This approach to integer programs (the Land-Doig algorithm [30]) does not enlarge the problem, as for example is the case with the Gomory algorithm [31]. In fact the problem is considered as a series of new problems and a smaller number of variables. The computational procedures are therefore generally more efficient. Another advantage claimed for the branch and bound technique is that it is often possible to obtain a 'good' integer solution long before the time required to obtain the optimum and in most cases to have some knowledge of how far this is likely to be from the true optimum. This can be a considerable advantage in very large problems.

Heuristics

Before leaving computational models it is important to consider some aspects of computation known as heuristics. In this case, by an 'heuristic' approach we usually mean finding a method that works by any practical means. As a rule, the mathematics is too complex to allow a direct, or simple, solution in some types of problems and we have to resort to simple 'common sense' approaches. Heuristics rarely yield an optimum solution, but with the help of a computer they yield generally satisfactory solutions which are not very far from the optimum values. In operational research terms such solutions tend to be known as 'robust' solutions or 'rugged' solutions. There is, within the meaning of the last two expressions, an implication of a certain insensitivity to small changes of the input variables. Problems which have been solved by heuristic methods can have very robust solutions since a large number of changes to the variables are often necessary to arrive at a 'solution', and the variability of the outcome can be observed during the solution process.

'Hill-climbing' techniques can be an example of heuristics. In this method a solution is initially obtained and this is compared with the values obtained when the variables are changed slightly. The process is continued and changes which make the most favourable changes to the objective function are investigated further and so the solution progresses up the steepest road to improvement until an optimum value is obtained. The disadvantage of this approach is that the 'optimum' may prove to be a local optimum, so the method is best applied by seeking a solution at a number of different points, probably over a spaced network of values.

Useful solutions can sometimes be obtained by a mixture of methods. These often begin by carefully observing the procedures used by people currently obtaining the best results. For example, in attempting to produce a suitable algorithm for scheduling the movements of lorries, a number of despatchers were carefully studied. It was noted that some features tended to be common to the best despatchers, such as, say, scheduling those lorries which had the longest journeys before all others. By observation of these features and combining them with the high-speed combinatorial assessment power of the computer, it was possible to produce a scheduling algorithm which gave a 10 to 15 per cent improvement over the average despatcher. These improvements are very significant in terms of fuel and road usage, and the Ministry of Transport has encouraged hauliers to make use of such programs.

This figure of 10 to 15 per cent is fairly typical of the kind of improvement which can be expected when a good heuristic approach is applied with the aid of a computer. There is, however, no guarantee that any solution of merit is going to be obtained, and many experiments may be necessary before a satisfactory algorithm is found.

In the relatively early days of computing there was a great desire to develop a software package for job scheduling in workshops. Many packages were produced. Few were successful, and some were dramatically unsuccessful, including the software of at least one very large computer manufacturer. The UKAEA, however, had in the meantime produced an effective package (WASP) which they had used extensively for their own purposes before marketing it. The method used to develop this package is noteworthy as a good example of heuristics. Although combinatorics also eventually featured heavily in the final solution, the basic ideas were initially obtained by playing with blocks, rather like children's building blocks. Lengths of wood were cut in proportion to the time required for each job and coloured according to the machine on which the job had to be processed. By 'playing around' with these blocks the programmers quickly found a process which tended to load all the machines to capacity and at the same time complete the total tasks as quickly as possible. They were then able to combine this with some combinatorial methods in order to produce a program which found a 'robust solution' fairly quickly.

6 Simulation models and system dynamics

Simulation models are not confined to computers, but these days they are most frequently discussed in this context. In some senses any model is a form of simulation of a situation which exists, or could exist, in the real world. In this chapter some of the methods of model building more normally referred to as 'simulation' are briefly described.

Computer simulation

A simulation model used on a computer belongs to a CA (computational analogue) class of models and is also sometimes described as a heuristic approach. The latter, strictly speaking, is probably not correct. A simulation model, as used on a computer, is a model in which an attempt has been made to describe events realistically on the basis of past knowledge. We can, for example, build a computer model which will show us the probable waiting times for ships queuing to enter a port, provided that we know probable unloading times for different types of load, given certain facilities such as cranes, trucks, etc. The model is basically a CA model and the heuristic approach enters when we try to argue the probable effects of adding another crane for loading and unloading.

Automatically repeated simulation runs may then be made to discover the results over a period of time. This procedure moves the activity towards the CM and heuristic group, although the model itself, without these procedures, stays firmly in the CA category.

Simulation models are often disliked by mathematicians, and frequently this is not without a certain amount of reason. A simulation model with enormous complexity which is only half understood is a dangerous tool. Its behaviour is not only unpredictable but also often found to be apparently irrational in the post-analysis assessments. It is at this stage that disillusionment with simulation models tends to occur. This is unfortunate because such models can be extremely valuable if used properly and the correct procedures adopted in constructing the model.

Computer simulation models can have great value because they can be used when the mathematics is difficult or impossible at the present time, or when experiments on the full scale would be dangerous. They can incorporate a complexity beyond the scope of current mathematics and can be extrapolated to run in situations which on the full scale would involve great cost or danger, and possibly even both. For example, it is possible to simulate the behaviour of nuclear reactors on a computer with considerable accuracy and in complete safety.

The kinds of problems which are handled by simulation methods are often defined as falling into three main categories. These are:

1 Problems with many *interactions*, such as are often experienced in process control, and in which the outcome of one process may affect the outcome of another. This gives rise to the rather general expression *interactive* models for work of this nature.

2 Problems involving *variability* such as the shipping problem outlined earlier in which the shipping arrived at various times in a more or less random manner despite scheduling.

3 Problems which require *complex decisions* also fall into a rather different category. A typical example would be the model of an airport, e.g. the arrival times may be near to the scheduled times, but delays on the ground may introduce a stochastic variable and rather complicated decisions may have to be made about the necessary actions if an aircraft is temporarily unserviceable.

In one sense all these three categories of problem are similar, since they all envisage a multiplicity of possibilities and it is usually the likelihood of a variety of events happening which is the objective of the investigation.

Top-hat or Monte Carlo methods

Let us return briefly to the example of shipping arrivals at a port, and consider how this kind of problem was originally studied using a method known as 'top-hat' simulation. It will then be seen how this method subsequently came to be used with computers.

Suppose that a dock has a number of berths and unloading facilities available for shipping. It is found that despite theoretical scheduling, as a result of weather, tides, etc., the ships arrive in a completely random fashion. It is also found from experience that the unloading times are distributed in a known manner. For example, with every 60 arrivals it is found that 20 take half a day to unload; 20 take one day; 10 take two days; 8 take three days; and 2 take one week. Similar figures are also available for the loading times.

Now supposing that it is known that the amount of shipping using the docks is likely to increase by 20 per cent over the next three years, and management wish to know to what extent the dock facilities should be extended if there is a penalty imposed on them for keeping shipping waiting for longer than say a period of six hours.

At first sight we may be inclined to conclude that this is a rather 'messy' problem with a lot of complicating features, and a solution may be rather difficult. In practice it can be solved very simply by merely 'trying out' the system. Let us deal with the events as they might occur in real life. In the first place we will assume that a ship arrives at the docks and is unloaded. This may take any of a variety of times to complete. However, we probably have data on unloading times in the past and all that is necessary is to sample one of the unloading times at random.

Thus we would simply take 60 cards and write half a day on 20 of them, 1 day on another 20 cards, 2 days on 10 cards, 3 days on 8 cards and 7 days on a further 2 cards. The cards are then placed in a hat and one card drawn at random. Hence the expression 'top-hat' simulation. ('Monte Carlo' simulation is another name which is used for this kind of approach.) The unloading time listed on the card may then be written down and the card returned to the hat. The loading time can be established in a similar way by having another hat containing cards giving probable loading times and drawing one of these at random in order to establish a typical time.

The periods between ship arrivals can also be observed from past data. The arrival intervals can then be decreased by 20 per cent, since the arrival rate is expected to increase, and by writing these times on cards and placing them in yet another hat, it is possible to draw times at random which will be typical of the new situation.

By selecting cards from the arrival times, the berths can then be used up until the docks are fully occupied and the facilities totally absorbed. A queue of waiting ships will then start to form, and the time can be advanced until either the next ship arrives or a facility is released by the departure of one of the vessels. This can be determined quite simply by keeping a calendar and by comparing the sum of two randomly selected loading and unloading times with the time of arrival of the next ship. It will then be clear whether a ship will arrive before further resources and dock facilities are released.

The process is continued for a period of time equal to six months, or even possibly several years, so that it is possible to be certain that a fairly stable situation exists and that the true average length of the queues have been determined. The dock facilities can then be changed. For example, it may be decided to add extra equipment such as cranes, etc. or extra berthing facilities. The simulation can then be re-run and the effects of the changes noted. In this way it is possible to obtain a realistic decision as to whether or not the additional expenditure will be worthwhile.

The answers to the problem are usually not immediately obvious after only a few ship arrivals, and some systems may have to be run on paper for the equivalent of five, ten, or fifty years in order to be sure that no serious problems arise over the period. This really depends on the nature of the problem and the seriousness of any difficulties which may occur.

In this particular instance the problem is relatively small and the simulation can probably be carried out by hand, although the process may be rather tedious. In cases where the problem is larger it can usually be handled on a computer without too much difficulty. Most computers have facilities for sampling from distributions and this facility is the principal requirement of any simulation. The languages available, although important, are usually a secondary consideration.

Problems in computer simulations

One of the difficulties with simulation models is the task of ensuring that there are no mistakes in the program. Some errors are easy to detect, but the ease of detection often depends on the plausibility of the events. Thus in an East–West battle model we are unlikely to end up with 200 British cruisers docked in several enemy ports. In the first place we know that we are unlikely to have 200 cruisers and second they would not be docked in enemy ports.

The problems arise in cases where the errors are not obvious. To some extent this can be tested by putting special trace features into the model which enable a particular input to be traced through the various stages. It also sometimes helps if the simulation is gradually built up in steps so that the effects of added variables can be separately assessed. Often this kind of assessment check has to be qualitative rather than quantitative, but it can prevent gross errors.

It is usually necessary to ensure that there are no 'end' effects due to starting or finishing the simulation. The estimation of the time required to remove end effects, before monitoring for data record purposes, is not easy from the theoretical aspect, but is usually not too difficult to determine from a practical examination of the results. In the majority of simulation problems it is desirable to avoid sampling from the very extreme ends of the distributions, so that in very large models which have many distributions, extreme values are often eliminated in order to obtain a reasonably accurate result within a reasonable computer run-time.

The use of mathematical, rather than electronic, methods of generating random numbers means that a series of random numbers can be easily regenerated; and so the storage of the series of numbers used, when comparative runs are required, is not necessary. This means that distributions can be studied to ensure that abnormal results are not obtained with relatively small samples. The disadvantage of this approach is that most mathematical methods tend to fail eventually when the sample of random numbers required is extremely large. There is, however, little doubt that mathematical random number generating functions are often very valuable during the early stages of 'debugging' simulation programs, even if they are not used for the eventual output runs.

A checklist for CA modelling

The following are considered essential for good simulation or CA modelling:

1 A good knowledge of the system states (usually known as the state of the world).
2 A careful formulation of the important factors for use in the model.
3 Isolation of events when changes of states occur.
4 A knowledge of changes of states when events occur.
5 Methods of time advance.
6 Tests and evaluations on the simulated system.

It is worth spending a fair amount of time deciding the minimum number of variables required for a reasonable description of the 'state of the world'. Furthermore, care should be taken with regard to the best methods of time advance, and whether the event or activity approach is the most suitable arrangement. Some assumptions may have to be made about events as possibly the required information may not be available. Where these data are important every attempt should be made to obtain them so that the model will show realistic reactions to events. The methods of time advance need to be considered carefully before deciding whether to use the event- or activity-based approach. A few notes showing how a model was developed (32) to simulate the life and death of business organisations is given in Appendix 5 and some notes on computer simulation languages are given in Appendix 6.

Simulation models can sometimes be helpful in rather unexpected ways. Attempts are often made to solve stochastic problems by deterministic means, often simply because the stochastic nature of the problem is not appreciated. As a consequence no 'realistic' solution emerges despite the effort expended.

Take, for example, the case of defence planning. We can decide to program this problem in a variety of ways. If we believe that we can estimate the 'effectiveness' of various types of weapon or force mixes, then we might choose to use some form of integer linear program, a dynamic program, or branch and bound methods. But do these approaches carry the realism of our normal state of the world? It is possible that in many situations they do not. For example, the word 'effectiveness' is a strange term. How do we measure it? Effective against what? How does effectiveness change if the enemy introduces a new weapon?

In fact we might well be advised to express 'effectiveness' in probabilistic terms, even if these are quite simple. For example, we could talk about a weapon having an effectiveness of say 10 units with a probability of 0.5, and of 3 units with a probability of 0.25 (in the event, perhaps, of the enemy producing a new weapon, or certain other, only partially foreseen, circumstances).

Indeed, the probabilistic elements in planning defence expenditure are quite large. In the first place there is the question as to whether any ideas or a new design will be successful. Research and development costs and time schedules are notorious for their unexpected variability. Second, there is the question of factory implementation of the design, and although this is much more deterministic, there is still an appreciable variability in this part of the problem. Finally, there are the delays due to production failures and implementation of the weapon in service.

These are the arguments the simulation expert would put forward for constructing a model which incorporates probabilistic features. Thus for defence cost planning we would sample from a distribution of delays found in typical research and development programmes, and from a distribution of delays for factory implementation and for service implementation. This should more clearly indicate the likelihood of any long-term forecast being achieved. Such a model might well incorporate some deterministic mathematical programming features, or may more effectively use a form of mathematical programming mentioned earlier known as chance-constrained integer linear programming, but this requirement should emerge naturally from the needs of the simulation model rather than as a rather artificial paper study, however technically excellent this may be.

Another group of models which can often fall into the computational analogue category are models employing what is frequently known as the 'system dynamics' approach. In this method the problem is first visualised in terms of a control theory network. The network is sometimes described in an analogous manner to an electrical circuit, when electrical control theory may be applied; but it is also often expressed in terms of hydraulic flow by using a series of valves and reservoirs to describe different parts of the system.

Once the problem has been satisfactorily formulated in this way it may be immediately soluble using pencil and paper and normal analytic methods. More usually, however, the problem is sufficiently complex to require the use of either a digital or an analogue computer. Analogue computers are rarely used these days as for most problems they are not as accurate as digital computers. They consisted of a series of small electrical generators and potentiometers which controlled the current flow through an electrical circuit network. Measurements of voltage levels and current flow were taken under various conditions in order to determine certain quantities in the original problem, such as, for example, the effect on stock levels given a sudden change in purchasing policy.

The rapid growth in digital computers has meant that system dynamics problems are now largely tackled only on digital computers using special software packages such as Dynamo, Dysmap and Microdyn. The technique is briefly outlined in Appendix 7.

Perhaps one of the principal weaknesses in system dynamics is the difficulty in finding the real world values for the coefficients in the rate equations which control the flow in the circuits. After all, one does not go along to a storeman and say to him 'What is your dy/dt please?'. You say to him 'When do you decide to reorder?'.

It is possible to overcome some of these difficulties by using 'spreadsheet' software for problems of the system dynamic type. If a conceptual diagram of the model is redrawn in a time-sequenced manner, as described in Appendix 8, it will be found that many solutions can be formulated very quickly. Moreover, with this technique it is possible to use the actual decision rules employed by people in real life, such as, for example, storemen, heads of department, etc. This method provides a powerful tool for solving problems quickly, and a number of examples using the software package 'Visicalc' are included in the appendix.

Simulation using research games

These games are used to study the response of individuals to different situations. Basically, the idea is to use real people and carefully instruct them in a typical situation. New information is gradually fed to the participants in the game and their behaviour and interactions in this situation are observed. Some games (see Cooper [33]) are heavily confined and the freedom of choice is strictly limited, whilst with others the outcome is uncertain and depends upon the interactions–which may be many and complex.

Complexity of interactions often means that many outcomes are possible and this is one of the difficulties with research games. What we can learn, given a certain situation about which a relatively few simple decisions have to be made, is approximately how many people will react in one way and how many in another. We can even subject people of different nationalities to these tests in order to ascertain whether they will respond in a fundamentally different manner.

Simple two-player games have been used to determine the response pattern of people to gaming situations. Thus some players are found to maximise their own gain, others the relative differential between their own pay-off and that of their competitors, whilst a number were also found who tended to maximise the communal good. [34,35] It is difficult to know exactly how representative these studies are of real life situations. In the first place the rewards are usually relatively small and not particularly significant; and, second, reactions under social observation may be different from behaviour where opponents are not drawn from the same group of students and anonymity may be preserved more easily.

Highly sophisticated gaming studies are carried out in military operational research establishments in most major countries in order to develop military tactics. The method used is quite simple in that the rules of the 'game' are first carefully drawn up and the armies of the two sides deployed on maps set on large tables. The opposing field commanders sit in different rooms with various aides and supporting officers as necessary for realistic decision making. Some games have even included sound effects for realism. Each commander has a map showing the deployment of his own forces, and such information about the enemy forces as he has been able to glean. An umpire then notes the various moves of the field commanders and informs them when they are in contact. Then, on the basis of the forces they have deployed and past statistics which indicate the outcomes of similar situations, the umpire and assessors inform the commanders of the outcomes of the different skirmishes and battles as they arise.

It was found that one of the disadvantages of this kind of gaming simulation study was the length of time it took to work through a simple three-day battle. Indeed the considerable effort involved meant that, originally, only a single simulation of a particular battle situation could often be carried out. Nowadays, it is possible to conduct this kind of simulation experiment much more quickly with the help of a computer, even where the direction of the game itself is not fully automated. The calculation of the thousands of movements involved during the play can be carried out very rapidly by means of a computer. This can in turn improve the realism of the game by allowing a larger command structure with the greater number of minor decisions.

Some research games are used for business studies. Business games such as Lansus [36] are designed to study commercial situations. In this particular game the student, or trainee executive, forms a member of a team of senior officers responsible for running a company producing several different products. The trainee is assigned a role such as sales manager, buyer, production manager, etc. and has to make some of the kind of decisions that might have to be considered in real life. For example, the buyer has to bid for raw materials in the open market at auctions and the company officials, as a group, have to decide just how much of the various types of products should be produced and to decide the appropriate prices.

It seems probable that these games provide useful training although no-one will suggest that a player is likely to become a potential company magnate overnight. What is certain is that the games are played with great intensity and that participants become very involved. Indeed I have seen 'buyers' walking up and down the corridors afraid to go back to report to their company the prices they have had to pay for raw materials!

I am also aware of a simple First World War simulation played in a summer school at the Open University where, despite the variation in outcomes due to a wide range of possibilities, the players cast in the various roles tended to respond in a behavioural sense in much the same way as the leaders of the different countries who were concerned with the outbreak of the war, had done. The stress on the man cast in the role of the Kaiser (known by the British at the time as the 'Mad Dog of Europe'), for example, was such that he eventually walked out of the game, never to return to it.

The majority of these kinds of research games using people are in the PP (pure physical) or IA (interpretive analogue) category. They can often have a useful part to play if properly conducted. One of the principal difficulties is that unlikely results are not believed when they are obtained. In several cases where real political situations have been 'gamed out', the results have been discarded because of their improbability, only for the players to find that these events have, indeed, come to pass.

Another serious danger must also be mentioned. In certain cases, whilst abroad, political officials have been subjected to games with their opponents in order to gain an insight into another country's point of view. There is a disturbing correlation between the number of times these officials have returned to their own country and been executed for expressing views not in line with those of their colleagues.

One thing seems fairly certain, however, and that is that research games can at least broaden our understanding of a particular situation, and at best provide a surprising insight into some types of problem. An example of the results of a game designed to investigate the conduct of negotiations is given briefly in Appendix 9. The outcome of this particular game was rather sad in that it provides a vivid illustration of how aggression and bitterness could have been avoided if both sides had been completely honest from the start of negotiations.

The physical as well as the mental response of an individual can be studied when the environment is also simulated. For example, the so-called 'battle courses' which involve the use of explosives and firing bullets close to the trainee are designed to provide experience of battle conditions. Stress interviews, which involve near torture methods, such as loss of sleep, incessant interrogation, etc. have been used in the last war to train overseas agents who were liable to capture by the Gestapo. The whole cabin of an aircraft, complete with controls, may be built in some aircraft flight simulators used to train aircrews. Stalling conditions may occur in complete safety on the simulator and the aircrews not only save valuable air training time but can undertake manoeuvres with the simulator which are not possible in practice. This kind of environmental simulation therefore often provides experience cheaply, quickly and with virtually no danger.

Having briefly discussed research games it may be useful next to mention a little about the theoretical aspects of conflict 'games'. There is some doubt whether the methods used have ever fulfilled the original expectations, and quite often analogous situations in real life are perceived only after the event.

7 Models using game theory

A standard mathematical and operational research method of dealing with some aspects of conflict situations is by the use of 'game theory'. In practice game theory, in its present state, has proved to be of limited use both to industry and to the armed services, although it has given some insight into behaviour patterns and also some explanation of global actions.

Game theoretical models

One classical game theory situation is referred to as the 'prisoner's dilemma'. However, such dilemma situations, when they do occur in real life, are often not quite so readily soluble as textbook methods imply. In real life there are many variables which can directly, and indirectly, influence decisions to a very considerable extent.

Let us briefly examine the way in which game theory deals with certain conflicts so that the limitations of the method will be clear.

In many mathematical game models of conflict the contestants are regarded as having exactly similar information. This is a very different condition to card games such as poker or bridge, where a player can only hazard a guess as to the precise cards his opponent is holding. Games in which all contestants have exactly similar information, in which all possible strategies and outcomes are known to both competitors in advance, are sometimes described as 'games of perfect information'. Note in passing that some purist experts would tend to disagree with this definition.

The effects, or pay-offs as they are usually called, resulting from the various possible strategies (or moves) by the competitors are normally displayed in the form of a matrix, as in Figure 7.1.

		Competitor No 2	
		Strategy 1	Strategy 2
Competitor No 1	Strategy 1	0	20
	Strategy 2	30	10

Figure 7.1 A typical game matrix showing the pay-offs and choices available to the competitors

In the diagram if competitor no. 1 plays strategy 2, and competitor no. 2 simultaneously adopts strategy 1, then the gain to competitor no. 1 is 30 units. On the other hand, if competitor no. 2 decides on strategy 2 the first competitor gains only 10 units.

Some games are described as 'zero-sum' games and in practice it will be found that not all strategies need to be considered. It is probably easiest to indicate what is meant by a dominant strategy, or by a zero-sum game, by means of an example.

Two semi-hostile (cold war) countries each have fairly similar weapon systems, both of which are roughly matched for efficiency and effectiveness. Both countries could decide to develop new and more effective replacement systems, and may be able to recoup a considerable part of the cost from overseas sales of the new system. In addition they may gain in prestige and this could prove to their material advantage in the future.

Development programmes take a long time and if one side decides secretly to develop a new system and the other does not, then the new system will take a larger share of the defence market and the developer will acquire appreciable prestige and advantage. If both sides decide to undertake the development, or both decide to do nothing, the situation will remain the same with regard to defence sales market share and also prestige. The pay-off matrix, with all values in millions of pounds, is as shown in Figure 7.2.

		Country No 2: Undertake new devel. programme	Country No 2: Do nothing
Country No 1	Undertake new devel. programme	0	+ 10
	Do nothing	– 10	0

Figure 7.2 An example of a 'zero–sum' game

The values in the matrix in Figure 7.2 represent the pay-off or return the first country receives as a result of various strategies. The pay-off to the second country is obviously exactly the same but the positive and negative signs are reversed.

The best pay-off the first country can obtain is by adopting the first strategy and undertaking the new development programme. Moreover, since the pay-off strategies are greater or equal no matter what strategy the second country adopts, this strategy is known as a *dominant strategy.* Hence, by classical game theory the country should always adopt this strategy.

Similarly, the second country should also undertake a development programme in order to reduce the pay-off to the first country as much as possible. It will be noted that the sum of the pay-offs to both countries adds up to zero. Such games, where one side loses as much as the other side gains, are known as *zero-sum games* and are relatively few in real life. Indeed in this particular situation, when all the special advantages and disadvantages to each country are carefully assessed, it is very doubtful whether such a game would prove to be 'zero-sum'.

In this example we assumed that the intangible asset 'prestige' could be suitably transformed into a tangible asset—cash. In practice this can often be very difficult indeed, and White [37] has dealt at some length with the matter of ordering choices. Generally, it is an easier matter to arrange things as a series of choices. Thus we have: first choice, second choice, etc. Moreover, once a series of such 'choices' can be expressed in relationship to each other, it often becomes more easy to transcribe a particular choice into monetary terms, since at certain points in the scale there may be known cash equivalents. This means that offering fictitious choices can be sensible where monetary equivalent values are known in order to assist in the scale conversion.

The estimated monetary equivalent of an item such as 'prestige' is likely to vary quite considerably over the years. It depends upon a conglomeration of attitudes present at a particular time and on the economic resources of a country. A country will tend to respond to monetary equivalents in much the same way as it is suggested in utility theory (see Appendix 10) that an individual will partake in lotteries, and this clearly depends on the financial situation of the person at a particular time. Thus the value a country attaches to items such as 'prestige' is likely to change with time.

Already we can see some of the problems involved in using conventional game theory in real conflict situations and how conventional theory presents a very much oversimplified view of real life problems.

Let us look rather briefly a little further into some of the basic theory of games.

Suppose we have an $n \times m$ game matrix for two competitors, *A* and *B*, with the pay-offs denoted by a_{11} a_{12} a_{13} etc., as in Figure 7.3.

In matrix algebra the letter i is generally used to denote the row and j the column, so that competitor *A*'s strategies are given by the i label and competitor *B*'s strategies are indicated by the letter *j*. The pay-offs for each strategy are indicated by the a_{ij} values.

Competitor B

	Strategy	1	2	3	.....	n
	1	a_{11}	a_{12}	a_{13}	.....	a_{1n}
	2	a_{21}	a_{22}	a_{23}	.....	a_{2n}
Competitor A	3	a_{31}	a_{32}	a_{33}	.....	a_{2n}
	..	..	..	..	.....	..
	..	..	..	..	.....	..
	..	..	..	..	.....	..
	m	a_{m1}	a_{m2}	a_{m3}	.....	a_{mn}

Figure 7.3 Pay-off matrix for *n* X *m* game

Whichever strategy competitor *A* selects, provided that competitor *B* behaves sensibly, *B* will always try to minimise *A*'s gain (i.e. the a_{ij} value). Hence if we consider the ith row; and *B* guesses that *A* will play this i^{th} strategy, then *B* will respond with a strategy which selects the smallest a_{ij} value for this row. Let us call this α_i, i.e.

$$\alpha_i = \min_j a_{ij}$$

Thus at the end of every row in the matrix we can write an α value, which is the minimum pay-off for the row (Figure 7.4).

Consequently, if competitor *A* wishes to play with great caution, he will choose the strategy with the highest α value, since no matter what move *B* may make he is assured of α. This is expressed as follows:

$$\alpha = \max_i \alpha_i$$

and since

$$\alpha_i = \min_j a_{ij}$$

$$\alpha = \max_i \min_j a_{ij}$$

Competitor B

Competitor A

Strategy	1	2	3	.....	n	i
1	a_{11}	a_{12}	a_{13}	.....	a_{1n}	α_1
2	a_{21}	a_{22}	a_{23}	.....	a_{2n}	α_2
3	a_{31}	a_{32}	a_{33}	.....	a_{3n}	α_3
..	...	...	...	.....	...	..
..	...	...	...	.....	...	..
m	a_{m1}	a_{m2}	a_{m3}	.....	a_{mn}	α_m
β_j	β_1	β_2	β_3	.....	β_m	

Figure 7.4 The α values represent row minima and the β values column maxima

This is called the *maximin strategy*, and α represents the *lower value of the game*, or the *security level* for competitor *A*.

Similarly, competitor *B* is interested in minimising the pay-off to *A*, and can find the maximum pay-off to *A* for each of his strategies β_j, i.e.

$$\beta_j = \max_i a_{ij}$$

Obviously if competitor *B* finds the least β_j value and plays this strategy, it is the worst that *A* can inflict on *B*. The smallest β value is therefore:

$$\beta = \min_j \beta_j = \min_j \max_i a_{ij}$$

Since β represents the worst that can happen to competitor *B*, provided he plays this strategy, it is known as the *upper limit* of the game, or the *minimax strategy*.

When the upper and lower values of the game are equal (i.e. $\alpha = \beta$) the game is said to have a *saddle point*. In other words the smallest number in the row is also the largest number in the column.

Many games do not have a saddle point with the result that if the game is played a number of times it may be necessary to vary the strategy in order to maximise the average pay-off. The various strategies are then employed in a random manner but with definite frequency. This is described as a *mixed strategy*.

A typical mixed strategy game is as follows. The Blue army has two important installations and can successfully defend either of them, but not both. One of the installations is twice as valuable as the other and the Red army has only sufficient forces to attack one of them. What should the strategies be for the two armies?

If we suppose that the value of the least important installation is 1, then the value of the other installation is 2. If both installations survive the pay-off to Blue is 3 and if the lesser survives the pay-off is 1.

Defining the defence or attack of the lesser installation as strategy no. 1 gives the pay-off matrix in Figure 7.5.

		Red 1	Red 2
Blue	1	3	1
	2	2	3

Figure 7.5 Pay-off matrix given two installations and ability to defend only one

For this kind of twin strategy game it is possible to find the optimum strategy by subtracting the numbers in the second row from the numbers in the first. Thus for Red we have:

1 and -2

Similarly for Blue we subtract the second column from the first to obtain:

2
-1

This gives the 'odds' or the frequency with which the strategies should be played in the reverse sequence (see Figure 7.6).

Hence the Blue army should defend the more valuable installation twice as often as the less valuable installation, and the Red army should attack the lesser installation twice as frequently as the large one.

		Red 1	Red 2	Blue Odds
Blue	1	3	1	1
Blue	2	2	3	2
Red Odds		2	1	

Figure 7.6 Estimating the 'odds' indicates the frequency of installation defence

There are many ostensibly similar instances of this problem, e.g. whether to locate a megaton weapon in a bomber or a fighter bomber when the two are flying together, and the bomber is more defensible than the fighter; or the question facing a security courier company as to whether it should hide the jewels in the armoured car or the more lightly protected escort vehicle, etc. These are all similar problems and can have various other complexities.

8 The use of statistics

Statistics is playing an increasing part in model building and model testing. Sensibly applied, statistics has great value; but, incorrectly applied, it can cause much damage, partly because it gives a strong impression of authenticity.

Random difficulties

Only too frequently these days statistics is used in a kind of 'blanket' manner. We pour into the computer a mass of variables and eventually a statistical software package produces correlations between a number of these variables. What is not realised, although many users of statistical packages may claim to be aware of it, is the relatively high probability of certain random correlations.

Consider the following series of 0's and 1's which was generated without thought by simply tossing a penny. There was a 50 per cent chance of either a head or a tail, this being denoted by a 0 or a 1:

010110101110000010111010001101010100101011001 0011

Now at certain times of the year in the UK there is approximately a 50 per cent chance of fine weather and 50 per cent chance of rain. If we call the 0 a fine day and 1 a wet day just such a series could have been generated by the weather itself.

Let our imagination take us a step further and let us also suppose that we are, indeed, visitors from another planet who have been sent to investigate the weather on the planet Earth! We arrive secretly in the UK at just the time when the above weather series is taking place. We have no time to work out translations of the Earth languages, but having the ability to deflect light we possess powers of invisibility and sneak into an office and find a calendar. We rapidly decide that there is a certain pattern about Earth men which is connected with the calendar. We carefully note this preoccupation with the calendar but we are primarily asked to investigate the weather on planet Earth and to find out any immediately obvious deterministic patterns.

As explained earlier, the data series generated was obtained in an entirely random manner, and is typical of what might happen given a 50 per cent chance of rain. However, as we have learned by discovering the calendar, Earth men seem to think and work in terms of weeks, so let us now arrange our data in this way. Thus we have:

0101101
0111000
0010111
0100011
0101010
0101011
0010011

Our excitement increases. Clearly it does not rain on the first day of the week. We get out our statistical package and become enormously excited. The chance of this happening on one day of the week is only once in 128 times! So this is why Earth people do their washing on a Monday–because it is always fine. But look; the significance is even greater if we consider only Mondays so we must multiply by seven since there are seven days in the week, so the chance of all this happening on a Monday is only once in 896! Unquestionably, this business of washing on a Monday is due to the weather. This also raises some interesting speculations back in the space ship as we depart as to whether these primitive Earth people are able to control their climate. If only we had stayed a little longer we would have discovered it rained for the whole of the next week! (This happened when, after writing the foregoing, the penny was tossed for another seven times.)

Here then is the problem: ingenuity in looking for answers in areas where no answer is to be found can cause the data to be analysed in various ways until a correlation with something is found. If the fine days had occurred on Saturdays it would have been possible to hypothesise that this was why the Earth people had holidays and played sport on Saturdays.

If we look at a sufficiently large set of data we can usually find one variable which will correlate with another. Thus it has been found that over certain periods the birth rate correlates with the migrating habits of storks, and even that the numbers of pregnant unmarried females correlates with the numbers of clergy ordained.

One data set which was frequently quoted was the series 1860, 1880, 1900, 1920, 1940, 1960 from which the next value, 1980, could be predicted. One was then asked would you accept the US Presidency if it was offered to you in 1980 since these are, in fact, years in which Presidents who have died in office were elected. I am happy to report that, despite one or two near misses, President Ronald Reagan managed to survive.

It may be thought that this point is being emphasised rather heavily, but it is astonishing how very many bad models have been created which have relied on trying to make data fit a particular model of the world and then, on finding difficulties, all sorts of additional variables have been shovelled into the computer in the hope that something will correlate with the objective function, or output, which it is intended to predict.

Statistical models

Having sounded these cautionary notes let us look at some of the statistical methods available. One of the simplest forms of model is the case where experimental data are available on the relationship between two variables x and y. By using a statistical approach a mathematical curve can be fitted through the points which satisfies the requirements that the sum of the square of the distances of all the points from the line is a minimum. This is known as the 'least squares' method of curve fitting. One of the biggest temptations, and hence continual sources of error, with this method is the desire to produce a 'perfect' curve fit by using high order polynomials. Great care should, therefore, be taken not to use polynomials of a higher order than is justified by the data. Indeed, if anything, one should try to work with an order lower than that which might be indicated on the basis of the known experimental accuracies.

The accuracy of the regression coefficient may be important and some assurance, or measure of confidence, of the accuracy of the data may be required. This can be established by using what in statistics is known as the t-distribution.

Only a limited number of problems can be modelled using two variables x and y. Many problems require the use of numerous variables whose significance may be different in every case. For example, in advertising problems, it is unlikely that the return will be the same for all media, so that newspapers, television, billboards, etc. will each have a different effectiveness and value. Even between newspapers there is likely to be a different reponse per unit advertising cost if we try selling Rolls Royce motor cars in 'Top People Newspapers' compared with, say, 'The Daily Excitement'. Thus if y represents the total return, x_1 is the cost of advertising in 'Top People Newspapers' and x_2 the cost of advertising in 'The Daily Excitement' we might find some formula for a particular 'spend' such as:

$$y = 2.5x_1 + 0.6x_2$$

If the 'spend' is increased then the return per unit expenditure may initially increase in an almost linear manner over a very small range, and it is quite easy to visualise the problem being similar to that outlined for a single variable linear regression except that in this case the values represent points in three-dimensional space.

It is not possible to visualise the physical relationship of four or more variables, but the same problems arise as with the linear regression for two variables. The regression method when more than one variable is involved is known as *multiple* regression and although it is moderately complex it can be handled fairly readily by a competent statistician using determinants and matrices, or by a novice who knows how to apply the appropriate packaged program and data to a digital computer. Despite the dangers inherent in the application of statistical computer programs to raw data by non-statisticians, there is little doubt that for routine analysis these software packages remove a great deal of the routine hard work from statistics.

In many cases the x variables may not be related to y in a linear manner and for these situations polynomial, or non-linear *multiple regression*, is possible.

The techniques of multivariate analysis have come to be employed to a much greater extent in recent years due to the increasing use of computers which can save a great deal of time and drudgery in executing the necessary calculations. The principal methods are *component analysis, factor analysis, cannonical correlation analysis, discriminant analysis* and *cluster analysis.*

These methods require a good knowledge of matrix algebra. Green and Carroll (38) have provided an excellent mathematical introduction to multivariate analysis.

Consider the case where there is a large number of variables and n sets of observations. They can be arranged in matrix form thus:

$$
\begin{array}{ccccc}
x_{11} & x_{12} & x_{13} & \cdots\cdots & x_{1n} \\
x_{21} & x_{22} & x_{23} & \cdots\cdots & x_{2n} \\
\vdots & \vdots & \vdots & & \\
\vdots & \vdots & \vdots & & \\
x_{p1} & x_{p2} & x_{p3} & \cdots\cdots & x_{pn}
\end{array}
$$

As an example Kendall (39) considers the case where the yields of 10 different crops were measured for 48 different counties. Hence, in this case n is 48 and p is 10. In *component analysis* the basic objective is to reduce the number of variates and still provide a good model. The method is to obtain a transformation such that some new, and relatively few, variates T which can be expressed in terms of certain of the x variables and which account for a large part of the variation. This is achieved by matrix algebra and an analysis of the correlation matrix of the variables.

In the example just quoted the objective was to have only one variate which could be regarded as an index of productivity. Kendall found a suitable index in this case to be:

$$I = .39x_1 + .37x_2 + .39x_3 + .27x_4 + .22x_5 + .30x_6$$
$$+ .32x_7 + .26x_8 + .35x_9 + .34x_{10}$$

In component analysis therefore we start from the data and see what emerges. In a sense we let the data see what variables they require in order to model the situation. A special test known as Bartlett's test, which uses a statistical distribution known as χ^2 can be employed to determine which variables are significant. Unfortunately this test sometimes tends to allow too many variables and the suggestions by Jolliffe may be rather more useful in reaching a decision in this matter.(40)

In *factor analysis*, on the other hand, a structure is proposed beforehand and the purpose of the analysis is to see whether the data fit this structure, and to use the data in order to estimate the parameters. In this case, we again have p variables $x_1 x_2 x_3 \ldots\ldots x_p$ which are to be related to say a number of underlying variables or factors $T_1 T_2 T_3 \ldots\ldots T_m$, where T_j are $N(0,1)$. Thus

$$x_j = \sum_{k1}^{m} l_{jk} T_k + E_j$$

where E_j is the error term, and l_{jk} are the coefficients of the principal components.

Some of the l_{jk} values are zero and having found the factors T_1 T_2 etc. the matrix can be transformed and rotated in order to make other l_{jk} zero.

Holzinger and Harman in 1941 (41) used data obtained by Gosnell and Schmidt (42) to determine two factors affecting political allegiance to the USA. In the 1934

Presidential elections, Chicago was divided into 147 voting areas and 17 variates were studied. Eight of these were selected by Holzinger and Harman as follows (i.e. $n = 147$, $p = 8$):

1 Percentage Democratic and Republican vote for Democratic Candidate, Lewis.
2 Percentage Democratic and Republican vote for Democratic Candidate, Roosevelt.
3 Percentage of straight party votes to total votes.
4 Median rental.
5 Percentage of families owning own homes.
6 Percentage of gainfully employed workers over 9 years of age in 1931.
7 Percentage of families living more than one year at present address.
8 Percentage of population over 17 who have completed at least 10 grades at school.

Using two factors Holzinger and Harman found that the first factor contributed about 62 per cent of the variance and the second factor about 14 per cent. The first factor depended strongly on variates 1, 2, 3 and 6, listed above, and negatively on 4, 7 and 8. The second factor depended positively on 5 (home ownership) and negatively on variates 2 and 7. Hence a home permanency factor appeared to be involved.

There are many other examples where factor analysis has been used. Rummell, in 1970, [43] carried out a comparison of 82 countries based on some 230 different characteristics. He wished to discover the underlying factors accounting for trade levels, memberships of international organisations, foreign conflict, etc.

The interpretation of the physical meaning of the factors can be difficult and although many studies have been undertaken, relatively few appear to provide factors capable of physical interpretation.

Cannonical correlation analysis is concerned with groups of variables, and the estimation of whether there is any connection between the groups. In this method the data are transformed into new sets of variates so that any variable in one set is uncorrelated with the variables in the same set. All correlations between new variables in one group and the new variables in the other group are zero, apart from limited correlations directly between the first of the variables in each new set and then between the second in each new set and so on.

All this possibly sounds relatively complicated and the reader may be relieved to learn that software packages are available which enable the process to be carried out relatively easily by using a digital computer. Hotelling's T-test is also used in *dispersion analysis*. This test is a generalisation of Student's t-test for sample n from the multinomial distribution.

Discriminant analysis is concerned with the analysis of data which divide into two groups, but in which there is usually an area of overlap so that there may be doubt as to which of the groups some future data shall be ascribed. This problem can be studied by using likelihood ratio tests and the probable chance of misclassification can also be estimated.

Cluster analysis is normally used to obtain a measure of the closeness between variables. One method uses the correlation r_{ij} between the variables x_i and x_j. The matrix of correlations is then examined and if say r_{23} is large, variables 2 and 3 form a cluster. Then the next highest correlation coefficient is obtained and if it involves a variable in an existing cluster then the variable joins that cluster, otherwise a new cluster is formed and so on. *Dendograms* may be used for this purpose. Other methods have also been developed by Kendall and Beale. (44)

A dendogram provides a method of eliminating redundant variables. The correlations between variables and the principal components are usually translated into a matrix. Then the first pair of variables with the highest correlation is taken, and these are usually called 1 and 2. Let us assume that they have a correlation coefficient of .95. They can then be drawn in the 'family tree' form as shown in Figure 8.1. The next highest correlated variable is then found and we will suppose this is variable 3 and it is correlated with variable 2 at the .87 level. Join this on to the tree diagram at the .87 level.

Now suppose that in descending of order of correlation we find that the variables 4 and 5 are correlated at the .85 level and these are drawn separately. In the next stage it is found that variables 3 and 6 are correlated at the .76 level. All the other variables exhibit correlations of less than .7 and, since it is advisable to cut off at about this level, only variables 1 to 6 are included in the analysis. The remaining lesser variables are then associated with the highest ones as explained earlier.

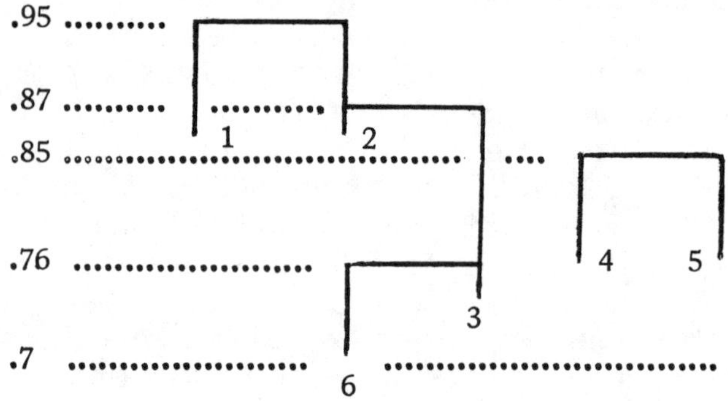

Figure 8.1 A dendogram showing which variables are highly correlated

Other tests can be carried out to see whether the latent roots of variables are significant. One of these tests is known as Bartlett's test and it depends on the χ^2 distribution. It is not altogether satisfactory and often tends to show rather more significant roots than can be justified in practice.

Multivariate techniques provide a very powerful method of data analysis, but the answers are not always as clear-cut and definite as one might hope. There is always a danger that, given sufficient variables, one of them will yield a good, but purely by chance, correlation with the objective variable. However, at the best, multivariate methods can yield good information about the existing data and sometimes even go some way to providing an explanatory model.

Computer software is available for most forms of analysis, but its use occasionally leads to over-elaborate print-outs of data since the software packages are liable to include everything and have inadequate suppression facilities. Even so, the use of computers normally saves an enormous amount of time and effort. Problems can arise in component analysis if the same data are inadvertently presented to the computer twice in a very slightly different manner, i.e. if the same variables are effectively used twice.

9 Decision theory and more about games

The field covered by the expression 'decision theory models' is not very easy to define. Some decision analysts include in this topic almost everything that involves making a decision. Some include decision trees and certain types of gaming and statistics, whilst a few others mean simply decision trees. In this report we shall assume that it includes decision trees and some types of gaming, since this seems to be the most common meaning of the term, although it may not necessarily be the most logical.

Decision trees

Normally decision theory is used for problems where there is a probability, or chance, of certain things occurring rather than for definite, precise events. Under these circumstances the variety of possible outcomes can make the problem appear quite complicated and difficult to resolve.

To take a very simple example: supposing you have £1,000 to invest, and there are only two companies in which you could possibly invest the money. Company A is a blue chip investment, whose price behaviour is relatively stable. You estimate that there is a 50 per cent (or 0.5 probability) chance that the value of company A's stock will increase by 20 per cent over a given period. You also think that there is a 25 per cent (or .25) chance that it will increase by 50 per cent and that the chances are one in four that it will decline in value by 10 per cent.

Company B, on the other hand, is a high-flying glamour stock. You think there is a one in ten (or 10 per cent or .1) chance that B's stock will go up by 400 per cent. You estimate that the chances are four in ten that the stock will double. But you also think there is a 50 per cent chance that you will lose your entire £1,000 investment.

This can be expressed very simply in the form of a diagram known as a decision tree, as shown in Figure 9.1. Starting at the left-hand side of the diagram at point D we are at the position of having to decide whether to invest £1,000 in company A or B. If it is assumed that we decide to invest in A, we progress across the diagram to the point A (ignore the figures in the brackets under A and B for the present). At this point a variety of events over which we have no control can occur with varying probabilities. Hence at this stage, which branch we will travel along is entirely a matter of chance.

Assuming that it was possible to take hundreds of thousands of similar chances at point A, about half of the time we would progress along the uppermost branch to the 200 pay-off but about a quarter of the time we should travel along the lower branch and lose 100.

It is therefore possible to put an average value on the action to buy into company A by multiplying the probabilities by the pay-offs and summing them in order to obtain the expected return which would be obtained if this chance could be taken many times. This is known as the 'decision value' and is the figure shown in brackets under the points A and B. It is calculated as follows:

Investment A

(£500 x .25) + (£200 x .50) - (£100 x .25) = £200

Investment B

(£4000 x .10) + (£1000 x .40) - (£1000 x .50) = £300

In this case the risky investment shows the best chance of gain if the probabilities have been correctly estimated.

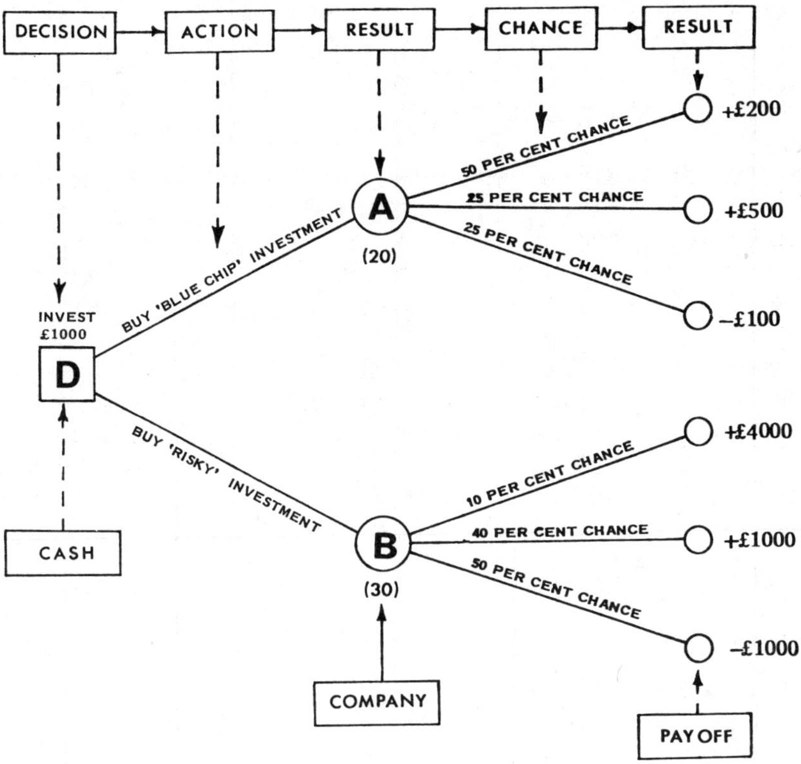

Figure 9.1 Simple decision tree example

Whether we can afford to take an investment such as this depends on our risk aversion. For example, a large company might be only too pleased to take this sort of risk if a few thousands of pounds were involved, but it might be much less willing to take the risk when hundreds of thousands of pounds were at stake.

Most of this was reported in the *Financial Times* in October 1970. (45) The merits and value of 2½ per cent Southern Rhodesian bonds were discussed in relation to decision theory and an investment at that time would have shown an almost fivefold increase in value within a few years. Even allowing for inflation, this must be regarded as very satisfactory.

Limitations of decision trees

One of the difficulties with decision trees is that the analysis tends to be oversimplified, or alternatively the decision tree tends to develop an unmanageable complexity. Moreover, for simplicity it is usually necessary to use the 'present value' of the pound in the calculations by considering discounted interest effects and the possible inflation or deflation of values. Assessing the true pay-offs is therefore not always an easy task. Rapid changes in costs or commodity prices can have a startling effect on possible outcomes.

A more complicated example of decision theory using decision trees, which also shows how to deal with further uncertainty aspects in a politicial–military situation, is given in Appendix 11.

Games 'against nature'

Another type of game which is also often considered in decision theory is the so-called 'game against nature'. In the games so far discussed we have mainly considered situations where people are playing against each other and probably actively trying to prevent each other from achieving their goals. This is not always the case in real life. In the first place an opponent who is not viciously inclined may not mind others attaining their goals provided that it does not affect his own goals, and also provided that he does not feel threatened by their actions.

It is, nevertheless, true that in games against nature the strategy of the game may be different, since nature is unlikely to be playing the game especially to defeat you. Indeed, if that were so, as our American colleagues might remark, we would probably never get off first base!

However, since nature intervenes in most affairs, the so-called 'games against nature' form an important part of decision theoretic analysis. Moreover, even with an active opponent information may be lacking. This tends to create a similar situation to simply regarding nature as the opponent.

More about game strategies

Some of the conventional strategies used in games of strong contest and good information were discussed in the section dealing with game theory. One of the methods described was the minimax strategy, which involves consideration of the worst possible outcome for each strategy and then selecting the best of these (i.e. the least worst). In games against nature the minimax strategy may not be the best, since nature cannot be regarded as an opponent seeking to minimise our gain. Consider the pay-off matrix in Figure 9.2.

	B1	B2	B3	B4
A1	200	4	1	3
A2	195	180	0	177

Figure 9.2 The minimax strategy is not ideal for some games against nature

The minimax strategy for player A is number 1, giving a lowest pay-off of 1. Yet if player B has imperfect information of the pay-off values, or makes an impersonal random choice of strategy, it would be foolish for A to choose strategy 1 unless he cannot even take the differential risk between 1 and 0 pay-offs, even though that risk is small. If this game is played many times and

B plays each strategy with equal frequency, then the average pay-off to A of strategy 1 is 208/4 = 52, whereas if A plays strategy 2, and B selects each available strategy the same number of times, then the average pay-off to A is 552/4 = 138.

Another possible strategy is described as the *maximax* strategy. This involves selecting the strategy giving the maximum of the maxima. In this example, player A would consider the maximum in row 1 (200) and in row 2 (196), and would then select the strategy giving the greatest of these. In this case the maximum of the maxima is 200 in row 1 and this again involves strategy 1. This method is clearly as wildly optimistic in games of uncertainty as the minimax criterion is pessimistic.

Hurwicz has suggested using a weighted average of the minimum and maximum pay-offs for each strategy. Thus if the minimum pay-off is given a weight equal to $\frac{2}{3}$ and the maximum pay-off is weighted by $\frac{1}{3}$, according to Hurwicz the strategy A1 is valued as:

$$1 \times \tfrac{2}{3} + 200 \times \tfrac{1}{3} = 67.33$$

Similarly, the strategy for A2 is rated as:

$$0 \times \tfrac{2}{3} + 195 \times \tfrac{1}{3} = 65$$

It will be noted that the strategy A1 is again chosen in preference to strategy A2 due to the fact that only the extreme pay-offs are used in the calculations.

For this reason the so-called *Bayes*, or *Laplace*, strategies usually give a better criterion. If nothing is known about the possible outcomes, then it may be best to attach equi-probability to nature's various strategies. Thus strategy A1 would be evaluated as follows:

$$\tfrac{1}{4} \times 200 + \tfrac{1}{4} \times 4 + \tfrac{1}{4} \times 1 + \tfrac{1}{4} \times 3 = 52$$

Similarly for strategy A2:

$$\tfrac{1}{4} \times 195 + \tfrac{1}{4} \times 180 + \tfrac{1}{4} \times 0 + \tfrac{1}{4} \times 177 = 138$$

It is clearly possible to assign fairly accurate probability figures to some kinds of natural phenomena, and in this case nature can be regarded as an opponent playing mixed strategies with a given frequency, although not necessarily in the optimal manner to defeat an active opponent.

Another method suggested by Savage is generally known as the *minimax regret* strategy, but is is also sometimes referred to as *Savage's* strategy. In this method the figures in the matrix are transformed into the cost of an incorrect decision and a new matrix showing the cost of mistakes, which is regarded as a function of regret, is obtained. If the new matrix is required to give a better measure of regret in quantitative rather than qualitative terms, the figures in the original matrix should be the utility values of the various outcomes.

Let us suppose that the original figures were expressed in utils, then the new matrix is as in Figure 9.3. The figures are calculated by considering each move made against the best that could have been employed. For example, if nature follows strategy B1 and the player employs strategy A1, he has no regrets (i.e. the pay-off matrix value is zero), since he has obtained the maximum possible. On the other hand, if he employs strategy A2 he only obtains 195 utils and he might have obtained 200 utils if he had followed strategy A1. Hence, his regret in this case is 5 utils.

	B1	B2	B3	B4
A1	0	176	0	174
A2	5	0	1	0

Figure 9.3 The application of Savage's minimax criteria to the data in Figure 9.2 gives a matrix of new values

Similarly, if nature follows strategy B2 the player has no regret if he follows strategy A2, since he then obtains the maximum pay-off possible under the circumstances. It will be seen that the maximum regret for strategy A2 is 5 utils, compared with a possible 176 utils for strategy A1. Thus, using Savage's minimax regret criterion, we would also select the second strategy.

There are, however, a number of disadvantages with the minimax regret criterion, one of which is that only the largest regret figures in any row are considered. It is also obvious that the probability of obtaining any particular regret is not considered, and this will undoubtedly psychologically affect the real regret. For example, player A may regret strategy A2 to the extent of 5 utils if nature plays strategy B1. Now suppose that A has 50 other strategies available to him, all of which yield 0 utils if nature follows strategy B1. Far from regretting anything if he follows A2 rather than A1, or the 50 other zero value strategies, he may be exceedingly pleased that he has chosen so well. In other words, satisfaction with his decision, based on his probability of making a good decision and the pay-off value, is a better guide.

One method which can be used to obtain a 'satisfaction' matrix is to describe satisfaction as the gain in utility divided by the expected utility multiplied by 100 to express it as a percentage. In the same way anything less than the expected utility is likely to give rise to dissatisfaction.

	B1	B2	B3		Bm
A1	a_{11}	a_{12}	a_{13}	......	a_{1m}
A2	a_{21}	a_{22}	a_{23}	......	a_{2m}
A3	a_{31}	a_{32}	a_{33}	......	a_{3m}
⋮	⋮	⋮	⋮		⋮
A	a_{n1}	a_{n2}	a_{n3}	......	a_{nm}

Figure 9.4 A sample game matrix

Consider the pay-off matrix in Figure 9.4. The expected gain from strategy A1 depends on the frequency

B plays B1, B2, B3 etc. Let the probability of these strategies be q_1, q_2, q_3, ... q_m etc. Thus the expected gain E_{ij} from strategy A1 is:

$$q_1 a_{11} + a_2 a_{12} + q_3 a_{13} + \ldots\ldots q_m a_{1m}$$

$$= \sum_{j=1}^{m} q_j a_{1j}$$

The expected gain e_{2j} from strategy A2 is:

$$\sum_{j=1}^{m} q_j a_{2j}$$

The average expected gain from all the strategies A1, A2, A3 ..... A_n, etc. is determined by the frequency with which these strategies are played. Let these be equal to p_1, p_2, p_3 ..... p_n, etc., so that the expected gain for optimal play is:

$$p_1 \sum_{j=1}^{m} E_{1j} + p_2 \sum_{j=1}^{m} E_{2j} + p_3 \sum_{j=1}^{m} E_{3j} + \ldots\ldots p_n \sum_{j=1}^{m} E_{nj}$$

$$= p_1 \sum_{j=1}^{m} q_j a_{1j} + p_2 \sum_{j=1}^{m} q_j a_{2j} + p_3 \sum_{j=1}^{m} q_j a_{3j} + \ldots\ldots p_n \sum_{j=1}^{m} q_j a_{nj}$$

$$= \sum_{i=1}^{n} p_i \sum_{j=1}^{m} q_j a_{ij}$$

A measure of our satisfaction for any particular value in the matrix is therefore given by:

$$\frac{a_{ij} - \sum_{i=1}^{n} p_i \sum_{j=1}^{m} q_j a_{ij}}{\sum_{i=1}^{n} p_i \sum_{j=1}^{m} q_j a_{ij}}$$

Naturally the solution to this new matrix of satisfaction is exactly the same as the original problem, since given a large number of plays the optimal strategy will be equal to the value of the game.

Decision theory limitations

There are a number of difficulties which may be encountered in practice when using decision theory which apply equally to military conflict problems and to business situations. In the first place it is often difficult to enumerate all the possibilities. If a large number of outcomes can be envisaged, then as Raiffa states, the decision tree 'becomes a bushy mess'. It is therefore first necessary to define the important policies which provide turning points or state changes. Then one should conduct a preliminary data analysis, and perhaps finally carry out a more detailed analysis of the various ways to implement a proposed policy.

Another difficulty is the sheer problem of envisaging all the possible outcomes (and sometimes, in retrospect, even the most probable). Consequently in practice unexpected things appear to happen because only a restricted range of variables has been considered. Variables which hitherto seemed unimportant in modelling a situation may suddenly become critical to the solution to a problem (see Appendices 12 and 13).

This difficulty is not peculiar to decision theory and other statistical treatments can suffer in the same way. For example, forecasting models based on multivariate analysis solutions in which dendograms are constructed showing the importance of the variables, may be completely unreliable if data used to construct the original model does not indicate the need for all function variables required in the near future. Unless an explanatory model is provided, in parallel with this kind of multivariate analysis model, forecasts usually run into trouble sooner or later as a consequence of the exclusion of an explanatory variable which has previously been quite unimportant.

In using and applying decision theory it is necessary, as with all statistics, to use one's head at the same time, and to consider very carefully indeed whether the analysis is thoroughly practical from all aspects. The accurate estimation of probabilities is, in itself, not necessarily an easy task. A variety of tricks can be employed in order to try to ensure an objective approach to what is frequently a subjective situation.

It is worth noting in this connection that in most cases nature is more reliable than man. We can, for example, estimate the probable rainfall in a country over a year fairly accurately since we have been able to accumulate past information by using measuring devices which we have developed. On the other hand, men unaided by instruments, and seeing exactly the same situation, will often estimate the probabilities quite differently–especially where discomforting situations are involved. Thus in the example just given, a man who dislikes rain will tend to estimate the rainfall higher than a man who is indifferent to rain, assuming that neither have any form of measuring device. Both could be incorrect for different reasons.

There is usually little redress on proffered probabilities which turn out to be incorrect and which have been given in good faith. Even the racing tipster who claims his tip is a 10 to 1 certainty is likely to reply to the failed punter 'Whotchaknow Guv, the 10 to 1 chance turned up. Bad luck!'

The only solution is to examine every facet of the information very carefully, and whenever data is available from apparently similar situations which have occurred in the past, to use this as a guide.

Whatever the disadvantages of decision analysis, using also game theory and the statistical approach, it often provides the only way of splitting up extremely complicated situations into less difficult problems which are capable of analysis.

It is also worth noting just once more that time often changes the pay-off values in the matrix, and in conflict situations against a human opponent information about the proposed moves of one's opponent, and also what he expects you to do, can be important. This unfolding of the future has not really been treated in depth in classical game theory, and decision analysis can tend to gloss over the more difficult aspects. Since both the problems mentioned in this paragraph can be important in practical situations, in the next chapter we will consider these aspects in relation to the apparently simple 'prisoner's dilemma', and then turn briefly to some interesting comments by Howard on metagames.

10 Repeated game play and deep strategy

The repetition of play in most competitive games can lead to very different participant behaviour in comparison to games which are only played once. The game entitled 'the prisoner's dilemma' has been used quite frequently in repetitious play situations and so it merits further attention. It appears to provide a model of many real world situations.

Movement in conflict games

The 'prisoner's dilemma' game is described as a non-zero sum game in that the pay-off to one player is not collected from the other. In non-zero sum games the pay-off may be influenced by a third person, sometimes in the guise of nature, circumstances, or environment, surrounding the game.

The story on which the 'prisoner's dilemma' game is based is as follows:

> Two criminals are arrested and are placed in separate cells. If neither confess there is only sufficient evidence to secure their conviction on a minor charge, and each will receive a six months' jail sentence. The police point out to the criminals the advantage of turning 'Queen's Evidence', and if this is done the

confessor will receive a sentence of $1\frac{1}{2}$ years provided the other does not also confess. In this case the one who fails to confess will receive five years, but if both confess they will receive three years.

If we label the criminals A and B, their pay-off matrices are as in Figures 10.1 and 10.2 respectively.

		B1	B2
		Not Confess	Confess
A1	Not Confess	– ½	– 5
A2	Confess	– 1½	– 3

Figure 10.1 Matrix for the prisoner's dilemma

		B1	B2
		Not Confess	Confess
A1	Not Confess	– ½	– 1½
A2	Confess	– 5	– 3

Figure 10.2 Matrix for the other participant in the prisoner's dilemma

It is usual to present the data for non-zero sum games in a single matrix, as in Figure 10.3. The first value in any cell is the pay-off to A and the second is the pay-off to B. A simple method for finding the solution to non-zero sum games would appear to be a development of the technique of optimal play simulation. The method in this case would be as follows.

	Not Confess	Confess
Not Confess	−½, −½	−5, −1½
Confess	−1½, −5	−3, −3

Figure 10.3 Combining the pay-offs for both players in one matrix

A decides to follow strategy A1. Write down B's opportunities for strategy A1 underneath the matrix (i.e. $-\frac{1}{2}$ and $-1\frac{1}{2}$). Choose and encircle the maximum of these, which is B1. Now write down A's opportunities which are given in the first column of A's pay-off matrix. Choose the maximum which is A1 and add the pay-off values for the A1 row of B's pay-off matrix to the row we first wrote down (see Figure 10.4).

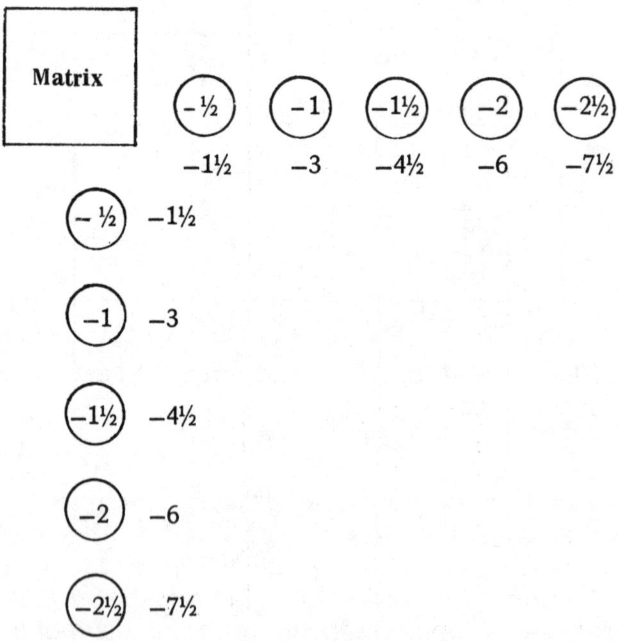

Figure 10.4 The technique of optimal play simulation

Select the maximum and encircle this value. Add the values of the column in which the circle occurs of A's pay-off to the first column which was written down on the right-hand side of the matrix. Continue until the strategy can be determined.

Now suppose that A had chosen strategy A2. In this case the first row would have been -5 and -3, and so B would have chosen -3 which involves strategy B2. A's choices are then also -5 and -3 with the result that he will again choose strategy A2. Continuation of the play as shown in Figure 10.5 does not require any change in strategy, and so A2/B2 is also an equilibrium point.

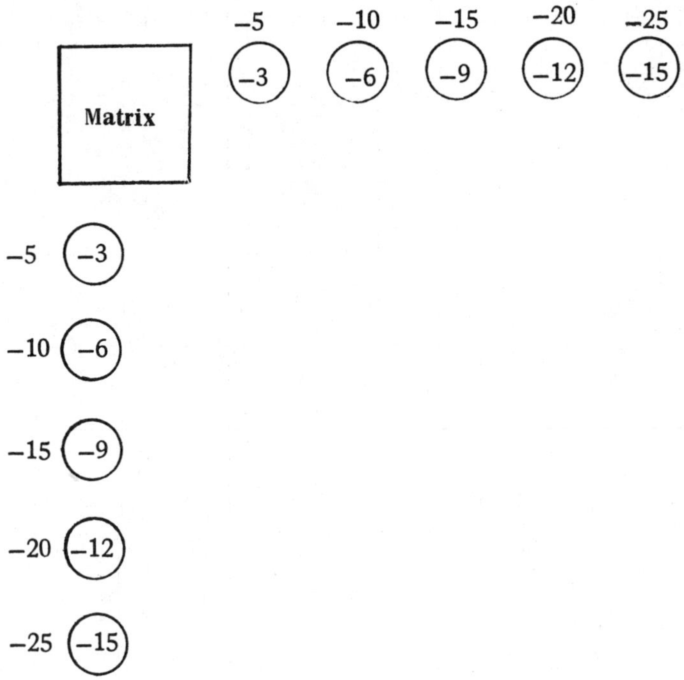

Figure 10.5 The choice of a different strategy gives a different equilibrium point in the prisoner's dilemma game

The only real difference between the equilibrium points A1/B1 and A2/B2 is that the latter has a lower value. In repeated or continuous play A may move to A1 to see whether B responds with B1. If he fails to do so, A has no alternative but to return to strategy A2. This game therefore contains a kind of altruistic optimum which can be realised by both players given goodwill on both sides. If one side seeks to 'punish', or minimise the other's pay-off, there is no alternative for the player than to return to the lower equilibrium point.

The effects of pay-off changes

The problem of the 'prisoner's dilemma' is very important and has various practical features and so it is worth spending a little more time considering the motivating factors in causing a migration from one equilibrium point to the other. Although it is clearly ridiculous, just suppose that Prisoner A had been faced with the matrix in Figure 10.6.

		B1 Not Confess	B2 Confess
A1	Not Confess	–½	– 100
A2	Confess	–98	– 99

Figure 10.6 The values in the matrix influence actions in 'prisoner's dilemma' situations

Knowing that B's choices were similar he would feel with some certainty that B would make the 'correct' or sensible choice, and so he would probably choose strategy A1. Now suppose that A's matrix had been quite different and as in Figure 10.7.

In this case A would be quite worried, and although he might feel that B would probably try strategy B1, it is likely that having weighed the possibilities he would decide to play strategy A2 in order to eliminate the thought of what would happen if he played A1 and B played strategy B2. A has therefore two motivating influences, greed and fear. The greed, or desire to maximise his pay-off, is obvious but he may also be fearful of B's actions. This has political implications.

		B1	B2
		Not Confess	Confess
A1		−½	− 100
A2	Confess	− 1	− 2

Figure 10.7 Most people faced with the choices available to player A would choose strategy A2 in this case

It would be useful if we could find some decision rule so that in these circumstances an attempt could be made to predict the probable moves of the competitors. In the first place B will certainly weigh up the possible prize against the possible penalty, and he will probably seek the smaller and more certain prize (albeit a small penalty in actual fact) rather than risk a very high penalty. Unless, and even if, he has some knowledge of statistics, he may behave apparently irrationally at times in terms of expectation (even allowing for such factors as risk aversion). It may be possible to deduce something about the probable response to risk situations from previous behaviour and in some respects this is easier in the case of companies or governments, who tend to behave as a group of people with slowly changing policies, rather than an individual who may have sudden spells of optimism or pessimism.

In the context of the general problem, and in the absence of information to the contrary, B may assume that A will follow the Laplacian assumption of attaching equal probabilities to A1 and A2, and if he has information about A's pay-off matrix he may well assume that A will play in the same manner as himself.

Consequently, if we consider the utility U_B of the pay-offs to B and assume that he expects A to play A1 or A2 with a probability of 0.5 (the meaning of utility is explained in Appendix 10), then, given the pay-off matrices a_{11}, a_{12}, b_{11}, b_{12}, etc. it is possible that B may assess his expectations as in Figures 10.8 and 10.9.

	B	
A	a_{11}	a_{12}
	a_{21}	a_{22}

Figure 10.8 The actions of player A depend on the utility of his pay-offs a_{11}, a_{12}, a_{21} and a_{22}

	B	
A	b_{11}	b_{12}
	b_{21}	b_{22}

Figure 10.9 The real pay-offs for player B depend on his utility for b_{11}, b_{12}, b_{21} and b_{22}

The expected utility to B of strategy B1 is:

$$E\ \{U_B(B1)\} = .5U_B(b_{11}) + .5U_B(b_{21})$$

Similarly the expected utility for strategy B1 is:

$$E\ \{U_B(B2)\} = .5U_B(b_{12}) + .5U_B(b_{22})$$

If it is assumed that the most likely strategy for B is that which maximises his expected utility $E_L(U_B)$, then:

$$E_L\ U_B(B_j) = \max \begin{bmatrix} E\ \{U_B(B1)\} \\ E\ \{U_B(B2)\} \end{bmatrix}$$

More generally this can be written:

$$\max_{0 \leqslant j \leqslant m} \quad \frac{1}{n} \sum_{i=1}^{n} U_B(b_{ij})$$

Hence the most likely best strategy for B can be determined. This makes it possible to find the most likely best strategy for A.

It is also possible to write the value of B's strategies in a more general form which considers the desirability of punishing an opponent by using probabilities, expected values and utilities. Thus the expected value of strategy B1 is:

$$E_B(B1) = p_{A1}E_B(b_{11})+p_{A2}E_B(b_{21})+p_{A1}U_B(a_{11})+p_{A2}U_B(a_{21})$$

Where in addition to the symbols defined p_{A1} is the probability of A playing strategy 1, etc.

Price wars

This model has a number of practical applications in industrial conflict situations. For example, suppose that two companies are the principal competitors in a particular field and are seeking to maximise their profits. They could maintain their prices at the present level or reduce them in order to improve sales and hopefully also their profit by obtaining a bigger share of the market.

			Company B	
			B1	B2
			Maintain Price	Reduce Price
Company A	A1	Maintain Price	0	– 5
	A2	Reduce Price	2	– 3

(All values in £0000)

Figure 10.10 The prisoner's dilemma game provides a model for some types of price war

A pay-off matrix for such a situation might be as in Figure 10.10. Both companies may be tempted to play their second strategy and reduce prices. This may result in a loss to them and they both may reduce prices still further in the hope of recovering some of the ground by increasing their market share. The continuance of this policy will cause changes in the values of the pay-offs and penalties in the matrix.

It is likely that the penalty for A2/B2 policy will increase rapidly, in terms of utility values, as the company profit margins are squeezed. Then again if one company reduces prices and the other does not do the same, certain sections of the market such as 'impulse purchasers' may be readily attacked, and further reductions may not produce similar *pro rata* benefits. These changes mean that the first strategy will eventually have a better utility expectation for both of the companies, and the system will stabilise at A1/B1 once again.

A similar non-zero sum problem will exist with regard to price increases (Figures 10.11 and 10.12).

Payoff Matrix for Company A

		B1 Maintain Price	B2 Increase Price
A1	Maintain Price	0	3
A2	Increase Price	–2	2

Figure 10.11 Eventually in price wars the perceived pay-offs in the matrix change. Prices stabilise and often gradually increase again.

Payoff Matrix for Company B

		B1 Maintain Price	B2 Increase Price
A1	Maintain Price	0	– 2
A2	Increase Price	3	2

Figure 10.12 If company B perceives pay-offs as above it will follow company A and also increase prices

The cold war and computer tournaments

A situation analogous to the 'prisoner's dilemma' exists in the cold war conditions between East and West, when each side would like to reduce armament expenditure but feels that the other might be able to acquire a considerable advantage if it followed such a course of action. Eventually, economic conditions may force a 'detente' situation, in that the utility expectation of the different policies has changed. Conditions will then tend to drift to a new level of stability until one of the opponents observes a sufficient migration from this point, or a change in international affairs alters his utility expectation for different policies, and as a consequence his strategy may change quite suddenly to the disadvantage and surprise of his opponent.

An interesting computer tournament based on iterative play of the prisoner's dilemma game was arranged by Axelrod under the sponsorship of the Institute of Public Policy Studies of the University of Michigan. (46-48) A program submitted by A. Rapoport entitled TIT for TAT proved to be one of the most robust. This program responded in the manner suggested by its name, and it was effective against most policies.

The importance of the role of communication in non-zero sum games, and in multi-equilibrium games repeatedly played, as a means of rapidly attaining stability, is obvious.

Let us now briefly consider Howard's work (49) on meta-games. This kind of conflict study is concerned with an expectation of moves by an opponent and with deviations from the expected reactions.

Meta-games and reflexive control

In order to develop the basic theory of meta-games it is only necessary to know the order of preference of the players' choices and these can be listed in a single matrix, player A's preferences being listed first (see Figure 10.13).

In the event that B can predict what moves A will make, he has the choice of four different policies depending on A's strategy. B can, for example, follow strategy B1 or B2 regardless of A's intended moves, or he can deliberately make a similar move or deliberately make a different move. This state can be represented in an expanded matrix and is known as a *meta-game*. Hence, for example, Figure 10.14. It will be noted that (2,2) again occurs, as in the original, as an equilibrium point.

	B1	B2
A1	3, 3	1, 4
A2	4, 1	2, 2

Figure 10.13 The values to be considered in a meta-game

		Competitor B			
		B1 Regardless	B2 Regardless	The Same	The Opposite
Competitor A	A1	3, 3	1, 4	3, 3 (B1)	1, 4 (B2)
	A2	4, 1	2, 2	2, 2 (B2)	4, 1 (B1)

Figure 10.14 If player B can estimate the intended moves of player A, he has an increased number of decisions

Now consider the situation where A believes that he can predict which of four possible strategies B is likely to adopt. Each possible policy of B gives rise to a further four possible moves by A, so that there is now a total of 16 feasible strategies. This is known as the second meta-level and the new matrix is as shown in Figure 10.15.

It will be seen that this matrix contains three equilibrium points, and it can be shown that further development of the strategies beyond the two levels, as described above, will not indicate any further equilibrium points for games involving only two competitors.

Reflexive control is not quite the same thing as a meta-game. It is discussed in an excellent work by the Soviet operational analyst Tarakanov, (50) in relation to Soviet policy. Basically, it involves making a competitor or enemy do what you want him to do. A good military commander may be able to achieve this on the battlefield but reflexive control can be even more cost-effective at the political and policy level. The Soviets appear to place some emphasis on this approach.

COMPETITOR B

COMPETITOR A	B1 Regardless	B2 Regardless	The Same	The Opposite
1, 1, 1, 1	3, 3	1, 4	3, 3	1, 4
1, 1, 1, 2	3, 3	1, 4	3, 3	4, 1
1, 1, 2, 1	3, 3	1, 4	2, 2	1, 4
1, 2, 1, 1	3, 3	2, 2	3, 3	1, 4
2, 1, 1, 1	4, 1	1, 4	3, 3	1, 4
1, 1, 2, 2	3, 3	1, 4	2, 2	4, 1
1, 2, 1, 2	3, 3	2, 2	(3, 3)	4, 1
2, 1, 1, 2	4, 1	1, 4	3, 3	4, 1
1, 2, 2, 1	3, 3	2, 2	2, 2	1, 4
2, 1, 2, 1	4, 1	1, 4	2, 2	1, 4
2, 2, 1, 1	4, 1	2, 2	3, 3	1, 4
1, 2, 2, 2	3, 3	2, 2	2, 2	1, 4
2, 1, 2, 2	4, 1	1, 4	2, 2	4, 1
2, 2, 1, 2	4, 1	2, 2	(3, 3)	4, 1
2, 2, 2, 1	4, 1	2, 2	2, 2	1, 4
2, 2, 2, 2	4, 1	(2, 2)	2, 2	4, 1

Figure 10.15 The complexity of meta-games depends on the depth of perceived actions. Points of stability tend to occur in the matrix.

A very simple example of the way in which we could employ reflexive control would be a decision to build one or two very large hovercraft capable of operating in the North Sea and off the Northern coast of the USSR or in the Black Sea. Such a threat would be very costly to counter and would probably require the development of special weapons which would need to be deployed over a very wide area. This may involve a potential enemy in considerable expenditure for relatively little expenditure of our own.

Soviet operational research literature contains many references to the importance of the gross national product of a country as a rough measure of its military capability. Clearly, therefore, reflexive control might reasonably be expected to be concerned with damaging the GNP of Western countries. This matter remains something of an enigma since Communist doctrine, and also that of the Soviet Ministry of Economics, strongly encourages world trade.

Soviet operational research literature also suggests the possibility of military actions being based on subjective estimates of the probability of success within a given time frame. Certainly such a policy would help to explain their actions in Afghanistan, Czechoslovakia, Hungary and Poland. Having been attacked from the West many times, it is natural for the Soviets to want a bulwark of friendly neighbours; but in the long term such activities are hardly likely to build lasting understanding among nations or even reliable allies.

Option analysis

The method entitled 'Analysis of options' as proposed by Richard Sharpe (51) is an extension of meta-game theory.

It will be recalled that in meta-games the decision maker begins by putting himself in his opponent's place and formulating a series of alternative policies which could be used. The player then decides, according to some decision rule, which policy his opponent is going to use, and selects his own strategy accordingly. This process can be continued to various deeper levels where the opponent guesses the strategy you will use and decides to act somewhat differently again. This process is said to have not only the advantage of improving the quality of the decision maker's analysis of the situation, but also to enable the detection of positions of stability for both parties, which would otherwise not be apparent.

In option analysis a similar process is used to detect best strategies, or positions of stability, which are in general too complex to be satisfactorily modelled in a game theoretical manner.

The first step, as in game theory, is to identify all the players and to try to determine all the possible options (or moves) available to them in a particular situation. When this has been done, a particular player is selected and the options available to him examined.

From this consideration, strategies, consisting of options, singly or in combination, available to the player, can be determined and evaluated. In this process, it is assumed that all the other players make no changes to their policies. In such a case each strategy can be classed as giving results better than the starting position, when they will be 'preferred', or worse, when they will be 'not preferred'. Some strategies will contain options which are irreconcilable (e.g. 'become a cannibal' and 'remain a vegetarian'), in which case they are described as 'infeasible'.

The group of strategies which have been classed as 'preferred' here are known as the 'unilateral improvements on the starting position' for the player concerned. Before they can be described as good moves, they must in turn be scrutinised to see if any of the other players, or combinations of players in the game, possess any options which might prevent the move from taking place. Such options, which might, according to circumstances, either nullify the action or make the resulting position less advantageous than the starting one ('the deterrent effect'), are known as 'sanctions'. The existence of a 'sanction' is not in itself, however, sufficient to prevent a move from taking place, since the decision maker may feel that the 'sanction' will probably not be used. Thus the 'sanctions' themselves must in turn be examined to evaluate whether or not they are credible.

After all the sanctions available have been reviewed, all the strategies which are subject to credible sanctions must be removed from the set of unilateral improvements.

The remaining strategies, which are advantageous for the player concerned whatever the actions of the other players, are the inescapable improvements for the player concerned. If no such strategies exist, the situation is 'stable' for that player or coalition.

It can be argued that the whole process functions as a kind of 'filter' for possible strategies (see Figure 10.16). Any strategy passing through the filter will represent a potential improvement, and although the technique has not attempted any form of optimising, the decision maker will have managed to satisfy himself.

Stage 1 & 2	OPTIONS	
Stage 3	------------------	Classification by Preferences
	UNILATERAL IMPROVEMENTS	
Stage 4 & 5	------------------	Evaluation of Sanctions
Stage 6	INESCAPABLE IMPROVEMENTS	

Figure 10.16 Illustrating the various stages in options analysis

The process of selecting and evaluating strategies should continue (in theory, at least) until the position has been examined from the point of view of each of the players, and all possible coalitions. When this process has been exhausted, the inescapable improvements of all the players will be known, and thus their probable moves can be assessed.

Detecting such positions of stability is of crucial interest to all players involved, for while they may not be intrinsically attractive to all parties (unconditional surrender brought stability to the Germans in the Second World War), they represent potential 'final positions', which once arrived at will be maintained by everyone.

Hence, the option analysis technique procedure listed below:

1 List all the players involved.
2 List the options available to each player.
3 Select one player and attempt to find strategies leading to unilateral improvements.
4 Investigate each of the improvements and see if any of the other players or coalitions have 'sanctions', which will prevent the improvements from being realised.
5 Evaluate the sanctions and decide whether or not they are credible.
6 List all the inescapable improvements (i.e. the unilateral improvements for which no credible sanction exists).
7 Select a new player or coalition, and repeat the process until all possible combinations of players have been exhausted.

Sharpe goes on to describe in some detail some of the weaknesses of option analysis and in part this appears to be due to the failure of one side to perceive options which seem quite obvious to the other. He shows that the First World War resulted from a tragic misperception of options. The correct interpretation of available information and the need for a flexible 'model of the world' are both essential if options are to be properly perceived.

I am of the opinion that although game theory can be of use, option analysis is generally much closer to the real world. Indeed in the real world we seem to climb through a multi-dimensional array of options with some options only being revealed as time progresses. Thus looking at the prices being charged for oil in the early 1970s it is possible to develop a multi-dimensional matrix and see how the addition of certain options in the late 1970s suddenly enabled the oil-producing countries to climb to a completely different part of the matrix and thereby charge very greatly increased prices.

11 Forecasting and time series models

The desire to predict future events has probably been with man from his earliest days. In order to survive we must have some kind of model of the world. Such a model must be predictive by nature. It is possible that the better we are able to construct this kind of model, the better the quality of our survival.

A variety of techniques

Virtually all models are designed with the object of being able to forecast something. What is being forecast may not be quite so obvious in some cases. For example, we may ask how an aircraft flight simulator fits into the pattern. The answer is simply that although we may be *using the simulator for training* purposes, the simulator is quite clearly *predicting* the *flight behaviour* of an aircraft given certain pilot responses. Indeed we can go further and say that the simulator is helping us to predict the ability of a trainee pilot to control a particular aircraft.

However, a number of techniques have grown up over the years in which a particular kind of mathematics is used to treat a series of data values gradually being obtained in an unfolding time series. The method attempts to establish the way in which the data have varied in the past and then to predict the way in which future data movements will occur according to some relatively simple formula derived from an analysis of the original data.

I would like to say at the outset that I am not very enthusiastic about the use of these techniques as an approach to model building in their own right. It is naive to imagine, merely because quite a large amount of mathematics may be involved in producing a simple formula in order to achieve an extrapolation, which an office boy with a good eye and a steady hand might achieve using graph paper in a few seconds, that a profound model has been developed which contains some deep inner truth. The techniques may, however, form useful adjuncts to the various other methods which can be employed and for this reason, if no other, they should not be neglected by the model builder.

It is generally necessary to know some of the 'in' expressions. Thus people not only talk about long- and short-term forecasts, the meaning of which is obvious, but they also talk about such things as *causal* and *statistical* forecasting models. A causal forecasting model is one which provides explanations, whilst statistical models may arise from multivariate analysis or similar statistical analysis. Occasionally *subjective* and *objective* forecasting is mentioned. By subjective forecasting one means that there is no real theory behind the prediction and the forecast is based purely on a 'hunch'. Objective forecasting means the reverse.

There are various ways in which time series have been analysed over the years and some of the very simplest methods still remain reasonably effective under certain circumstances. Staff at ICI [52] have considered various straightforward ways of forecasting, and the methods they describe have achieved some credibility in industrial forecasting. The methods include exponential smoothing, moving averages, seasonal adjustments, etc.

Brown [53] has developed various approaches, some of which are based on sine wave oscillations. It is sometimes said that his methods can be very satisfactory in models which involve oscillatory effects. In my view this is somewhat uncertain, and much depends on the circumstances.

Fourier analyses of oscillatory functions have been conducted with varying success. This process consists of resolving a function into a series of sine waves of different frequencies. Again a great deal depends on the type of problem, and it is necessary to look at the underlying variables. For example, if we wish to predict the torsional vibrations of a crankshaft, then a Fourier analysis of the cylinder pressures can be helpful in isolating pressure components which have the same frequency of harmonics as the crankshaft frequency. Although this is a practical tool for predicting the magnitude of the torsional oscillations, it is quite likely to be useless if applied as a predictor for a random walk time series.

The fact is that there is a considerable difference in the way we are attempting to apply the analysis. In the case of the torsional vibration analysis we are trying to develop a predictive formula from an explanatory model of the situation, even if this is rather vague. On the other hand, in the case of the straightforward application of Fourier analysis to any time series where we have no explanatory understanding of the situation we should not be surprised if things go wrong, for we have hardly the right to expect any kind of sensible solution.

The Box–Jenkins method

One of the approaches to forecasting which has received great attention in recent years is the so-called 'Box–Jenkins' method. (54) The way in which the predictive formulae are derived is by considering different values in the series and examining the differences between certain values. If the Box-Jenkins method is to be applied, a pattern must be observable. This involves taking the first or second differences between the various values in the series and is called 'the process of obtaining stationarity'. For a discrete process to be stationary, the joint probability distribution must be unaffected by shifting all the times forwards or backwards by an integer amount k, where k is known as the 'lag'. The autocorrelation coefficients are then calculated for the original series, for the first differences of the series, and for the second differences of the series. This process is described in Appendix 13. These autocorrelation coefficients are diagnostic of the type of model which should be used, these relationships also being shown in Appendix 13.

Let us consider a simple example of what not to do in forecasting and then see if the Box–Jenkins approach can be of assistance.

One of the difficulties in tracking the flight of projectiles by means of radar over the sea is that when the angle of the radar over the sea becomes small there are considerable reflections from the waves (clutter) which cause the shell to be lost from the radar screen. For certain trials the point of splash-down is an important parameter which must be measured. In order to obtain a point of splash-down from the radar data it was therefore necessary to extrapolate the trajectory over the last few hundred feet into the sea. On one particular range where this kind of work was carried out, the range scientists apparently wrote a small extrapolation program and all seemed to go well for several months.

One day, according to this extrapolation program, and in seeming defiance of the laws of gravity, one of the shells apparently never reached the sea despite being tracked by radar on a downward path pointing directly into the water until the reflection was lost in 'clutter'. The military officers on the range were somewhat reluctant to believe that their shell had, in fact, turned round and soared upwards when approaching its watery grave, and so the Central Scientific Group were called in to investigate. They found that the extrapolation program which the range scientists had developed involved fitting a high-order polynomial. This was a totally unsatisfactory procedure in this case. Indeed it is rarely advisable to fit high-order polynomials for extrapolation at any time. The problem was caused by an error at the end of the curve, possibly as a consequence of the radar beam passing through a piece of reinforcing material on the fibreglass dome and causing a temporary aberration.

What should, of course, have happened was that a simple approach such as a 'tied' low-order polynomial or a 'sliding arc' method should have been used. The 'tied' polynomial simply means that the curve is constrained in a particular direction, and in this case only downwards. Sometimes a combination of 'sliding arc' (which is simply sliding the curve along a bit) and tied polynomials are used in curve extrapolation. In this case it would also have been possible to use a simple ballistic trajectory model and fit the data to this in order to derive the necessary extrapolation.

A Box–Jenkins example

Let us also consider what happens if we try to use the Box–Jenkins method for this task. The theoretical trajectory had been fairly well established for this particular shell, so let us consider the value of the Box–Jenkins approach as a one-period predictor of this theoretical trajectory.

The data might be as follows:

Time (sec)	Range (metres)
0.0	0.0
2.0	1063.56
4.0	2021.29
6.0	2892.38
8.0	3691.77
10.0	4431.20
12.0	5120.03
14.0	5769.05
16.0	6394.15
18.0	7002.32
20.0	7596.03
22.0	8176.12
24.0	8743.12
26.0	9297.39
28.0	9839.17
30.0	10368.55
32.0	10885.55
34.0	11390.09

The autocorrelation coefficients were calculated for the various lags as explained in Appendix 14. These values have been plotted in Figures 11.1, 11.2 and 11.3.

It will be seen that the values decrease almost linearly. Figure 11.1 could be regarded as an exponential, particularly after the first lag. Since this diagram is based on first differences, this would imply from an analysis of the diagnostics that the model is an autoregressive integrated moving average model, or ARIMA (1,1,0). Alternatively, if the first lag is significant, it is an ARIMA (1,1,1). If we assume that the model is an ARIMA (1,1,1), which is the more complex of the two models and the model turns out to be close to an ARIMA (1,1,0), we will find the ARIMA (1,1,1) condition.

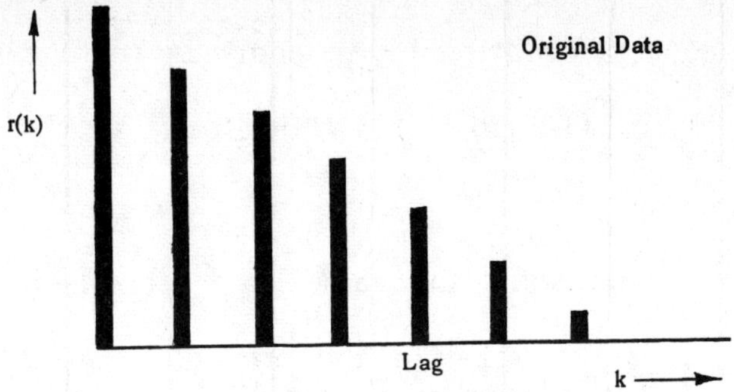

Figure 11.1

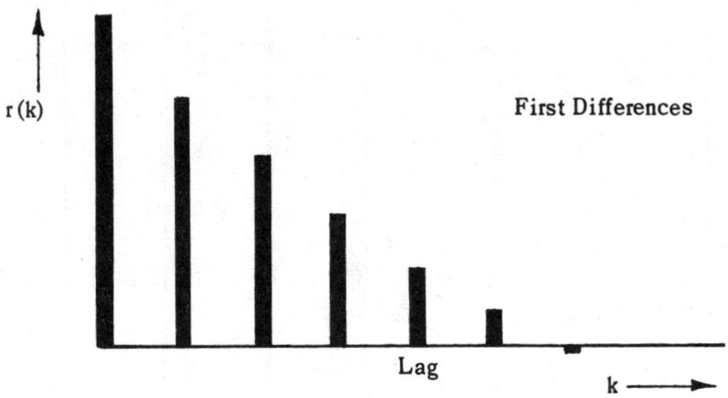

Figure 11.2

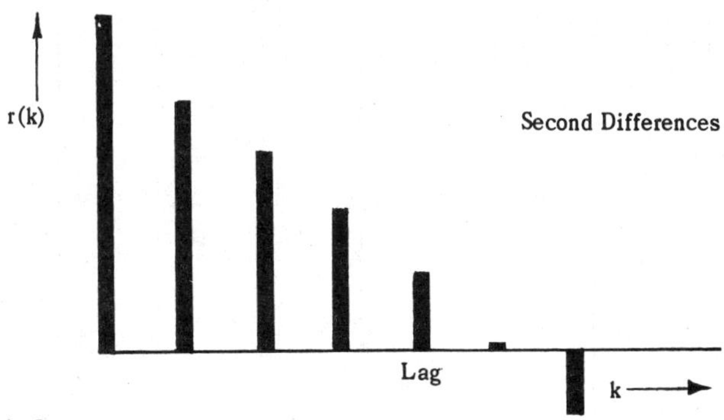

Figure 11.3

First order autoregressive integrated moving average ARIMA (1,1,1)

From the general equation in Appendix 14 we obtain:

$$(1-\phi_1 B)\ \nabla z_t = (1-\theta_1 B)\ a_t$$

where

z_t	represents the value of a variable at time t,
z_{t-1}	is the value of the preceding variable in the time series at time t-1
z_{t+1}	is the value of the succeeding variable in the time series at time t+1
$\bar{z}$	is the mean of the data series $z_{t-n} \dots z_{t-2}$, z_{t-1}, z_t
$\hat{z}_{t+\ell}(\ell)$	is the forecast of the value of $z_{t+\ell}$ at time t
$\dot{z}$	is the difference between z_t and the mean value of the series $\bar{z}$
a_t	is an error term and is the difference between the actual value at time t and the value which was forecast at time t-1, i.e. $z_t - \hat{z}_{t-1}(1)$
∇z_t	is the difference operator such that $\nabla z_t = z_t - z_{t-1}$
B	is the backward shift operator such that $Bz_t = z_{t-1}$
ϕ	is the 'weighting', or coefficient, given to the z values in the autoregressive process and
θ	is the 'weighting', or coefficient, given to the z values in the moving average process.

Removing the difference operator and the shift operator gives:

$$z_t - z_{t-1} - \phi_1(z_{t-1} - z_{t-2}) = a_t - \theta_1 a_{t-1}$$

So:

$$z_t = z_{t-1} + \phi_1 z_{t-1} - \phi_1 z_{t-2} + a_t - \theta_1 a_{t-1}$$

i.e.

$$z_t = (1+\phi_1)z_{t-1} - \phi_1 z_{t-2} + a_t - \theta_1 a_{t-1}$$

Thus, if ℓ is the period ahead of the required forecast:

$$\hat{z}_{t+\ell}(\ell) = (1+\phi_1)z_{t+\ell-1} - \phi_1 z_{t+\ell-2} + a_{t+\ell} - \theta_1 a_{t+\ell-1}$$

Hence, for one period ahead forecast:

$$\hat{z}_{t+1}(1) = (1+\phi_1)z_t - \phi_1 z_{t-1} + a_{t+1} - \theta_1 a_t$$

But a_{t+1} must be expected to be zero at this stage, and so:

$$\hat{z}_{t+1}(1) = (1+\phi_1)z_t - \phi_1 z_{t-1} - \theta_1 a_t$$

Similarly, for a two-period forecast:

$$\hat{z}_{t+2}(2) = (1+\phi_1)z_{t+1} - \phi_1 z_t + a_{t+2} - \phi_1 a_{t+1}$$

But a_{t+2} and a_{t+1} must be zero at this stage, and so:

$$\hat{z}_{t+2}(2) = (1+\phi_1)\ z_{t+1} - \phi_1 z_t$$

Also the only value available for z_{t+1} is the forecast of this value, viz. $\hat{z}_{t+1}(1)$.

So:

$$\hat{z}_{t+2}(2) = (1+\phi_1)\ \hat{z}_{t+1}(1) - \phi_1 z_t$$

or more generally:

$$\hat{z}_{t+\ell}(\ell) = (1+\phi_1)\ \hat{z}_{t+\ell-1}(\ell-1) - \phi_1 \hat{z}_{t+\ell-2}(\ell-2)$$

where $\ell > 2$.

With some algebraic manipulation it is also possible to rewrite this equation in the form:

$$\hat{z}_{t+\ell}(\ell) = z_t + \frac{\phi_1(1-\phi_1{}^2)}{(1-\phi_1)} (z_t - z_{t-1}) - \frac{\theta_1(1-\phi_1{}^2)}{(1-\phi_1)} a_t$$

This is a very general model and can be used in many situations. Clearly some other models are subsets of this model, and represent the case where certain of the coefficients are zero.

Preliminary estimates of θ_1 and ϕ_1 can be obtained from the equations given in Appendix 14.

It will be found that preliminary estimates indicate that the value of ϕ_1 is expected to be about .75 and the value of θ_1 about -.1.

The forecast for one period ahead for this model is given by:

$$\hat{z}(1) = (1+\phi_1)\, z_t - \phi_1 z_{t-1} - \theta_1 a_t$$

It is important to establish a 'settled' a_t value for the model in making the one period ahead forecast. This can be done by starting at the beginning of the data series and calculating the a_t term for the known data.

Thus:

$$\hat{z}_{t-15}(-15) = (1+.75)z_{t-16} - .75z_{t-17} + .1a_t$$

i.e. $$= 1.75 \times 1063.56 - .75 \times 0.0 + .1a_t$$

$$= 1861.23 + .1a_t$$

But we have no knowledge of the previous error so initially we must assume a_t to be zero. Thus our forecast of the -15th value is 1861.23. However, we know from the available data that it proved to be 2021.29 metres, so the error (i.e. the next a_t value) is 2021.29 minus 1861.23 or 160.06. Thus in making the -14th value we would use this a_t value in the formula.

If we work through the series in this way, we will be able to find the current a_t value, and by using the formula we find that the forecast for the next 'range' value is 11,779.09 metres. It proved to be 11,881.99. This is not too bad considering the simplicity of the model. If we are interested in a particular portion of the trajectory, it is likely that we can improve the model still further by adjusting the constants.

Indeed if ϕ_1 is taken to be .98 and θ_1 to be -.8 it will be found that the forecast is 11,881.89 metres as compared to the actual 11,881.99. This error is within the limits permissible. With this particular set of ϕ and θ values the forecasts fluctuate in the earlier part of the series, and then settle down to a small error of up to 3 metres. Thus, in making predictions from a radar, it may be adequate to smooth the data using a second-order polynomial curve fit over the latter part of the trajectory and to then apply the following simple formula:

$$\hat{z}_{t+1}(1) = 1.98\ z_t - .98\ z_{t-1} + 0.8a_t$$

The reason for using a 'tied' polynomial for smoothing purposes is quite simply that the accuracy with which the data are obtained using a radar is variable. Servos, reflections, etc. can all cause an individual measurement to be in error, and in using a very simple formula of the type shown above, it is necessary to ensure that the z values are as accurate as the known errors in the measuring system will allow, i.e. we should try to enter the most accurate estimate of the z values that we can make.

It will be evident from these remarks that the Box–Jenkins technique must be applied intelligently and the more we understand the physical processes which are involved the better we are able to apply the Box–Jenkins method. There is, for example, normally little point in excessive departure from the calculated ϕ and θ values, unless as in this example we know that the data curve is changing for purely physical reasons.

It is, of course, possible to write a fairly simple computer program which develops the Box–Jenkins formulae by virtually 'trial and error' methods. In all probability the 'branch and bound' approach, combined with simple 'hill-climbing' techniques should result in a fairly rapid solution with the minimum error. The estimation of the ϕ and θ values for the ARIMA (2,d,2) model, for example, is not a simple task without at least some form of computational assistance. I have, in fact, developed a simple program for a programmable pocket calculator which finds the coefficients automatically for all models up to the (2,2,2) case. It can take several hours to run but since it involves the user in no mental effort his mind is not distracted from the main task.

The chartist approach

It is not possible to leave forecasting models without mentioning the work of the chartists. Chartists are not early trade unionists or medieval freedom fighters, they are people who plot graphs in different ways and attempt to deduce the future from consideration of whether their charts produce good or bad vibrations. As in many pseudo-scientific areas, theories which have been 'proven' and 'falsified' alike, abound. The activities of chartists are deemed by many mathematicians to lie within the witchcraft category. Similarly chartists, who are often closely involved with practical aspects, tend to regard mathematicians as useless, especially since one or two mathematicians claim to have 'proved' certain things cannot happen when anyone with practical experience of trying to forecast what will happen in a market knows that they both can and do take place. Mathematicians therefore tend to be regarded rather badly in the London Stock Exchange and in the Commodity Exchanges. As a consequence market and security analysts have turned to chartism.

I became interested in chartism in 1956 when Ellinger [55] was virtually the sole disciple in this country. In 1972 when chartism was very strong on Wall Street it was still barely used here; although it had been pointed out [56] that, even if one had doubts about certain of the chartist activities, some chartism could be profitable. It was therefore astonishing that by 1979 most stock and commodity brokers in the London Exchanges should have their own tame chartist, and yet surprisingly enough, no tame mathematician. In the USA Wall Street seems to have partly passed through this phase and various senior investment advisors have been known to describe chartism as 'snake oil' which is about as rude as they can be on the subject.

Let us take a brief look at how chartists work since the subject appears to have gained such credibility amongst the City institutions. Whole books have been written on this matter so it is very hard to try to provide a useful summary in a few lines. Basically, the original idea was to give the analyst a picture of the movements in market price, and volume, of an individual issue. After studying thousands of patterns the chartist technicians decided they could forecast when the price was about to go up, and from some other patterns they could decide when the price was about to go down.

Two types of charts are mainly used: 'vertical bar' and 'point and figure' chart. In the first the trading range of a given stock or commodity is indicated by the length of a small vertical bar. The volume of trading is also often shown at the foot of the graph. In the case of the 'point and figure' chart the x-axis does not represent time but indicates a change in direction of the movement of the stock or commodity. The vertical axis represents a unit of price movement. For example, a unit on the graph may represent a 5p movement of a £2 stock. Only unit moves are recorded, and when a reversal occurs, the unit movement is made along the x-axis and the movement recorded with a different symbol. Usually upward movements are recorded with an 'x' and downward movements with an 'o'. Dates are also shown on the chart so it will be appreciated that this chart is, in effect, recording three variates. Figures 11.4 and 11.5 show typical charts.

Chartists talk about patterns such as triple tops, triple bottoms, bearish triangles, bullish triangles, head and shoulders, inverted head and shoulders, saucers, flags and pennants, etc. The value of some of these patterns is highly dubious, to put matters charitably, but there are mathematical reasons why some of the patterns recognised should have definite significance.

It will be appreciated that there is a considerable amount of 'noise' in any free market system. This is due to all sorts of things such as rumours, political speculation, etc. This noise often masks the underlying trends. In addition there may be reasons for certain short-period oscillatory actions in the market. A market which has some positive feedback, and most markets do, will tend to oscillate more than once. If a large financial institution starts to invest, then others may join in until the shares or commodity being bought has been driven to an abnormally high level and far beyond the price the institution deems reasonable. At this level either the

Figure 11.4 An example of a point and figure chart

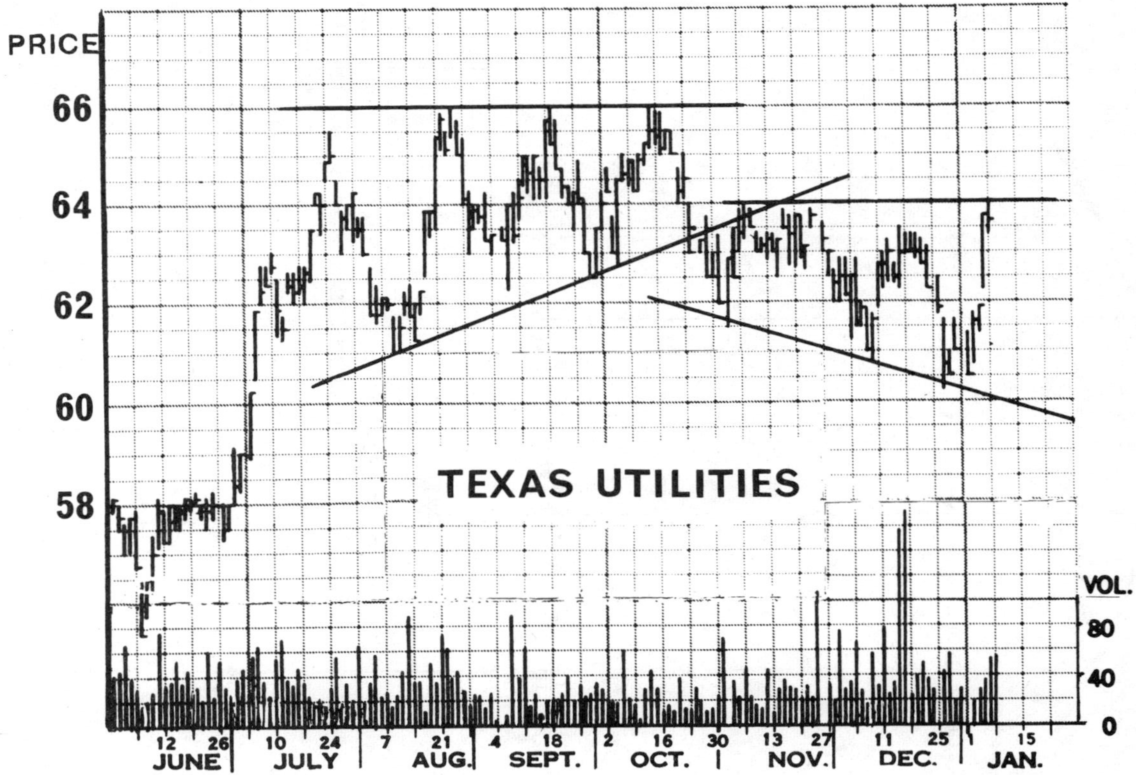

Figure 11.5 An example of a vertical bar chart

institution, or more likely some of the early investors, will sell and the dealers will mark prices down rapidly. While the decline continues at a pace few want to buy and so it continues until there are no more sellers. At this stage a few buyers may creep into the market and the whole process may repeat itself.

Figure 11.6a Underlying trend

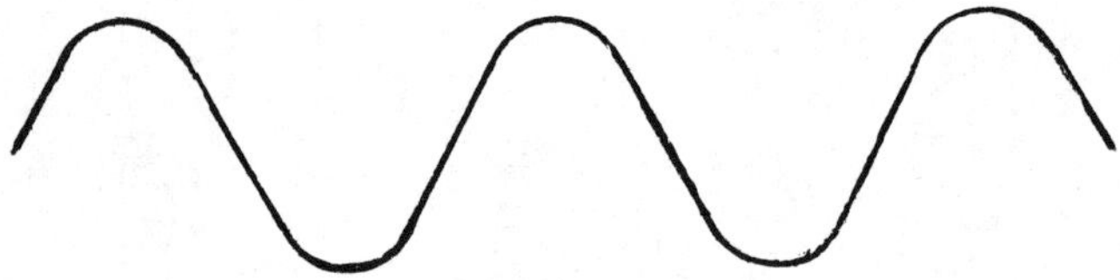

Figure 11.6b Natural market trading oscillation

Figure 11.6c Day-to-day noise

Suppose that we have a time when the economic conditions have reached a peak and the market is about to fall. Let us show this as Figure 11.6a. In addition the market has a natural trading oscillation due to over- or under-buying as shown in Figure 11.6b, and there is a certain amount of 'noise' in day-to-day sales as shown in Figure 11.6c. The latter effect, as explained earlier, may be due to rumours, political speculation or good or bad trading results from a company generally regarded as a market leader. If we combine the underlying economic trend, the natural market trading oscillation, with the day-to-day sales noise, as shown separately in the three graphs, we get the result shown in Figure 11.7.

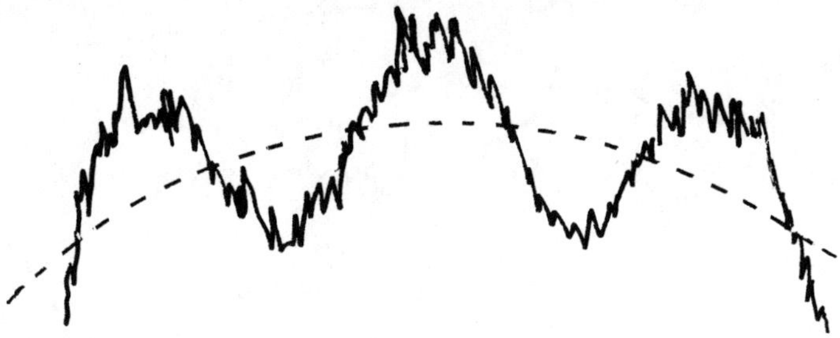

Figure 11.7 The result of combining market trend, oscillation and 'noise'

This is exactly the formation which chartists refer to as the 'head and shoulders' formation and which is said to occur when the market is about to fall. If we invert this graph we obtain the graph shown in Figure 11.8. Chartists draw two lines on their graphs and if the line breaks through then a strong upward movement is signalled. This could, in fact, equally well be deduced by mathematical analysis by simply removing the higher-order oscillatory motions. Thus, a number of chartist models can have a sound mathematical basis but others almost certainly do not. I have never seen these comments or explanations given elsewhere. Perhaps this is since mathematicians and chartists each show little sympathy for the opinions of the other.

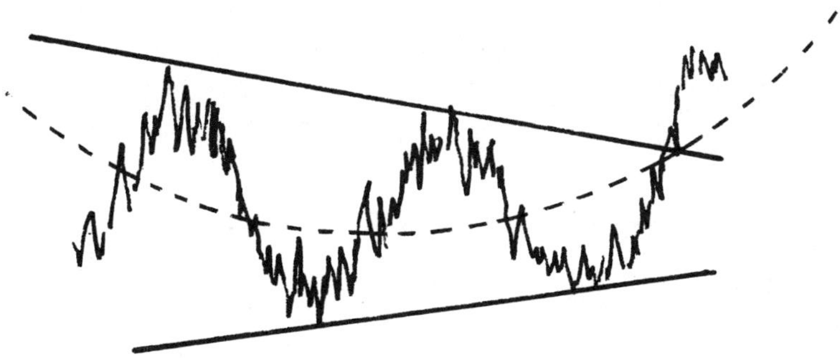

Figure 11.8 A 'breakthrough'. The converging lines drawn by chartists may be simply due to a combination of natural oscillations

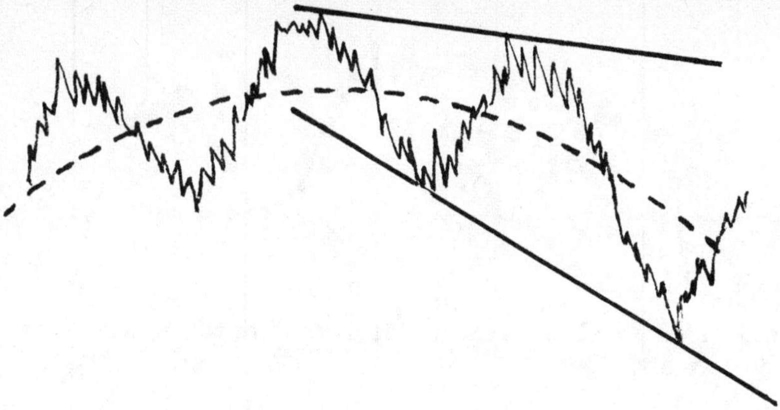

Figure 11.9 An oscillation superimposed on a commencing downbend gives the 'expanding wedge' formation

As explained earlier a pattern of the type shown in Figure 11.9 gives an expanding wedge and this wedge shape also suggests a downward trend. This is, however, only to be expected from the observations I have made already. A wedge in the reverse direction can also be achieved if the resistance of the media in which the vibrations occur increases with the amplitude of the main oscillation. Thus frequencies hitherto regarded as 'noise' will be distorted and will narrow as shown in Figures 11.10a and 11.10b. This particular shape is again an indication that a firm market is developing and since things are rarely static for long in free market conditions a movement out of this condition in the upward direction is quite likely.

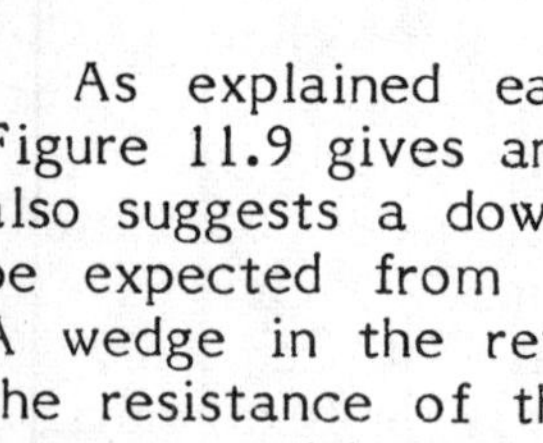

Figure 11.10a Increasing resistance

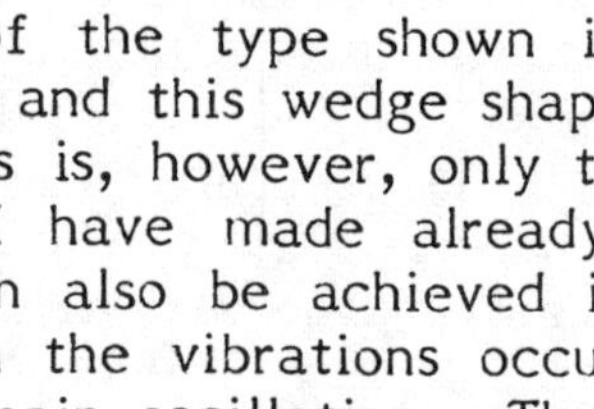

Figure 11.10b Complete mechanical resistance

All these signals can be completely masked if the appropriate frequencies and phase differences are not present and this is often the case. A properly developed model of the market with an occasional examination of market behaviour patterns can often be much more effective than a total recourse to 'chartist' activities alone.

12 Another method for building models

A first step frequently taken by consultants or operational analysts when attempting to analyse problems is the construction of 'conceptual diagrams'. A conceptual diagram indicates which variables affect other variables, although it does not give a precise mathematical description of the interactions between these variables. Such a diagram is also sometimes called an 'interaction' diagram, or a 'structure' diagram, since it helps to display the general structure of the problem. Difficulties tend to arise later when the analyst tries to quantify his views and to construct a formal model.

The conceptual diagram and beyond

The following method is suggested as a general method for building models. It is intended for use only where direct analytic methods are not possible. It should not be regarded as a formal method of analysis since it is subject to the kind of difficulties which tend to occur with dimensional analysis to which it is slightly similar.

The process is commenced by listing the variables and then noting which of them interact directly or indirectly with each other. It is also necessary to appreciate which of the variables are to some extent under our control, and also any restrictions or constraints.

Let us consider, as an example, the simple Wilson lot size formula which is sometimes used in inventory and stock control. Clearly we would not use this method to find the formula but since it is well known it will serve as a good example to demonstrate the method. The formulation makes various assumptions about constant demand rates, replenishment costs, etc. and the following variables are considered:

λ is the demand rate
I is the inventory rate
A is the cost of placing the order per item
C is the cost of the item
Q is the quantity ordered and
K is the annual cost.

It will be seen from Figure 12.1 that when the stock falls to a certain level it is replenished with a quantity Q, and it can be shown quite simply that:

$$\text{total cost } K = \frac{A\lambda}{Q} + \frac{ICQ}{2}$$

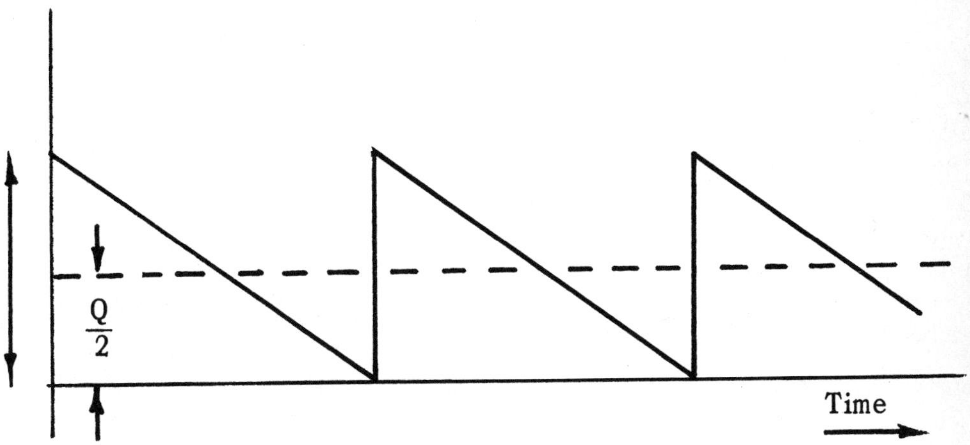

Figure 12.1 Stock quantities with the Wilson lot size model

Let us temporarily forget this formula and try to construct a model that will be suitable for stock control by building up a 'cause and effect' network which will show the way the different variables interact. Probably

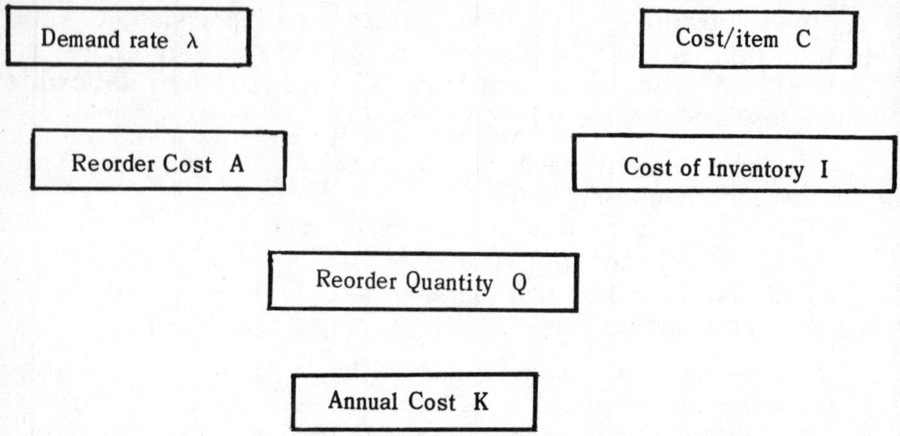

Figure 12.2 Variables in the Wilson model

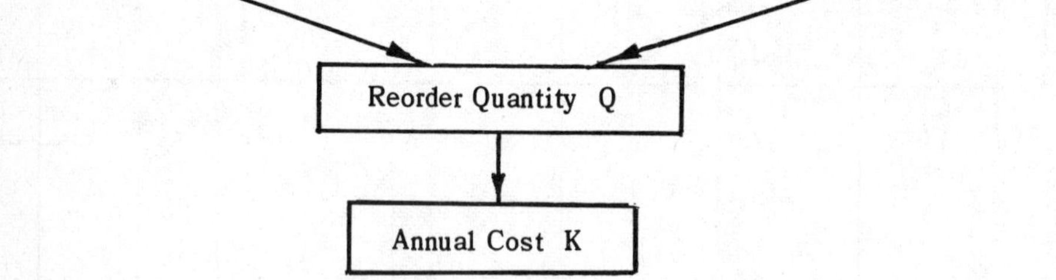

Figure 12.3 Conceptual diagram for simple inventory model

the demand rate λ is a good variable to consider first. Write this down inside the small box as shown in Figure 12.2. In the same way write down all the other known variables. By virtue of our earlier definitions the reorder cost is proportional to the demand rate so it is possible to draw an arrow from the 'demand box' to the 'reorder box' in order to show that the latter is affected by the former. Careful thought will show that the cost of inventory may be affected by the cost per item and that all these factors either directly or indirectly influence the reorder quantity and the total annual cost K. The linkages are illustrated in the diagram in Figure 12.3.

Rules for the derivation of formulae

The next step is to try to build up a formula from the inter-linkages. At present there do not appear to be any rules for executing this formulation so let us try to indicate what appears to be a reasonable method:

1 Find the box to which all the arrows lead. This is usually, but not always, the objective of the formulation. Write this down on the left-hand side of the page (it will become the left-hand side of an equation).
2 Find all the separate branches leading from this and write these down separately on the right-hand side. A plus or minus sign will occur in the formula to join each branch. Thus:

$$K \quad \pm (\lambda, A, Q) \pm (C, I, Q)$$

At this stage it would be possible to write the formula as:

$$K = \pm (a\lambda^g . \; bA^h . \; cQ^k) \pm (dC^l . \; eI^m . \; fQ^n)$$

where a b c d e f g h k l m and n are constants. Clearly it is desirable to try to elucidate this formula further. In order to do this the following stages are useful:

3 Start by considering the left-hand branch of the diagram. Does an increase in the flow along this branch cause an increase in the value of the left-hand side of the equation? If the answer is 'Yes', this term

is positive. Similarly for the next branch to the right-hand side of the one just considered and so on until all the branches have been treated. Thus the formula can now be written:

$$K = (a\lambda^g . bA^h . cQ^k) + (dC^l . eI^m . fQ^n)$$

Is it possible to evaluate the formula any further? The following procedure will indicate whether the rest of the constants $g\ h\ k\ l\ m\ n$ are positive.

4 Start at the end of the left-most branch. If this variable is increased will the left-hand side of the equation increase or decrease? If it increases place the variable in the numerator, otherwise place it in the denominator (i.e. the power is negative).
5 Progress to the next variable on the branch and consider whether this increases or decreases the left-hand side having regard only to the preceding variables in this branch.

In other words in the example, from Rule 4 above, the term $a\lambda^g$ will appear in the numerator. Taking the next variable in the branch, as per Rule 5, it will be seen that bA^h will also appear in the numerator. Now the effect of reorder quantity Q on this branch can be seen and it will be noted that the cost K on the left-hand side will be increased by decreasing Q since the reorder costs will be more. Hence, in this case the Q term is in the denominator.
Thus this branch can be written:

$\dfrac{a\lambda^g . bA^h}{cQ^k}$ where the k is now positive in the formula.

Similarly by developing the other branch it is possible to write the more complete formula:

$$K = \frac{a\lambda^g . bA^h}{cQ^k} + dC^l . eI^m . fQ^n$$

Now consider the values of the coefficients *a b c d e* and *f*. This evaluation of constants may or may not be possible by argument. The following forms the basis of this decision.

6 How is the left-hand side likely to change with the weighting of the variable? If the constant is doubled and the other branches are removed, the left-hand side will be doubled. Does this seem right? Should the term be unity or is there some reason for another value?

In the example this form of argument shows that *a b c d* and *e* should be unity but f is equal to 0.5 since the cost *K* on the left-hand side will depend on the average stock. Thus the formula can now be written:

$$K = \frac{\lambda^g A^h}{Q^k} + C^l I^m \frac{(Q)^n}{2}$$

Finally consider the power of the variables as follows.

7 How will the left-hand side be likely to change with the power of the variable? If the other branches are removed and the variable to be considered is changed in value, will the power constant induce the correct changes in the left-hand side? Should there be any other changes?

This part of the problem is often rather difficult and computation and statistics may be necessary for the evaluation of both the coefficients and the power terms; the latter may be especially difficult if they cannot be visualised as whole numbers. The solution to this example is relatively simple and after some thought it will probably be appreciated that in this case these terms are likely to be unity. Thus the final answer can be expressed as:

$$K = \frac{\lambda A}{Q} + I\,C\,\frac{Q}{2}$$

This is the same solution as before and has taken a great deal more effort to obtain than by the normal direct method but the object has been to show the way in which complicated situations can be resolved and quantified using this method of conceptual diagram development.

Problems which are extremely complicated often become much clearer when set out in this manner. As mentioned earlier, it is also possible to use conceptual diagrams in conjunction with statistical analyses for model building. The principal methods include component analysis, factor analysis, cannonical correlation analysis, discriminant analysis and cluster analysis. These techniques have come to be employed to a much greater extent in recent years as a consequence of the increasing use of computers, since the latter can save a great deal of time and drudgery in executing the necessary calculations.

Occasionally, the approximate mathematical relationship between variables may be known and it may be desired merely to study the effects of changing the constants or coefficients in the model.

It is often possible to use 'spreadsheet' software for this purpose, but in order to do so it is necessary to know not only the mathematical interactions between the variables, but also the sequence of actions between them. Once this has been achieved the mathematical relationships can be entered into the spreadsheet matrix in the correct sequence. Most spreadsheet software will sequence sideways and downwards and this is usually the most natural sequential arrangement for the variables. The effects of changes in the values of the coefficients or constants then become immediately obvious by entering the new values in the matrix and allowing the automatic recalculation procedure to take place.

13 Modelling human behaviour

Human behaviour forms one of the most difficult aspects of model building. In the physical sciences reaction and interaction can be predicted with accuracy and the repeatability of an experiment is relatively assured. Unfortunately, in operational research, actions and interactions are rarely deterministic and not necessarily ever statistically repeatable. We are normally much more concerned about the response of individuals, and any one individual may well react differently in circumstances that appear to be the same. However, the influences on a person's actions are very complicated, and so we could say that in any situation most of our difficulty lies in identifying the values of variables as perceived by that person.

Variability in human reactions

Experiments in game theory at Sussex University and elsewhere (34,35) have shown that in games between two players the behaviour of players and their intentions can be very different. Some players seek to maximise their own gain, some to maximise the gain on both sides, some seek to maximise their gain relative to the other player no matter how damaging this may be to themselves. There can be little doubt that such experimental games can be very uncertain. Attempts to quantify the number of people behaving in a particular way can only be related in broad general terms to the response of an individual to a particular experimental situation. Variations can occur as a consequence of the way in which people are briefed, the competitive nature of the environment (many of the tests are carried out using students) and so on. Even so, the experiments show that there are very real and wide differences in the way human beings may try to respond to certain set situations. (This, of course, extends some of the comments made earlier on the problems of using games as models for predictive purposes.)

Such differences are reflected in real life. Not everyone will steadfastly refuse orders to abandon a sinking ship and continue to work the guns in order to minimise the relative gain of the enemy. No doubt some differences are necessary for the survival of the human species, and in order to understand this, one may have to study the behaviour of other animals. For example, in the case of monkeys it is known that several may be prepared to risk almost certain death from a predator, in order to harass him and to save the tribe. We are therefore asking for trouble if we assume that the human reactions in any model we are building are all likely to be the same, in either a deterministic or a probabilistic sense.

It is true that we may be able to represent the macro-reactions of a large number of individuals fairly well, but nations have repeatedly failed to predict the behaviour of other nations, and this is quite often because they have different models, or pictures, of the world. Thus, in the gaming experiments mentioned earlier, it would be interesting to study the social background and immediate environment of the players.

Towards quantification

There have been many attempts to measure personality and motivation. The so-called 16 Personality Factor Scale (16PFS) was devised in an attempt to try to quantify human personality. It can be said of this approach that the measurements can sometimes prove useful in certain problems. Cattell (57) has spent much time in studying the scientific analysis of personality. His methods and those of many psychologists are interesting, but are generally not very easy to apply to the kind of problem with which the average operational analyst is faced. For one thing the analyst is not usually a trained psychologist, and for another there is usually something rather special about each individual situation.

The translation of personality and environment into the attitudes and beliefs which influence behaviour is even more uncertain. The whole process of attitude, belief and cognitive interaction is very complicated and it is doubtful whether simplistic models, such as those of Fishbein (58) are very helpful other than to provide a general definition of a possible explanation of decision processes. His approach is rather similar to that suggested by Newcombe in 1946. (59)

Fishbein suggests that an attitude can be expressed as follows:

$$A_o = \sum_{i=i}^{n} B_i \, a_i$$

where:

A_o is the attitude towards object O

B_i is the strength of belief i about O (i.e. the probability that O is related to some other concept X_i)

a_i is the evaluative aspect of B_i or the evaluation of X_i

and

n is the number of beliefs about O.

This approach to decision analysis assumes that the deductive structure of decisions is evident, and that the consequences of each move can be studied and evaluated. The final move therefore depends on the total scores obtained for the various alternatives.

It is worth noting at this point some of the general implications of these comments to the operational analyst. The attitudes of management to any model, or proposed solution to problems, is therefore likely to be heavily affected by their individual belief structures. This is why many analysts, who have constructed excellent models, become frustrated when their models are not implemented. The answer is simply that quite often the belief structures of the analyst and the managers are totally different. The analyst must therefore carry the managers along with him during the time he is working on the problem so that belief structures become exposed during discussions and fallacies or additional problems are fully revealed.

There have been many attempts to study and model group interactions, and perhaps a good general reference to this work is the book edited by Cartwright and Zander (60) which contains numerous papers by some of the foremost workers. Many of the papers also contain details of experiments in motivation, leadership and decision making.

The limited experiments I have undertaken on human interactions have been very rewarding and have shown that it is unwise to expect too much of human organisations. The first discovery came almost by accident when trying to find the optimal way for a group leader to allocate his time. During this analysis it was discovered that the cooperation between individuals who, from the nature of the management structure were supposed to cooperate with each other, was positively correlated to the time they spent talking to each other. Individuals who disliked each other, whom one would normally expect to find spending at least a couple of hours a week resolving their work problems, would perhaps spend only two minutes in discussion on the telephone in a whole week. Thus, by simply asking individuals how much time they spent talking, and how this was shared, it was possible to build a picture of the cooperation existing in an organisation. Still further, it was possible to discover the state of cooperation between the organisation itself and the other organisations with which it had to deal. (61,62)

Some years later a much more elaborate experiment was conducted into the rate at which information is lost or distorted in the course of normal face-to-face verbal communication. This showed how some 30 per cent of verbal information was lost at each transfer from one person to the next after a two-hour interval. It also showed that as the information was lost it became necessary to adopt rationalisation procedures in order that the remaining information should make sense. This often involved 'inventing' suitable information, so that sometimes, after several transfers, the new information was almost more believable than the original message. Only dates, times and numbers, which were usually written down, were not subject to the 30 per cent attrition process. Of the plausible additions, or 'inventions', approximately one-third conveyed almost exactly the opposite meaning to the original information. As a consequence, the original message eventually became very garbled. It was possible from these experiments to build a communication simulation model and to show the limits to which an organisation might be able to grow before encountering serious control problems.

All these analyses barely uncover the surface of the many problems the model builder has to face when trying to construct a model which will adequately represent the situation and will satisfy the human decision maker. Bowen (63) has been much concerned about the problems created by differences in perception, and Ackoff and Emery (64) have considered both this and the effect of attitudes on the ultimate solution which may be implemented. Various different studies have been carried out to try to understand how the decision maker sees, or recognises, the problem. Eden at Bath University has done excellent work on this problem, but some find it a little difficult to apply his approach. However, the whole matter of cognisance is critical so let us briefly examine some of the work in this field, at the same time remembering its inadequacies.

Cognitive modelling

It is often difficult to grasp the precise nature of the problem posed by the decision makers and cognitive mapping can sometimes be helpful. Cognitive mapping can be defined as a mental construct which enables people to acquire, code, store, recall and manipulate information about their spatial environment. Cognitive mapping or modelling processes are, in fact, part of a general process whereby an individual copes with information from his total environment.

The multidimensional cognitive maps which form the cognitive model are difficult to quantify. It has been said that they belong to a class of phenomena variously referred to as hypothetical constructs, intervening variables, or inferred states or processes. I tend to disagree with the view that they are within the same category as intervening variables except in so far as they are often difficult to quantify. Indeed, there are always difficulties in trying to quantify attitudes or mental concepts.

The quantification of mental maps can be attempted in different ways. Let us, for simplicity, consider that we are dealing with a particular area of a town and seeking to obtain people's views of that town, or area of the town. We could use: mapping which would include free-hand sketches; verbal descriptions obtained by interview or questionnaire; identification of photographs; observations of behaviour in the environment; and psychological tests which might include repertory grid, semantic differential, attitude scales, etc.

The results of such tests often indicate very different views of an area and seem to suggest that people generate different mental maps of their own individual physical environment. These tend to reflect the different uses they make of the environment.

It will, perhaps, be evident from the foregoing remarks that social psychologists have tended to view cognitive models in a rather limited way. One of the principal uses is the determination, in part, of the social environment of new towns, new blocks of flats, etc. In one sense there is nothing very novel in the idea of cognitive mapping, in that all we are often asked to provide is information which will yield a picture of part of our 'model of the world'. To some extent our model contains, or has a sub-model of, a set of attitudes. Attitudes are not always readily accessible directly, and it seems to me that this is one of the advantages of the cognitive search approach. In other words you do not have to say directly 'What do you think of Shepherd's Bush?' It becomes fairly evident from a person's remarks and descriptions when asked some question such as how to get from A to B, whether he sees the area in a purely functional way, such as a stage in getting to work, or as a social amenity, in which cases the local pubs or bingo halls may get a mention (for example, 'Turn right at the Red Lion').

Shapiro and Bonham (65) have been particularly concerned with cognitive processes and foreign policy decision making. In order to study how different knowledge and backgrounds gave rise to different decisions they carried out a simulation using a crisis game invented by Steinbruner. In this game three participants with different backgrounds were read a scenario of a crisis and asked to advise the US President what his response should be. The major event in the scenario was the placing of nuclear weapons in Syria by the USSR.

The first participant was a specialist in international relations, the second was a specialist in comparative politics, and the third was a Middle Eastern specialist.

It was found that the first specialist tended to see the Middle Eastern situation in the broad context of an international power struggle featuring the USA, the USSR and various uncommitted nations. He asked questions not only about the Arab positions but also about the policies and recent involvements in the area of the USSR, the USA, the UK and France. He saw many explanations for the Russian behaviour, including a move to force Israel to surrender occupied Arab territory, and also as a wider global strategic interaction.

It is said that he also commented as follows: 'I still think that the basic question which faces the United States is an assessment of Soviet intentions. If, indeed, it is the Soviets' intention to simply force Israeli evacuation of occupied territories, that is one thing ... But we don't know whether that is the Soviet objective, and we don't know what other Soviet objectives there may be and how tenaciously they may be held.'

As a consequence he suggested beginning negotiations along the lines of the French proposal for a Middle East settlement. However, since the Soviet intentions might be more ambitious, he also recommended various public and private moves, the former to include US Naval training exercises and an approach to India to give nuclear technical assistance to Israel (the USA could not do so due to treaty agreements). The covert activities included aiming Polaris missiles at Syria and a threat to attack Cuba if the weapons were not removed.

The comparative politics specialist enquired about Soviet military strength and about relations between Arab regimes, and also about intra-governmental conflicts in Israel. He saw the Soviet move in rather less global strategic terms and felt that the Soviet behaviour signalled an eagerness to start negotiations. He said that the

Russians always liked to negotiate from a position of strength. He also pointed out that the missiles were still under Russian control. He recommended protesting to the United Nations, demonstrating military readiness close to the trouble spot and attempting to begin negotiations.

The Middle East specialist asked about Soviet intentions, about the strength of some of the Arab countries, and about the activities of the Palestinian commandos. He was inclined to see the move as a result of developments in the Middle East. He also saw the Arab–Israeli conflict as the major determinant of Soviet penetration into the Middle East and thought that Arab regimes turned to the Soviets for support because no other nation seemed sympathetic. He advocated an immediate start to negotiations and the use of international auspices to effect a settlement which would touch upon all relevant sources of grievance in the area.

The three participants therefore used different theories of the environment in reaching their conclusions. It is worth noting that the complexity of thought of the participants matched the complexity of their perspectives; and the more aspects to the theoretical basis on which they operated, the more complex their policy recommendations.

The participants were asked to answer questions about what they saw as the three major objectives for US policy in the Middle East and to indicate the extent to which a number of alternative policy proposals (including the one they had selected) would achieve or block each of the three objectives they had chosen (a technique probably due to Rosenberg). As in the case of their perceptual mapping, the cognitive processing of the participants was arranged in order of complexity. This showed the first participant as having the most complex process, and the third the least complex. It is therefore possible to obtain some measure of the complexity of cognitive processing by simply looking at the responses to these questions and noting the variance in their beliefs about the effects of the alternative policies on the objectives.

The Middle East expert was invariant about his beliefs about the various policy alternatives. He scored +3 chance of attaining all his objectives (scale +5 to -5). Other policy alternatives were given scores of -5 for all three objectives, another -4 for all three objectives, and the other -1 for all three objectives. The other participants thought that each policy alternative would have a different effect on their chosen objectives. The first participant saw the greatest variation in his beliefs about the effects of different policy alternatives.

Shapiro suggests that beliefs represent the concealed experiences of the decision maker and his expectations about the environment. Beliefs act like templates for channelling information. Beliefs of decision makers, he states, can be represented as a map of causal linkages. *Affective concepts* refer to the policy objectives, or actors in the international system. *Cognitive concepts* concern beliefs about actions that occur in the international system, and *conative concepts* indicate possible alternatives from which the decision maker selects policy recommendations. The total *map linkages* between these concepts *form* the *cognitive map of the beliefs* of the decision maker. The linkages are represented by arrows carrying either positive or negative signs in order to distinguish between the direction of the causal relationships as seen by the decision maker.

According to Shapiro and Bonham four processes are invoked when a decision maker is confronted with a new problem which requires a response. These are: the amplification of beliefs; the search for an explanation; the search for policy options; and the choice of a policy.

In the amplification of beliefs stage, the decision maker attempts to place the situation in the context of his experiences. In this stage all possible relevant concepts are 'highlighted' and the information stored for immediate use. The search for explanations, the second stage of the process, is rather more complicated. If a set of causal linkages exists between two concepts an immediate explanation exists. The initial concept in an explanatory path is often a person's, or organisation's, intention; whilst the final concept may have been generated during the amplification process.

The search for policy options is restricted to those embedded in a decision maker's explanation of the situation. He looks for options of policy which will give him control to a satisfactory state, or will set off a chain of actions or events which will have a favourable impact on his policy objectives. This relates to the isomorphism concept of Asch (66) and comments by Campbell. (67)

The choice of policy options can employ various calculus methods. In the Shapiro and Bonham model, the decision maker is first concerned with his most important policy objective, and if the options affect it equally, he moves to the second priority objective, and so on.

The precise methods used by Shapiro and Bonham are quite complex in that the concepts are often expressed as matrices and diagraph theory is used to represent the cognitive structure and to calculate the cognitive processes. The concept of 'reachability' in diagraph theory is used in the amplification of beliefs and search procedures. If several explanations of a system exist, the model searches for the one with the most deductive support. This is the number of logically independent reasons which reinforce an explanation. It is accomplished by constructing and manipulating an arrangement called an 'adjacency matrix'.

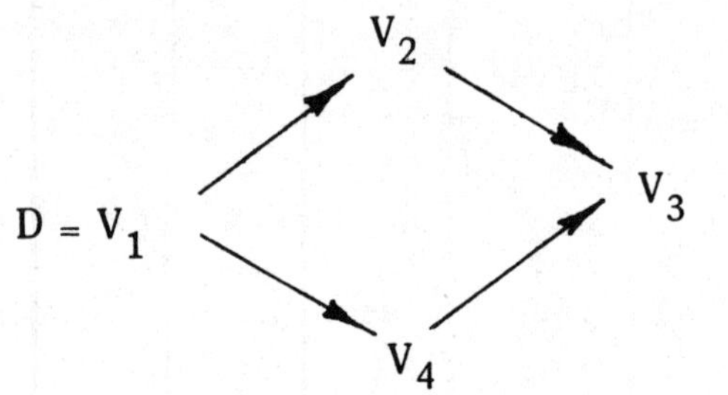

Figure 13.1 Simple diagraph

	V_1	V_2	V_3	V_5
V_1	0	1	0	1
V_2	0	0	1	0
V_3	0	0	0	1
V_4	0	0	0	0

Figure 13.2 Adjacency matrix

Consider the simple diagraph in Figure 13.1. The adjacency matrix for this is shown in Figure 13.2. Raising the matrix to the power of 2 causes the non-zero elements in the matrix to indicate the number of paths of length 2. This process can be continued to the nth power to discover paths of length n. The number of paths to highlighted concepts can therefore be counted. This counting process is deemed to be similar to deductive support. In checking the effect of policy alternatives on objectives, a valency matrix is used. This is a signed diagraph converted into an adjacency matrix which preserves the signs to show whether the concepts are positive or negative. With this type of matrix it is possible to determine whether there is a positive or negative path from a specific conative concept to one or more affective concepts.

The difficulty with most of the Shapiro and Bonham studies is that they are largely conducted after the event. Certain options, which, to some, might have seemed probable at the time, tended to get ruled out. In practice this is often less of a problem than the simple question as to whether a decision maker has perceived all the options available. In other words, one of the main difficulties is to input all feasible concepts and actions into the model. However, having made this comment, the Shapiro–Bonham model remains a most interesting approach for studying decision processes. It also involves some of the most advanced methods of cognitive mapping which, even in its simplest form, is worthy of further attention as a means of determining genuine attitudes and concepts.

14 Cybernetic models

The word 'cybernetic' is derived from a Greek word meaning 'the science of steering ships'. It was used in a wider sense by Ampere to mean the control of society. The concept was popularised by Norbert Wiener in 1948 in a book which dealt with control in machines and living systems. [68] It involves closed-loop feedback systems, the manipulation of information which guides these systems and the processes for filtering out random disturbances from information channels.

Cybernetic principles and feedback loops

Much is written in the literature about cybernetic principles and modelling [69,70] and some of this overlaps, or is simply a re-expression of well known principles of control theory. For example, descriptions are given of various means of sensing an output and feeding this information back to the input so that a new input signal is generated in the same phase as the already incoming signal. This is known as positive feedback, and not surprisingly usually results in an unstable arrangement which often tends to get out of control. In everyday life an example of this would be an inflationary situation where prices rise, leading to a demand for more wages, which in turn leads to a further increase in prices.

However, if the signal obtained from sensing the output value is fed back to the input in the opposite direction, then it will detract from the incoming signal. This is known as negative feedback and normally produces a much more stable system. Thus a government which threatens increased taxation if wages expand too rapidly is in effect apparently applying a form of negative feedback since it is virtually reducing wages.

Although this explains the nature of positive and negative feedback, it should be noted in passing that this kind of argument is much too simplistic as an economic concept. Unfortunately it appears to be often adopted by politicians, possibly as a result of economists talking about the need to reduce money supply. Since it is a lesson in the importance of isolating the causal variables and the multiplicity of feedback loops, it may be worth digressing from our main theme for a moment.

In the first place it is likely that dissatisfaction with wages is some function of the rate of change between prices and wages, although dissatisfaction is also likely to arise in an inflationary society as the result of temporary pay anomalies. Second, having obtained the additional tax money, there may well be some pressure on the government to spend it, e.g. strong demands from the social services. Many believe that value added tax provides a rapid means of controlling the economy more easily and seem to think that, when prices rise, value added tax should be reduced. Some monetarists might well argue that this will put positive feedback in the system, and what should happen is that, if prices increase, value added tax should be increased. This might be effective if the money were invested overseas or in purchasing commodity or power sources of long-term value. On the other hand, if you believe reducing value added tax to be a psychologically effective means of reducing wage claims, then it could be argued that a certain amount of positive feedback might be tolerated as a temporary measure.

It will, perhaps, now be appreciated that the whole question is quite complicated and that one may have all kinds of time lags and feedback loops in a complex system. Indeed, the economy is so complicated that it is not possible to deal with this subject in any greater detail at this stage. What should now be clear is the meaning of positive and negative feedback loops in a model, the way in which these affect the overall stability of the system and the multiplicity of loops which can exist in what, at first sight, may seem to be a simple model.

Many books on cybernetics also contain a great deal of information about the analogies between human and machine control systems. These are extremely interesting from the technological and learning aspects, in that they may greatly increase our understanding of systems. However, having briefly commented on the kind of things encountered in books on cybernetics, we should note that one or two books on cybernetic modelling become extremely involved in definitions and say relatively little about problem solving. Let us now turn to the kind of things which are usually intended by the words 'cybernetic modelling'.

A cybernetic model has the distinctive feature that it has the ability to 'think for itself' and to change its form as it is used over a period of time. This clearly involves some form of feedback and the comparison of expected answers with the actual answers obtained in practice.

The question then arises how do we make a model 'think for itself'? The answer is, of course, that we can only make it obey certain decision rules which we have provided. For example, in making forecasts about a certain variable in an unfolding time series, we may run a simple regression on the data and extrapolate the values into the future. If the points are not widely scattered about the regression line, the forecast may be reasonably valid and we may have a fair amount of confidence in our prediction. If the data are widely scattered about the regression line, then we are likely to have little confidence in the prediction, and we may well be advised to simply use the average of the last few values as a predictor of the value of the variable over the next time frame.

It is relatively easy to devise a simple test to check the data scatter when the data are being used for a model programmed on a computer. Thus, as new data are acquired from observation, and are fed into the computer, the program will test the scatter of the data, decide whether to use the regression extrapolation as a predictor, and may then feed this predicted variable into a more detailed model which in turn is designed to predict certain events. This is an extremely simple example of a model deciding for itself whether the input data are reliable or not. It is not very difficult to produce even more complex decision rules which are likely to be still more effective. Indeed a process of forecasting using what are known as Kalman filters, (71) which have a special ability to sort the signal from the noise by varying the filter characteristics automatically, has been found to be of value in adaptive models.

Felsen (72,73) has used cybernetic models extensively, and apparently effectively, to forecast the Dow–Jones Index of American stocks and shares. His approach is to feed all the relevant market data into his model and to let the model sort out the most important data from the rest, using past experience as a guide. In this way he can use the model to consider an enormous number of variables and their interactions, in a way which would not be possible for a single individual. Overall judgements are generally much more reliable when such a method is used. The precise way in which Felsen's model works has never been made absolutely clear, although he has written a great deal about it. Basically the idea is as follows. Let us suppose that we are trying to forecast the value of a commodity such as cocoa, at some time in the future. A cybernetic model might include a number of variables such as the price last year, demand last year, an estimate of the demand this year and next year, changes estimated in crop acreage, changes estimated in labour costs, changes in shipping and transport costs, etc. The model may also have a number of other variables to take account of the weather, government intervention, etc. It is this kind of random variable which can have a considerable effect on price, which may be very important in a cybernetic model. Thus we may have a model of some form:

$$\text{Price} = A\,f(x) + Bg(y) + Ch(z) + \ldots..$$

In other words the price of the commodity is a function of the variable x (which could be the weather) multiplied by the 'constant' A (which in a cybernetic model is not, in fact, constant), plus a function of the variable y (which might represent the demand) multiplied by a 'constant' B, and so on. Initially the most appropriate values of A, B and C will be chosen, but as new information becomes available which shows that the value of these constants might require revision, the sub-routines in the model are allowed to modify automatically the values of the constants.

Hence, if we see an article in the *Wall Street Journal* which reports some damage to crops from frost in a small producer country, we may feed a -1 into a formula for calculating the value of *A*. We may grade weather effects subjectively from -5 to +5. Reports of very favourable weather might give an input of +4. It is also possible to count the number of reports received and attach a weighting to this number, or to the reliability of the source. Sometimes a 'constant' may be allowed gradually to settle to its original level provided no further information becomes available, and sometimes not. Much will depend on the nature of the variable it is affecting.

The values of *A*, *B* and *C* may have been obtained originally from a multivariate analysis, but in the cybernetic model their relative importance will be reviewed continually. Thus not only may *A*, *B* and *C* be changing because as time unfolds new results are obtained, but the model will also be learning whether the weighting factors, used to calculate the constants, should be altered in order to make better predictions.

A large cybernetic model may therefore necessitate a considerable amount of time being spent in reading and assessing the relevant information and inputting it to a computer. A quite small amount of computing time will be required to adjust the various weighting factors in the light of new information received. Then, from comparison of actual and predicted events, the computer may decide automatically that it is necessary to review and revise the weighting factor calculation methods and, possibly, even the basic model. Ideally, only the collection of the data and their entry into the computer should ultimately involve human intervention.

Thus the model is initially operated with the existing data, and new data are fed into the model and compared with the predictions, as further information becomes available. A fully cybernetic model should therefore improve as time passes, in that the basic formula and coefficients should become more accurate.

As a consequence cybernetic models have certain special qualities and value. They have, albeit sometimes rather crudely, the ability to think for themselves, having been told in the first place the way in which they are required to think. For certain types of model this approach is particularly useful. I have found the method very helpful in developing certain economic models and have used the method to enable the model itself to apply some moderately simple tests and to decide whether the country was suffering from a certain amount of economic chaos. In the latter case it was required to function very differently from the way it did in situations of relative economic stability.

An additional and important advantage which cybernetic models often possess is that they can sometimes cope better than normal models with exogenous effects, provided their thinking logic has been carefully devised. In certain situations this can prove invaluable.

Various new names have recently appeared in connection with these kinds of model. The name 'IKBS', which stands for 'intelligent knowledge-based systems', and the name 'expert systems' have both recently come into use. An 'expert system' is simply a program which incorporates the knowledge of an expert and therefore allows an inexperienced person to operate it with considerable effectiveness. Intelligent knowledge-based systems are largely an extension of this concept but usually include a huge data base which needs to be searched before the appropriate action can be recommended.

Names in science, especially in the newer sciences, can sometimes be a matter of fashion and so whether these particular titles will survive is open to question. What is certain is that the pace of the automation of mental processes will continue to gather speed and will probably be one of the most important areas of activity in the coming decades.

15 Subjective judgements

It is surprising how many problems which initially appear to involve only quantitative measurements also rely greatly on hidden subjective judgements. It is also surprising how many subjective judgements, or value judgements, as they are often called, can be studied quantitatively if careful thought is given to the appropriate measurements to take. As mentioned earlier, methods have been found to measure unlikely parameters such as the extent of cooperation between individuals and between organisations, and the information content in verbal communication. (74,75)

Normalised expected marking

A beauty contest provides an interesting example of value judgements. There is no reason why the judges should be unanimous about the winner, since each judge will have a different set of in-built standards, or values, and these standards will have been modified by experience and environment.

Experience can be a great modifier of values, and a judge who comes from a tropical island and lives surrounded by beautiful women will have a different standard, or scale of values, from his twin brother who has spent his life incarcerated on another island with a group of rather elderly female witch doctors. This is an absolute, or level, difference. It is also probable that their discriminatory abilities will be different, since they will have been used to judging at different ends of the scale. This is a scalar difference.

Even so, and having regard to all these difficulties, judgements have to be made in beauty contests and also in many other situations. We could approach this matter in various ways. We could try a simple judgemental quantification by using a marking system where each candidate is awarded a number of points out of 100. Alternatively, we can ask each judge to rank the contestants in the preferred order from 1 to 10, since this is generally much easier to do. Points can then be awarded in an inverse ratio to the rank, i.e. 10 points for the one rated first, 9 for the second and so on. The judge's individual scores can be added and the candidate with the highest score named the winner.

A method of using indifference between choices has been developed by Rivett (10,76) in the process he calls 'structural mapping'. This does not produce probabilities, but shows the interlinkages and preferences where a number of choices exist. It is, of course, necessary with this method to have sufficient choices to link all the possibilities together and to view them in the proper perspective. We might try to use Rivett's system of selecting pairs of candidates whom we rate as equal and between whom we are indifferent as to which is selected. However, indifference appears to be hardly the right measure for a beauty contest!

As mentioned earlier, ranking is much simpler than awarding points and usually creates fewer anomalies in the marking system. If several judges are employed, then it is usually necessary to combine their assessments by awarding marks in the inverse order as explained above. However, the question arises as to whether this is a totally satisfactory method.

One method of developing a marking system is to consider the total quantity of marks to be awarded and the way in which we might expect them to be apportioned on average. Appendix 15 describes a method of dealing with random partitioning and ranking the partitioned pieces. Thus, if we were able to divide a hundred marks at random amongst ten candidates, then from the random break formula proposed in Appendix 15 we would expect the marks, on average, to be proportioned as follows:

Candidate No.	Expected random partition value	Normalised candidate no. 1 to 100
1	29.29	100
2	19.28	66
3	14.29	49
4	10.96	37
5	8.46	29
6	6.46	22
7	4.79	16
8	3.36	11
9	2.11	7
10	1.00	3

To obtain the normalised expected random values simply multiply all the expected random partition values by a factor such that the value for the top candidate is brought to a value of 100, or whatever is appropriate having regard to contestants in previous years.

Thus, if we had several judges and ten finalists, judges would simply rank the contestants in order, and the points appropriate to the rank (as from the table above) awarded.

Let us suppose that we have two contestants who have been selected from ten finalists. Five judges rank these two contestants as follows:

Miss A	Miss B
3 judges award 1st	4 judges award 2nd
2 judges award 8th	1 judge awards 4th

If we adopt the normal practice of awarding marks from 1 to 10 inversely with ranking we obtain:

For Miss A			For Miss B		
3 x 10	=	30	4 x 9	=	36
2 x 3	=	6	1 x 7	=	7
Total		36	Total		43

which makes Miss B the clear winner.

Several people were asked which contestant should be ranked first. The result was interesting. Most said that they thought Miss A should be ranked first but none were certain why, other than the fact that she had been ranked first by three judges. Being mostly OR men they immediately tried to quantify their beliefs by using a simple 1 to 10 point inverse rank marking system just described. This obviously resulted in a convincing win for Miss B but also left them quite unhappy about the outcome.

However, if we apply the random break approach we get a very different result. Awarding marks according to the normalised random partition gives for Miss A:

3 x 100	=	300	4 x 66	=	264
2 x 11	=	22	1 x 37	=	37
Total for 5 judges		322	Total for 5 judges		301
i.e.					
Average	=	64	Average	=	60

This makes Miss A the winner. It therefore seems likely that the expected random partition weighting will provide more satisfactory answers in this type of situation.

Difficulties can arise in trying to arrange different kinds of things in a preferred order. The problems are increased if one item depends on another, but it may be that the introduction of either one or two explanatory variables, for which values can be assumed, will make the original items conditionally separable and hence capable of assessment. If there is only one explanatory variable, its distribution can then be determined separately. If there are two explanatory variables, an examination may indicate that they are independent, or have algebraic combinations which are independent. If they are not, it may be necessary to introduce a new explanatory variable and so on.

It is possible to write a fair amount of axiomatic theory about this aspect, but it is all largely commonsense and may tend to confuse rather than clarify. In practice it is likely that the analyst will be able to sort out this kind of choice fairly quickly provided he gives the matter a little careful thought.

Using many judgements

In common with many judgement problems, it may well be helpful to split the beauty contest analysis into several parts in order to ensure a balanced assessment. Thus separate sets of marks may be awarded for figure, deportment, personality, etc. Some of the general techniques used for social problems are described by Kaplan and Schwartz. (77)

Another judgemental technique is the cybernetic approach developed by Felsen (73) for commodity forecasting. As explained earlier, newspaper articles and television news items carrying relevant information on the commodity are weighted according to their content and then fed into a large formula which is corrected as new data are acquired.

At least one of the major banks in Britain uses value judgements as a means of forecasting interest rates. A panel of 'experts' is asked to predict the movement of interest rates over the next month, and the average of these estimates may then be used as a prediction. Further improvements can be made by weighting the prediction according to individual past performance. The fact that such predictions are correct in forecasting the direction of movement for a period of one month ahead for better than 70 per cent of the time appears to be superficially a success for the value judgement approach. However, quantitative methods show that interest rates are highly correlated and, if we use a relatively simple approach such as a slightly modified Box–Jenkins technique, we can obtain equally good results without any experts at all!

The problem of time

Narrow considerations which seem critical in the short term can prove disastrous in the long term. To some extent the growth of large cost-effectiveness teams and working parties in the 1950s and 1960s may have been partly responsible for emphasis on the immediate rather than the future. Indeed, some might say that in this period value judgements were sacrificed on the altar of cost-effectiveness. Professor Ackoff (University of Pennsylvania, USA) in the early 1970s stressed what he called the 'teleological approach', i.e. the need to view the problem within a much broader framework so that the wider issues can be considered. The difficulty is, of course, that most people do not have the ability to look very far ahead. (Indeed, this is one of the main problems in using scheduling methods such as CPM/PERT as discussed in Appendix 16.)

A consensus of value judgements appears to have the greatest use for present-time assessments or extremely short-range forecasts. Indeed, the cost-effectiveness studies mentioned earlier are a form of present value judgement, since they tend to value situations in terms of present costs, or future costs as perceived at the present time. Hence, only time may show that valuing a local branch railway line in terms of its immediate operating profitability is a mistaken judgement in terms of the total long-term savings, when road traffic congestion, additional road building and energy conservation become increasingly critical factors as time passes.

The real difficulty with subjective or value judgements therefore tends to remain. Judgements which are likely to prove right about long-term future effects often do not represent the consensus of present opinion and are regarded as mavericks. This, at least in part, is due to the fact that the unexpected so often happens. One of the reasons why unexpected events occur so frequently is that a large, almost infinite, variety of possibilities exists. Many of these possibilities individually have a very low probability of occurring, but collectively they have a very much higher probability and so there is a reasonable chance of one of these unlikely events taking place.

It must also be accepted that, even in the case of short-term effects, attempts to directly quantify the opinions of individuals who are not numerate may often prove difficult. What is certain, however, is that value judgement is no substitute for reasoned analysis and some attempt at quantification. The latter should include, where possible, a study of the probable effects of time on the important parameters. If problems are fully explored in this way, value judgements and subjective guesses at probabilities may need to be applied only to relatively few intangibles.

Estimating subjective probabilities

The estimation of subjective probabilities often depends on guessing the true proportion of certain events. For example, suppose that you have been invited to visit an acquaintance and you plan to take a small present, what should you decide to buy?

In the first place take and carefully use every piece of information available. Let us call this primary, or stage one information. Suppose that all we know about the person we are visiting is that he is a retired Scottish ships' engineer. We can perhaps link this with another piece of information which we can recall from memory which is that, of the few Scottish ships' engineers we have known, none were averse to a wee dram of whisky. This is secondary or stage two information. The thought then occurs to us that perhaps we should take along a bottle of Scotch as a present. It therefore seems unlikely that our host will be completely teetotal, but what is the chance our present will be welcome?

Let us begin by asking ourselves out of every hundred retired Scottish ships' engineers how many, do we think, would be likely to welcome a bottle of Scotch. First, could the number be over 50 or under? Over 50 we say. Very well, if over 50, will it be under 99? You may say yes to this too. Now consider a number which you consider it is equally likely that the event will be above and below. Suppose that you say 90.

Next suppose that you were told that the number was known to be less than 90, divide the interval from 0 to 90 into what you would consider to give equal probabilities of the event. You may say 85. Now imagine (perhaps rather incredibly) that you have been told exactly the reverse and that the number is over 90. How would you divide this into equal probabilities of the event occurring? Let us suppose that you say 94.

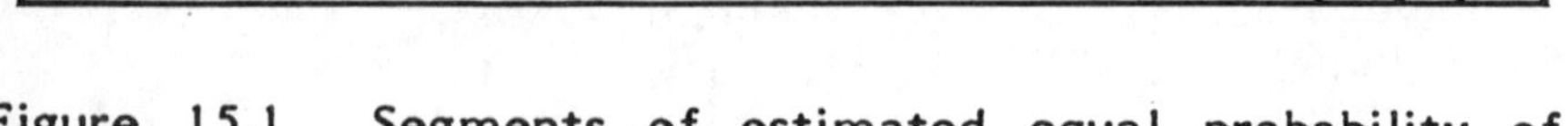

Figure 15.1 Segments of estimated equal probability of the occurrence of an event

Now show these points on a line as shown in Figure 15.1 and consider if asked to place a bet whether or not you would have any preference for placing it on any particular segment. If the answer is yes, then perhaps we have got the points in slightly incorrect positions, and it is necessary to carry out adjustments. At this stage just by a little further thought we may wish to change the first figure of 90, or we may wish to adjust the 85 or 94 values, which are the .25 and .75 fractile values denoted symbolically as $p_{.25}$ and $p_{.75}$ respectively.

It is now possible to try dividing the quartile sections into zones of equal probability (i.e. to obtain $p_{.125}$, $p_{.25}$, $p_{.375}$, etc.). Let us suppose that these values turn out to be 80, 88, 92.5 and 96 respectively.

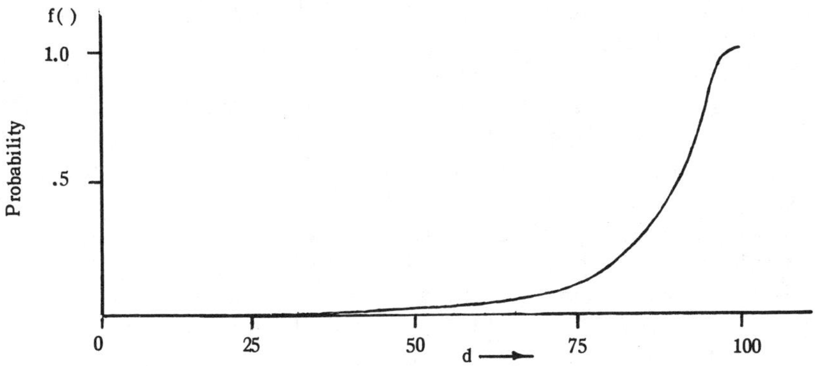

Figure 15.2 Subjective cumulative probability distribution

Taking these values it is possible to plot a rough cumulative probability distribution as shown in Figure 15.2, and to smooth out any obvious errors or anomalies in the curve.

It may sometimes be possible to obtain a few actual observations or measurements and see how these fit a subjective distribution. Confidence intervals with small samples of real information may not be very meaningful and the method of maximum likelihood is certain to be very wrong on occasions. Even so, the maximum likelihood method is usually worth trying, and in the example the maximum likelihood estimate of the probability that the gift is acceptable (*p*) can be obtained from the number of drinkers (*d*) in the sample and the number of non-drinkers (*n*) from:

$$\text{Max } L(p) = \frac{d}{n + d}$$

In making the estimate of the probability that the retired Scottish ships' engineer would like a bottle of whisky as a present it was implicitly assumed that he would not like it if he happened to be a non-drinker. This is not necessarily true and in effect this problem forms part of a decision tree on which new branches can be added to consider the possibility that he might like to have the whisky for his family, or for entertaining friends. Alternatively, it may have some utility value since he could probably always manage to swop it for a haggis.

It is also possible that after all this estimation we may have considered that there was just a one per cent chance that he was a reformed alcoholic and decided to take him a different present anyway, for fear that we might create problems. In this event we may, in effect, be unconsciously using the minimax regret criterion described earlier, but certainly we have discovered that one of the pay-offs in the decision tree is negative and as a consequence we are having to consider very much more carefully whether we can accept this kind of risk.

Subjective judgements enter into a great variety of problems and are frequently the reason why sound technical methods appear to fail hopelessly. Even apparently firm planning schedules using PERT and CPM (Appendix 16), which may have achieved the pseudo authority created by computerisation, are very vulnerable to failure in view of the fact that such schedules are frequently based on subjective estimates.

Many judgements

Provided the analyst has the time it is often worthwhile splitting down the problem to this extent. However, as mentioned elsewhere, decision trees can grow with great rapidity so that it is usually necessary to put some limitations on the problem and to consider certain possibilities simultaneously.

It is also necessary to provide some limitation to the time horizon otherwise the number of alternative branches and outcomes tend to infinity with alarming speed!

Saaty (78) has developed an interesting method of considering multiple choice and attributes. He argues that if difficult decisions have to be reached in complex situations then it is an observable human trait to attempt to break down the situation into smaller parts and that often these smaller parts form a hierarchy. Usually an attempt is then made to determine how the lesser parts of the hierarchy contribute to the goals that represent the apex of the hierarchy. He believes that this is a process to be encouraged and suggests a method for determining the contributions made by one level to elements at the next level up. His method relies on pair-wise comparisons of the relative contributions made by each of the lower elements to each upper element in turn, together with a similar assessment of the relative importance of the lower elements among themselves (possibly derived from an earlier application of the same method). From this information he is able to derive both a measure of the numerical consistency of the pair-wise comparisons and importance weighting factors for the upper layer. This process can be repeated till the apex is all that remains.

The pair-wise comparisons are in the form of a ratio lying between 1/9 and 9/1 and the matrix of ratios (a reciprocal matrix in which $a(i,j) = 1/a(j,i)$ is formed. The leading diagonal contains only *1*'s. If a group of comparisons were wholly consistent, then there would be linear relationships between the individual elements and the matrix would have only one non-zero eigenvalue equal to the sum of the diagonal elements. The corresponding eigenvector would provide weights in the given ratios. In real situations perfect consistency will not normally occur and some inconsistency will exist. A measure of the inconsistency is derived from the size of the maximum eigenvalue of the reciprocal matrix resulting from the comparisons as compared with the perfect consistency case. Again the eigenvector provides weights that most nearly represent the information in the comparisons.

Saaty claims, and cites evidence in support, that this process when dealing with situations where the 'answers' are known, produces very good results and in cases where answers are not known produces valuations that do not outrage those participating. He further claims that if entered into in a reasonable spirit the process provides a framework for discussion, or even negotiation, when several interested parties are concerned with the outcome of a decision.

Criticisms are made that if important elements are introduced into an evaluation at a late stage then some earlier weights might switch rank. The answer would seem to be that it is better to try to avoid omitting important elements rather than to abandon what does seem to be an intuitively useful process. Saaty's book contains a number of case studies and references to cases where the method proves of value or interest.

The estimation of subjective probabilities can, occasionally, be grossly in error. The use of Delphi techniques, (79) which are a kind of protracted 'brains trust' activity where a lot of people are asked for their opinions, can sometimes be helpful. It can also occasionally give rise to split viewpoints and bimodal distributions, one part of which is clearly in error. Unfortunately precisely which part is in error may not be at all evident from the discussions.

It is, perhaps, the subjective assessment of the probability of rare events which tends to cause the greatest difficulty. Usually, insufficient direct data to make an objective, quantitative, assessment of the probabilities are available, and as a consequence a number of people may be asked for their opinions. Some nuclear safety matters can fall into this category. We can, for example, assess the reliability of certain controlling safety valves as individual mechanisms and may have considerable data available on the subject. We may also have a great deal of information available on metal fatigue under heat stress. But how do we deal with a shift engineer on a reactor who fails to take the proper action on a red warning light and, because the circuit had been giving false alarms in the previous week as a consequence of circuit component faults, merely telephones the maintenance department; whilst round the corner a man doing the cleaning has dropped a Coca Cola can so that it rolled into a position which renders a shut-down valve inoperable. The effect of this is to cause a pipe to heat up, bend and crack so that the fluid spillage shorts out yet another automatic safety device.

Clearly this is an unlikely series of hypothetical events, but it is the kind of thing which can happen over a long period and when a lot of nuclear plants are in action. Attempts have to be made to try to visualise this kind of problem in dealing with nuclear safety.

To some extent the same kind of difficulties may be experienced by insurance underwriters. They are often able to assess risk on the basis of past information, but occasionally something unexpected occurs. The wise underwriter will fix his premiums such that, hopefully, he can still attract sufficient business, and yet at the same time keep his premiums high enough to cover the unexpected event whatever it may prove to be. The experienced underwriter possesses this ability as a kind of instinct; and yet is it instinct, or is he merely quantifying all the events or happenings about which he has first- or second-hand information?

Thus there may be sense in the inexperienced underwriter asking several of his more experienced colleagues whether he has made sufficient allowance for these effects.

It is necessary, before leaving the matter of subjective judgement and subjective probability, to mention the critical importance in military situations of assessing weapon effectiveness on the battlefield correctly. This often involves much more subjective judgement than is frequently appreciated–but it must be carried out in a careful manner (see Appendix 17).

One may well have figures for the explosive content of an enemy weapon, or for its range and rate of fire. A superficial analysis might simply compare these parameters whereas, in practice, much may depend upon how easy the weapon is to handle, its accuracy and the relative skill of the user in comparison to the enemy.

There is much evidence in history that the subjective assessments on the part of one side, at least, have frequently been incorrect. Thus it is said that the English won at Crécy with 12,000 against 40,000 Frenchmen. It appears that the most significant part of this battle was the 6,000 English longbowmen against the 5,000 French crossbowmen. Although both weapons had similar range, the English longbow appears to have had five or six times the rate of fire of the crossbow. Thus the English escaped with a score of casualties since they were able to decimate the crossbowmen, and at a later stage, the French men-at-arms were unable to close with the English archers in view of the fire power of the latter.

In the desert battle at Gabr Saleh in 1941 in the last war the British lured the Germans into a tank battle. The British had 110 tanks to the 79 German tanks. Although the results of this battle were rather inconclusive, it was the British who broke off the engagement. There is some evidence that the German tanks and crew were two to three times more effective than the British, and yet on a straightforward technical analysis of the weapons one would not have expected this to have been the case.

It is therefore important to remember that with weapons which seem superficially similar there can be a manifold variation in weapon effectiveness for a variety of reasons not the least of which may be the abilities and morale of the user. These variations seem likely to mean that the weapons of one side will be two or three times as effective as the weapons of the other. This is particularly true where the armies have not engaged in combat for some considerable number of years when the reliance on subjective assessment increases considerably.

16 Software for computer modelling

The choice of which language to use when building models on a computer, and the decision as to whether or not to use special software packages, is nowadays very much a matter of personal preference. The widely used languages, FORTRAN and ALGOL, are available on most machines. Quite often there are very slight variations in a particular language from one type of machine to another. Appropriately, these are known as 'dialect differences'.

Arguments have raged long and loud between the protagonists of the different languages regarding their respective merits. Quite frequently it is a simple prejudice that the language first learned, so painstakingly, must be defended at all costs. However, there are those who have gone to some lengths to familiarise themselves in detail with several languages and who are undoubtedly much happier with the structure of one language, or its facilities, in comparison to another.

Thus one might hear some proclaim that 'FORTRAN is a machine-type language more suited to machines than men', whilst others claim that: 'If, then, else, statements and some procedure calls in ALGOL are unnecessarily complex'. Such arguments are generally pointless, and most people find no serious difficulty in working with either of these languages. It is worth noting that the new language PASCAL is attracting users from both camps.

Software for large machines

Most of the older very large machines were amply supplied with special software packages. These packages performed with varying degrees of effectiveness, a number of them being quite frequently recalled to correct programming errors. The quality and quantity of the documentation varied considerably, and it was not unusual to have read a two-inch thick manual in order to learn the procedures necessary to carry out a simple multiple regression. This is somewhat alarming when one considers the ease with which this kind of calculation can be carried out on some of the better pocket calculators after a few minutes' perusal of the instructions.

As a consequence of this unnecessary complexity, many scientists using large machines tended to write their own special software, and a certain amount of trading in privately written 'user-friendly' packages occurred within universities and industry.

A number of special purpose packages and languages were generated by the large time-sharing computer bureaux in the late 1960s and early 1970s. For example, Automatic Data Processing (formerly Time Sharing Ltd.) developed a general purpose modelling package known as 'Oracle'. This was used for many years, although principally for financial modelling. A variety of simulation and statistical packages are available from this bureau, from others such as GEISCO, and from consulting groups such as SCICON, LOGICA and SIA. Many universities have now also produced software packages which will operate on a wide range of machines.

With the increasing power of micro-computers, many of the original time-sharing businesses have branched into slightly different fields. For example, GEISCO is now a subsidiary of the General Electric Company of America, and has widened its business into international networks. There are many bureaux still providing time-sharing services, but often now they also provide small computers to interlink (or 'front end') with their own very large machines.

New packages appear very frequently, especially for the smaller machines. Software can prove to be an expensive item when setting up a computer system, and it is often advisable to 'shop around' for suitable available software before purchasing the machine. On large machines the cost of bad software can be phenomenal, and a potential user should seek comments from current users before committing himself. This advice should obviously be sought when salesmen are not present, and some care should be taken to ensure that the way in which the software will be used is very similar.

The problems and future of large-scale software

Hired software must have a perceived pay-off and thought needs to be given with regard to how this may change as time progresses. In using bureau machines and software, consideration should be given to possible additional costs in the event that the model increases in complexity.

With some packages, perhaps especially those involving simulation, it is necessary to decide just how much time must be devoted to learning how to use the package. Quite frequently it is much better practice to write the simulation in a well known language which can be readily understood by other programmers, who may well have to unravel problems at a later stage, or may be required to make further additions to the package.

These remarks are primarily intended to instill a measure of caution with regard to what are frequently described as 'general purpose modelling packages'. Some, by their very size, are both cumbersome and costly to use. Large software packages have proved useful and effective to many people, but some may be unfortunate enough to discover the disadvantages just outlined.

This does not imply that users should attempt to write all their own software packages. It is, for example, probably unwise suddenly to attempt to write one's own integer linear programming packages. An activity of this kind may well be very instructive but it is also likely to be expensive in terms of time and effort. It is one of the purposes of this book to help the reader to decide precisely which technique is appropriate to his problem so that he can readily seek and select the most suitable specialist software packages. Specialist software packages are usually, but not necessarily, cheaper than general purpose ones. Moreover, since they are designed for a limited more specific purpose, they are also normally more effective.

It is likely that further development of the 'spreadsheet' software described earlier will lead to improved general purpose modelling packages. It is also important to remember that with the dramatic reductions in the cost of computation quite large software packages involving mixtures of 'spreadsheet' techniques, statistics and cybernetics, can be left to work for days to derive the 'best' solution to a problem by what is almost a process of attrition. This solution may, or may not, yield further insight into the problem and to more direct methods. Where the reasons for the answers cannot be explained the solution must be treated with caution since the results may be due to unexplained random effects.

The future development of software packages and their application to a wider range of problems remains an interesting and exciting prospect. General purpose packages will have to be written in such a way that they 'think' for themselves much more effectively than at present. In this way when large cumbersome statistical analyses are attempted probable random correlation effects can be avoided, or, where likely, the possibility highlighted for attention. The linkage of these activities with those in the artificial intelligence field is likely to be both interesting and important in the future.

Software for micros

The development of software for microcomputers is proceeding at such a pace that it is unwise to attempt to describe the available packages in excessive detail.

A wide variety of 'spreadsheet' software has been developed in order to answer simple accounting-type business problems of the 'What happens to profits if wages or material costs increase by say 10 per cent?'. Such packages include Visicalc, Supercalc, Plannercalc, Calcstar, Lotus 123 and Multiplan.

A number of slightly more user-friendly packages, which are suitable for staff with little computer knowledge and which have somewhat similar spreadsheet properties, are also available. A wide variety of accounting packages can also be obtained with a fairly wide range of cost and facilities.

Many different word processing packages can be used. Some of the earliest available for microcomputers were

Wordstar, Format 80 and Applewriter. The choice of software is, to some extent, limited by the operating system of the machine. Wordstar requires a CPM operating system whereas Format 80 requires an Apple DOS system. It is possible with some machines to run with either system providing arrangements are made when the machine is purchased to fit an extra processor card.

It is also possible to obtain a variety of statistical packages. Many of these are rather limited in their scope and so it is necessary to check the requirements carefully before making a purchase. A number of graph plotting packages are available. Some of these take a moderate time to load and run. A wide variety of database packages are available. Some are very flexible and can be adapted to a wide range of tasks. A number of packages are constrained in ways which preclude them from being useful for certain tasks, but makes them ideal in more limited roles. The number and size of fields, the number of records, the ease of making amendments and the probable nature of the searches should be given detailed study before deciding which package should be adopted.

There is a need for some of the best software packages to be more effectively tied together. The use of data interchange format files, known as DIF files, can enable data to be exchanged between different software packages, but this can be a slow process where a large amount of data is involved. Improvements in data handling facilities for microcomputers are continuing to take place quite rapidly.

Some microcomputer manufacturers have attempted to solve the problem by storing a variety of useful programs using semi-automatic dynamic memory management. Computers are available which use a 'mouse' or light pen to control the cursor to point to the appropriate program. The former is a movement-sensing block on a lead to the computer, and this causes the cursor on the screen to be moved in the same direction as the block. The so-called 'light-pen' is a rather similar device, but in this case the pen is used to point directly at the screen.

The rapid growth in the use of microcomputers has made microcomputer software big business. This can only result in better, and relatively cheaper, facilities for the user in the future.

17 Testing models and the use of data

A great deal of time is often spent in developing models, but in many instances only a surprisingly small effort is made to test them before they are pressed into service. In some respects this is to be expected, since once a representation of a system has been formulated it is believed to be accurate and there is a desire to obtain any benefits which might accrue from the use of the model as quickly as possible. Unfortunately, failure can often prove very costly and so the need to establish methodical testing is obvious.

Building and testing models

Perhaps it is not surprising that the model builder frequently tends to neglect the testing phase. He gradually acquires more confidence as be embodies new factors in the model to take account of various situations. What he may forget is that he could be elaborately tailoring his model to fit the existing data and that his explanatory reasoning may have become weak. It is frequently better to have a model which only approximately describes the real world, provided that it is known to be approximate, than to have an elaborate model which fits accurately over a wide range and then fails unexpectedly when confidence has been established and when much depends on the outcome.

One method of testing mathematical models is to divide the data into two halves and use one half to construct the model and the other half to test it. This is generally a good scheme but all too often there are insufficient data available anyway. In practice, it is frequently reasonable to use about two-thirds of the data to build the model and one-third to test it. Indeed, moderately elaborate mathematical exercises may be carried out to decide just how much of the data shall be used in the model construction and how much are required for test purposes. There is no magic number or scientific mathematical reason for taking two-thirds of the data to build a model, since everything depends on the quantity of information available and its probable error. However, as a general rule, if the model builder does not have sufficient 'insight' into the problem to construct a model having had access to this amount of data, he is usually unlikely to achieve a solution with the rest of the data.

Ideally, an attempt should be made to construct the first model on a purely theoretical basis. The minimum amount of data should then be taken and the model tested. Modifications should be made in order to explain differences between the data and the model, and the minimum amount of available data taken and the model retested. This process should be repeated again and yet again, until the differences between the predictions and the data grow less at each stage. Only in this way can one have any certainty that the model is performing realistically. This method is shown diagramatically in Figure 17.1. It is also obviously important to look out for any special situations in the data and to test the model over as wide a range of parameters as possible. If experimentation is feasible it is often advisable to exaggerate the controllable parameters and effects so that the model will fit satisfactorily over a wide range.

t-tests and analysis of variance

Statistics can play an important part in testing models (see Appendix 18). In the event that the model is attempting to predict the value of a certain variable we may be able to derive confidence limits which ensure that there is, say,

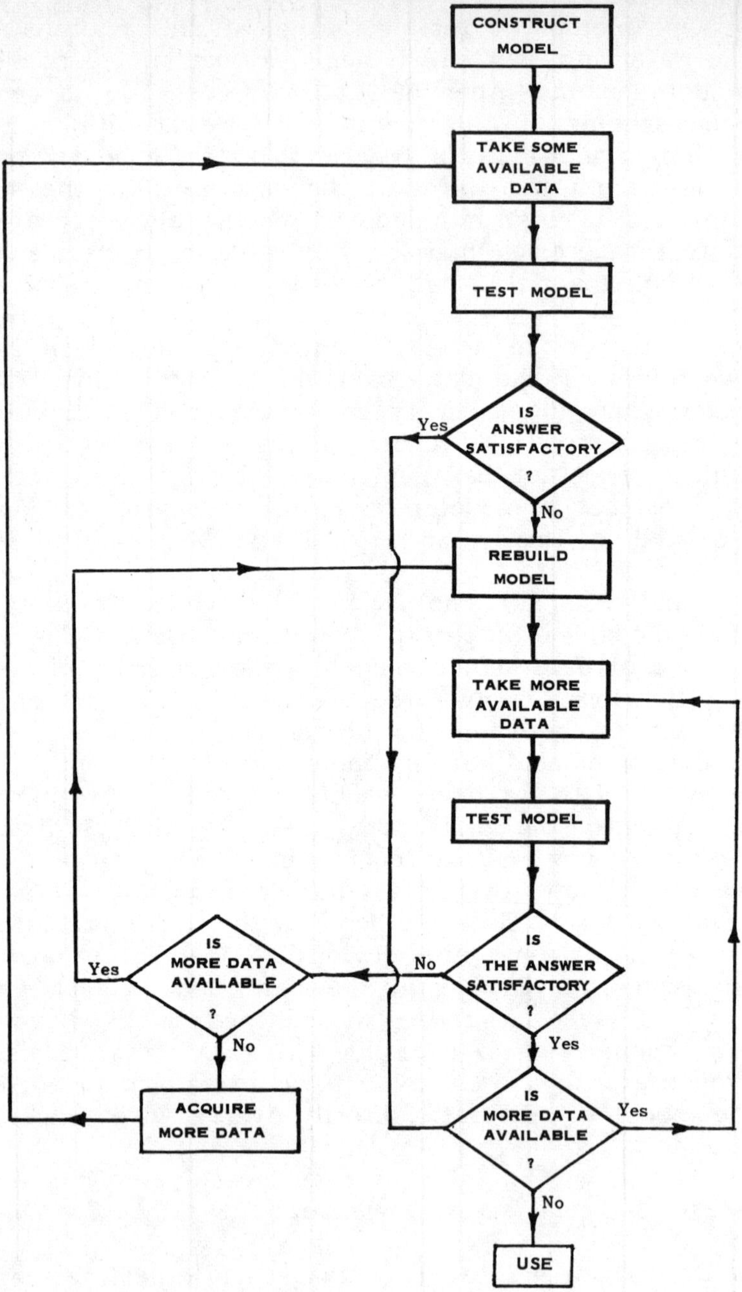

Figure 17.1 The algorithm for building and satisfactorily testing a model

a 90 per cent chance of the variable lying between a certain upper and a lower value. Similarly limits can be calculated for, say, the 95 per cent chance, or the 99 per cent chance. If only a small amount of data is available, confidence intervals can be established by using t-tables, and naturally, the smaller the sample, the wider the range of predictions. As a consequence it becomes more difficult to test the model satisfactorily.

It is possible, by using the appropriate formula and t-tables for one *degree of freedom*, to test whether the slope of the line for two plotted variables on which a regression analysis has been carried out, corresponds to some particular slope specified in a mathematical model. As a simple guide, the number of degrees of freedom corresponds to the number of parameters to be estimated, or if you prefer to express it another way, the number of different basic types of chance available. For example, in tossing a coin for heads or tails, there is really only one basic chance, because given a first chance frequency of heads then the frequency of tails is fixed automatically.

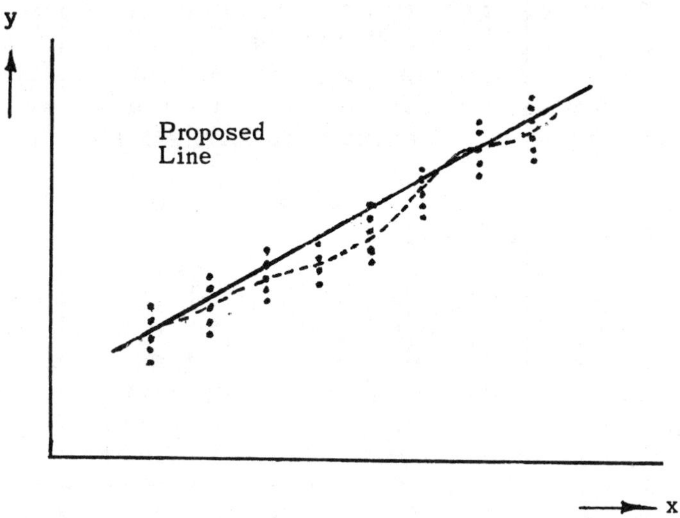

Figure 17.2 F-tests can be carried out in order to determine whether a linear fit is adequate

It is also possible, by a technique known as *analysis of variance*, to test whether a linear fit for a line derived from regression analysis of data is adequate. Consider the data plotted in Figure 17.2. In this test the sums of the squares of the differences between each group of data points and the linear regression line are compared with the sums of the squares of the differences from a mean within each group. A special test known as an F-test is carried out to compare these variances. If the sum of the squares of the differences between the group and the line, divided by the sums of the squares and the differences from within a group mean, is greater than a value determined from statistical F-tables, then the hypothesis that the fit is adequate is rejected.

Some of these tests are fairly tedious when 'hand' calculation methods are used. Computers are often employed for the task and special programs are available to carry out routine tests. Consequently, it is possible to execute fairly sophisticated statistical tests on models with only a relatively elementary knowledge of statistics. Naturally, in some ways this may be a rather dangerous procedure in that it is only too easy with the wide-ranging facilities available to apply the wrong methods to the problem. Furthermore, some of the more elaborate programs have a vast quantity of output, most of which may be unnecessary to the user, and this can sometimes be confusing. Perhaps this is partly due to the fact that surprisingly few statistical software packages have been written by experienced practical statisticians.

χ^2 and goodness of fit tests

It is often necessary to test whether a theoretical distribution provides an adequate representation of the actual results obtained. For example, we may wish to test whether our data fits a normal distribution* or a Poisson distribution, or some other kind of distribution.

In order to test whether the theoretical distribution adequately fits the data, the appropriate theoretical values are calculated and a 'goodness of fit' test may then be used. This involves the use of χ^2 (chi-squared) tables.

---

* There should be some a priori theoretical reason for the distribution selected.

The procedure in this test is to split the data values up into cells (i.e. to lump all closely similar values together) so that the quantities in each cell are not too small, and to compare the difference between the predicted data for each cell with the observed data.

Thus the χ^2 statistic:

$$\chi^2 = \underbrace{\left(\frac{\text{Observed--Expected}}{\text{Observed}}\right)^2}_{\text{Cell 1}} + \underbrace{\left(\frac{\text{Observed--Expected}}{\text{Observed}}\right)^2}_{\text{Cell 2}} + \ldots\ldots$$

The numerical result of the above calculation gives a χ^2 statistic for K degrees of freedom where K equals the number of cells minus one. This statistic is compared with the values of the $\chi^2{}_k$ in tables, and if it is less than the quoted value it is accepted that the fit is reasonable. In this case it means that the theoretical distribution provides a satisfactory model. Other types of distribution could be tested in the same way equally well.

Hypothesis tests

Another statistical method which can sometimes be used for testing models is known as *hypothesis testing.* It should not be supposed that the classical statistical theory for hypothesis testing is capable of dealing with more than a small proportion of the problems encountered in model testing. Even so, it can be extremely valuable for certain types of model involving probability distributions and it is therefore important that we consider it here, albeit fairly briefly.

Suppose that we have a normal distribution of events which occur with varying frequency as shown in the histogram in Figure 17.3. The number of times in an experiment some score, or parameter (-.2, -.1, 0, .1, .2, .3, etc.) is obtained, is plotted as a series of vertical bars. Clearly on the vertical axis we could just as easily have plotted the probability, or chance, of obtaining any of the values -.1, 0, .1, .2, etc. at any particular try. Now let us suppose that the variable parameter is not discrete (i.e. does not consist of separate steps) as in the graph, but is a continuous variable such as temperature and does not take on values in steps, but assumes intermediate values.

If this curve was plotted it would be of continuous form and might be as shown in Figure 17.4. In this graph the number of events has also been converted to the probability of the event occurring, and since it is certain that some x axis value must occur, the total area under the curve must add up to one.

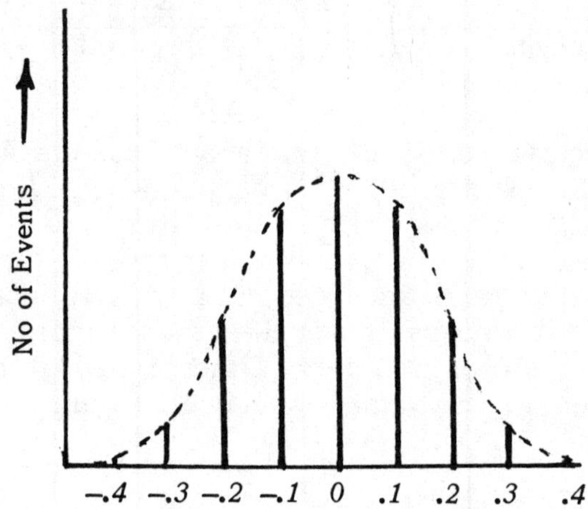

Figure 17.3 A discrete probability distribution

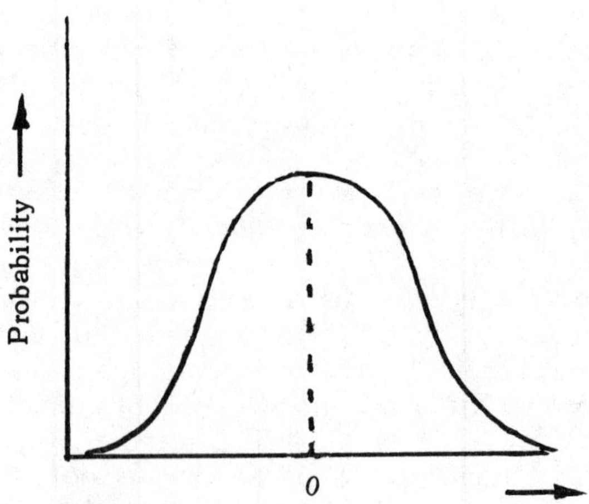

Figure 17.4 A distribution on probability for a continuous variable

Suppose that two sets of unmarked ingots are discovered in a metal store and from their weight and appearance they may be any one of two alloys used regularly. Unfortunately, the company does not possess the facilities to carry out a chemical analysis. As the alloys have different melting points, the foundry manager decides to sort out the ingots by measuring the temperature at which melting takes place. The pyrometers used to measure the temperature are not very accurate and the melting points of the alloys are fairly close to each other. Consequently, if a great deal of measurements on the first alloy are taken, many readings may be obtained around the true melting point (a) and in taking measurements on the second alloy, readings will be obtained which are distributed around the melting point of this alloy (b) (see Figure 17.5).

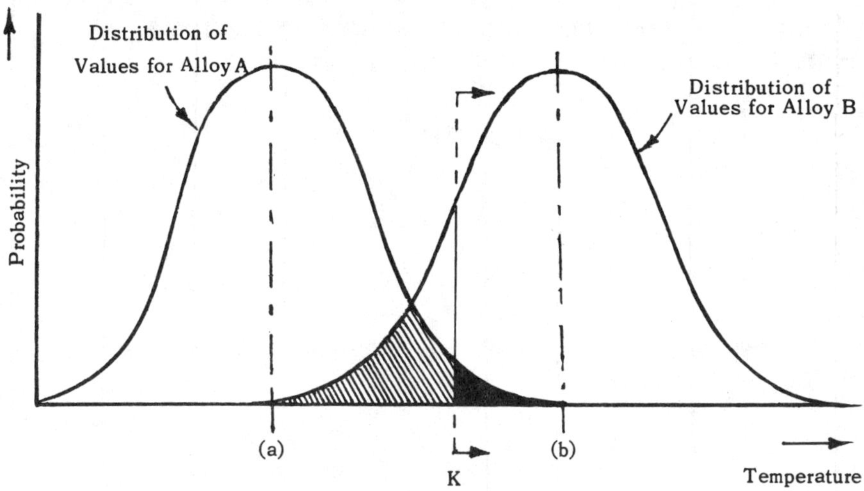

Figure 17.5 Illustrating the probability of a Type I or Type II error in hypothesis testing

Suppose we hypothesise that the ingots are composed of the first alloy type A. This then is known as the *null hypothesis* H_0. If the alloy is type B this is the *alternative hypothesis* H_1. Taking a single melting point measurement of the unknown alloy means that if it is type A the reading could fall anywhere under the A curve and with the greatest probability at the temperature (a). If the alloy is type B the reading would fall somewhere under the B curve and is most likely to occur at (b) since this is the greatest probability for alloy type B.

Thus it is possible to set up a test and accept that the alloy is type A if the melting point measurement obtained is less than some amount, say K, and reject it if it is more than K. Hence, the area shown in black represents the probability of rejecting the alloy as being A when it is really A, and is known as the *type I error*. The black area is generally known as α and represents the probability of this error occurring. Similarly, the probability of accepting the alloy as type A when it is really type B is given by the area shaded diagonally. This probability is usually denoted by β and the error is known as *type II error*. The region above K (i.e. on the B distribution side of K) is known as the *critical region*.

In the event that we choose to make α have a definite probability (e.g. 5 per cent) then a value which minimises β is known as a *best test*. This is achieved by using the *Neyman–Pearson lemma* and *likelihood ratio* tests, which in terms of the above problem state that:

$$\frac{L\ (\text{B})}{L\ (\text{A})} \geqslant K \text{ inside A,}$$

and

$$\frac{L\ (\text{B})}{L\ (\text{A})} < K \text{ outside A,}$$

where L means 'likelihood of'.

In the case of a normal distribution the value of K can be found from the formulae:

$$K - \text{a} = \frac{Z_\alpha \sigma_A,}{\sqrt{\text{n}}}$$

and

$$\text{b} - K = \frac{Z_\beta \sigma_B,}{\sqrt{\text{n}}}$$

where, in addition to the symbols already named, σ is the variance (or the square root of the sum of the variations from the mean divided by the number of observations, (n) and Z_α and Z_β are the standardised distances from (a) to

K and (b) to K respectively. For example, if $\alpha = 5$ per cent, then from statistical tables $Z_\alpha = 1.64$. Hence in the diagram

$$\frac{Z_\alpha \sigma_A}{\sqrt{n}}$$

is the distance from (a) to K. Another very simple example of hypothesis testing is given in Appendix 18.

Further tests and practical aspects

It is also important to consider some of the practical problems encountered in testing models, as well as the purely theoretical aspects. Various statistical techniques have been described which should assist in analysing data and in deriving relationships between variables. Whilst there can be no doubt of the value of these techniques, it must be remembered that they are basically 'unseeing' and merely relate the variables quite impartially on the basis of the data presented. Sometimes it is possible, with certain multivariate techniques such as principal component analysis, to isolate the variables which have the greatest effect on the model and to perceive their physical meaning, but this is perhaps the exception rather than the rule.

Hence, it will be noted that by employing such methods blindly the model has been constructed to fit the data. It is much better if careful logical arguments can be used to build the model and statistical techniques used to refine it and establish its value as a predictor. Under circumstances where the model has been derived using purely statistical methods, the importance of using a completely new set of data for test purposes is quite obvious, since it is quite likely that we have built into the model some purely random effects.

It is usually not difficult to develop a conceptual diagram of the problem which may have hundreds of variables and interactive features. Such diagrams are usually very pretty to look at, and undoubtedly indicate that the analyst has thought about the problem quite deeply but they rarely, in themselves, provide decision makers with a useful guide for making decisions. It would also be nice to think that by simply running a component analysis program it would be possible to extract all the important variables. As explained, this does not always occur in quite such a simple fashion. In practice, relationships between variables are only occasionally linear and yet component analysis is basically a linear technique. Consequently, unless it can be decided that a variable is likely to change in accordance to some known law, in which case it is possible to input this in order to see whether the residual variance is adequately reduced, component analysis may not reveal the nature of the situation.

In other words, although different techniques can help greatly, there is no substitute for deep and careful thought. In developing a formula using multivariate techniques, it is possible to use methods such as Hotelling's T^2 test to check that the means of the variables obtained from the data are significantly different. With this test the relative importance of the variables for any given set of data should be obvious from inspection, but the physical meaning of the principal components may not be deduced so readily. The elimination of redundant variables can be achieved by undertaking a full principal component analysis on all the variables and examining those with the smallest latent roots (i.e those having the least effect). Each of these variables can be associated with a variable having a larger component loading and thus they can be eliminated and made redundant. Theoretically, these variables should be eliminated one at a time, but in practice it is reasonable to take three or four at once.

Tests can also be carried out to see whether the latent roots are significant. Barlett's test depends on the χ^2 distribution, although this often tends to show rather more significant roots than can be justified in practice.

Although great emphasis has been placed on the need for simplicity in models, it is better to have too many variables in the initial stages rather than too few, provided steps are eventually taken to eliminate the unnecessary ones from the model. The omission of a relevant variable means that the model can never function properly until it is included, but insignificant variables can be removed at a later stage.

In cases where non-linear functions have been approximated by means of a series of linear ones, these must be checked over the full range to ensure that they operate properly, and that anomalous results do not occur. Constraints may need to be added or removed in order to ensure that the model operates satisfactorily. It is also usually worthwhile checking the constraints in any model, and removing redundant ones in order to make certain that no unnecessary computations or time-wasting procedures will take place.

18 Implementation, cost and performance

Eventually the model builder may be satisfied that his model not only fits the original data but has been adequately tested using additional data. Careful thought must then be given to the possibility of encountering an environment in the future which is dissimilar to that for which the model was constructed, and whether or not it is necessary to try to simulate such conditions.

The cost of implementation

At this stage, a decision has to be made whether to put the model into practice or to carry out further experiments and acquire more data. A continuous spectrum of possibilities of the model performance exists. For example there may be a probability of .6 of the model being satisfactory, .2 of its performing splendidly, .15 of its performing badly and .05 of it failing completely. It may be advisable to ask what will happen under each of these conditions. Let us suppose that if it performs satisfactorily we get a benefit or return of £2,000. If it performs splendidly, the pay-off is £4,000, whereas for a bad performance we lose £6,000 and for a complete failure £10,000.

Multiplying out these probabilities by the return, and summing, gives the average expected return. If we could go on making this kind of decision hundreds of times a day or a week, and indeed some large companies may well do so, then the average pay-off would be as follows:

.6 x 2,000	=	1200
.2 x 4,000	=	800
-(.15 x 6,000)	=	-900
-(.05 x 10,000)	=	-500
Total		£+600

Unfortunately, the decisions as to whether to apply a model in practice are usually much more complicated than explained above. It is not a simple matter to estimate the probability of failure and the costs of failure have also been seriously underestimated in the past. Several companies have had highly traumatic experiences in installing computerised information systems linked to inventory models, although most have proved highly profitable in the end. Incidentally, very few have proved capable of cost justification in advance. In part this is often due to benefits arising from unexpected sources and so, where attempts are made to cost-justify a model, particular care should be taken in trying to assess side-benefits.

Difficulties of implementation

However, there are even worse problems which may face the model builder. Occasionally, the decision maker requesting the study will favour a particular outcome, possibly for reasons he has not revealed to the model builder. The model builder may produce a robust model based on the information available to him, only to find that he has not produced a solution which pleases the decision maker. He may press for his model to be implemented on grounds which to him may seem almost sacred, such as unquestionable improvements in cost and efficiency, only to suffer irritation and frustration when no action is taken on the matter.

In these situations it is difficult to know exactly who is to blame; the decision maker for not revealing the whole truth and other relevant variables, or the analyst for not pressing the decision maker to disclose these fully. Of course, there are some variables so close to the decision maker's soul that he may choose not to reveal them. Thus a two-year resource allocation study supervised by some of the foremost management scientists in the UK may be brought to nought for no apparent reason. Only many years later, in a weak moment, when the dishonesty of the decision maker's soul has been laid bare by alcohol, may the truth be really revealed by a statement such as 'I didn't want anyone to know how overstaffed I was as I knew that a two per cent staff reduction was coming soon'.

One may say 'why not simply ask the decision maker what answer he expected to obtain and give him that answer?'. Quite apart from the doubtful value of such a procedure, there is often some difficulty in finding out what the decision maker really wants and who is the decision maker in the first place. It will probably turn out to be your boss, but, if weighty decisions are involved, it is more likely to be a collection of some rather senior individuals whose opinions have been canvassed either separately or in committee.

For these reasons, small commercial organisations are sometimes the most satisfying for the consultant model builder in that money and efficiency are clear motives. Moreover, the managing director is usually a person of some authority and if he owns half the shares in his company and pays money for a model to be built he is likely to implement it without delay if he thinks it has a reasonable chance of working satisfactorily.

Whatever the size and nature of the organisation, continual consultation with decision makers during the analysis and construction of the model is important. There should also be some agreement on how the model is to be tested and implemented. Detailed tests can be a lengthy process and precisely how costly they are can depend on the nature of the model.

In classical physics experiments are often conducted by holding all the variables except one constant (as far as possible) and changing this one variable in a predetermined manner. Average conditions may be taken for all the variables, and each variable is selected in turn and altered in a systematic manner throughout its range to see how sensitive the answer is to changes. This procedure is, in fact, more formally known as sensitivity analysis. Hence, in order to carry out a preliminary sensitivity analysis for the system with ten variables over a range of ten values, it would be necessary to carry out a hundred tests or experiments.

In practice, the experiment would then be repeated taking the values of the variables which have so far been found to be optimal, and then one or two readings on either side of each variable taken to find any improvement. The process is then repeated until there is no significant improvement in the optimum. It is possible to derive a variety of algorithms in which progress is made to an optimum by adjusting the variables in the direction which makes the greatest improvement. When this process is carried out on a computer it is described as a 'hill-climbing' technique and it is often undertaken when the general mathematical relationship between the variables is known but the exact values are not known precisely.

As already discussed, the existence of some sort of mathematical expression means that many fewer verification experiments need to be carried out and there is less chance of merely achieving a local optimum, as might be the case by using the broad sensitivity analysis method already described.

In setting up tests for the verification of an existing model, it is advisable to consider carefully the effects of the variables on the output (e.g. the output may be profit, costs, particle emission, etc.). Variables which have the greatest expected effect should be tested first, and fairly thoroughly, whilst those which are expected to have the least effect can be tested less rigorously provided the experimental data corrobrate these findings.

The need for a mathematical model as a predictor is obvious if the cost of conducting the experiments for a sensitivity analysis is large. The problem may not be serious if the 'experiments' consist of sending out batches of questionnaires, or market researchers at a cost of £20 a time, but if the experiment is concerned with the design of a warhead for which full-scale tests may cost more than £100,000 per warhead, then this alone is sufficient to preclude a complete full-scale sensitivity analysis and a part of the analysis may have to be carried out on the model scale.

Performance and 'robustness'

It frequently happens that one is concerned with obtaining a 'good' answer at a reasonable cost rather than the best result at an exorbitant figure. This is sometimes referred to as 'satisficing' because it is aimed to attain an acceptable level of efficiency. Alternatively, as mentioned earlier, it has also been described as obtaining a 'rugged' or 'robust' solution (i.e. one which will withstand attack).

We need to have a very clear idea of what we mean by the performance of the model. Some measurements of performance can be regarded as objective, because the output can be measured independently by a number of different people, possibly using some form of meter.

However, some measures of performance depend upon assessment by valuers and Roberts [80] suggests quantitative measures of people may be made with regard to performance in five different roles, but in one of these categories, the 'artist', the measurements are entirely subjective. Some would argue that it is difficult to obtain quantitative measurements in the other categories. For example, we may have a 'fuzzy' idea about what constitutes a good ruler and it is possible to have two rulers who are rated equally 'good' who have very different values in various attributes. This gives us a very vague or 'fuzzy' idea of what we mean by 'good'. Also who decides what is 'good'? The ruler's subjects, outside 'world opinion' and, on the micro-scale, what weights do we attach to the views of an economist in comparison to the opinions of a person concerned with human rights?

The same kind of difficulty in the assessment of performance can attach to models which have been newly constructed. It is for these reasons that management science is frequently concerned with satisficing. The kind of business problems which can be solved so elegantly in text books seem to occur relatively rarely in practice. The operational analyst frequently finds it impossible to apply pure mathematical techniques, and must use a semi-heuristic approach, or simulation methods, etc. These practical methods can lead to a solution which is quite adequate without necessarily being optimal and consequently, if definite targets can be specified, it is often possible to curtail unnecessary expenditure on model development (see Appendix 20 for example).

Tests to assess the quality or 'goodness' of this kind of model are often more difficult to carry out. Random tests and hill-climbing tests may be of some value. If the model is used for forecasting, then its quality will be evident from its accuracy as a predictor of events. Should the model be found to be inadequate, and given that it has performed well in the past, it is necessary to examine the structure to see what has gone wrong. It may be found necessary to incorporate another variable which allows additional actions to take place.

The modification of a model to account for exogenous variables is a much more difficult task. These variables are basically outside the normal problem but affect the situation in some chance way (often it is a Poisson-type process). Various examples can be given, and an instance would be what might happen to sales and prices as a result of a long dock strike. Although the probability of this kind of event occurring may be known, its precise timing cannot be determined. It is true that there may be some indications or predictors that the likelihood is greater at certain times; for example, high rates of rise in the cost of living, poorer payments relative to general community payments, absenteeism, drift to other jobs and fall in output per man, an increase in minor disputes, etc.

Unless exogenous effects are very important, it is generally advisable to neglect them, as the difficulty is not usually the exogenous variables which have been considered, but the ones that have not even been thought about. For example, suppose that a product being studied contained a sweetener such as cyclamate and it was suddenly discovered that this was a health hazard. The government could decide to act quickly and impose a ban on this constituent. It is feasible that virtually no warning of this hazard could be given and so it would not be possible to attach meaningful probabilities to the government action and the consequent effect on the product. Under such situations the best that an analyst might be able to do is to suggest diversifying to other products to avoid the risk of failure incurred by one-product companies. Appendix 15 contains some relevant comments.

We have discussed what operational research workers describe as 'rugged' solutions, but some would say that there is a class beyond this level which could be called 'really rugged' solutions. It is not suggested that at this further level the action of the variables within the problem itself should be researched to any greater extent, but rather that the scenarios and assumptions on which the problem is based should be studied more extensively than is usually the case. This means that during the testing stage of a model one should consider the effect of some variation in the underlying assumptions, including even the more obvious ones. Ackoff's 'teleological systems' concept, in which he considers the need to investigate at least one level deeper than the immediate problem, is very much in line with the foregoing remarks.

Many analyses fail because the scenarios on which they are based do not occur, and a completely unexpected scenario takes place. It is therefore necessary to consider, in as much detail as possible, the ways in which unlikely events can happen, and to state clearly the limitations of the model. This is important not only for the effective use of the model, but is also in the best interests of the model builder himself.

Concluding remarks

The way in which models are constructed depends on the individual approach of the model builder to problem solving. It is very much a personal activity. Individual experiences greatly influence one's approach. Personal experiences can therefore provide valuable guidelines. As a consequence I have tried to outline some of the things I have found to be important in building models, and to indicate some of my mistakes. A realistic method for classifying the various types of model has been suggested. This should enable the model builder not only to see clearly the class of models of immediate interest, but also enable him to consider more readily other types of model which might be used to obtain additional corroborative evidence at reasonable cost.

A simple method of converting conceptual diagrams into formulae has been outlined, and the ways in which the diagrams can be modified to build models using 'spreadsheet' software has been described. The latter includes an explanation of how this method can be applied to system dynamics problems. A wide variety of different types of modelling technique are described in the text and in the Appendices. These indicate many of the general methods used and in most instances should be adequate in themselves. A more detailed study may require recourse to some of the references and the reader should follow these up where possible.

However, one of the most important tasks in model building is to be able to sort out precisely which method is appropriate to the problem. Expert advice on a particular technique can then readily be sought if this proves necessary. The selection of the correct approach may be critical, and if the reader follows through all the examples in this text he should have a good overview of the available methods.

It is also worth emphasising again the need for a *completely impartial attitude to the model* and the need for careful testing after any amendments have been made. A model which has continually to be modified is likely to be based on unsatisfactory explanations and interpretations of the system and may best be discarded completely if the cost of failure is high.

Remember also that it is helpful if the limitations of the model are understood and its worst idiosyncracies known. Only in this way can serious mistakes be avoided, since the real world reveals a continually unfolding pattern of events and situations which are often only superficially similar to those which have gone before. It is important not to become mesmerised by the apparent desirability of using large numbers of variables but rather to understand fully the ones which you are using.

Models are intended to produce a better overall view of a particular situation and so consideration should always be given as to where the completed model fits into the total pattern. This relationship should be stated clearly and, if it is to form part of a series of models, the need for standardisation procedures must be considered carefully, so that the models can be linked together at a later stage if this becomes necessary. Good documentation is essential, and this can take longer than building the model in the case of some techniques such as simulation.

The wider the area that can be covered by a set of models, the better the overall picture of the situation, and the greater the chance of benefits being obtained by consideration of the interactions which may arise from the solutions to smaller individual parts of the problem.

People talk about certain scientists having 'intuition', but it is doubtful whether any scientists of whom this is said would regard this as being the case. More probably they would say that it is a matter of trying to look at things differently and to use a logical process of self-argument which goes on in their minds for days, weeks or perhaps even years. They try to fit the pieces of the problem together, and then perhaps decide to pull the existing pieces apart and start all over again. This kind of mental muttering, which reasons 'If this happens, then this cannot occur, and this must', etc. is as important in the development of explanatory models as a logic chart is to a computer program.

Above all it is worth remembering the need to keep models as simple as possible, but the balance between simplicity and complexity is seldom easy. Then again the need for adequate tests on the model using sufficient quantities of new data must be re-emphasised, since surprisingly this sometimes gets overlooked in the excitement of the completion of the task. The use of corroborative models, such as IA models in pure physical studies, or the use of mathematical models to check points in simulation studies, should always be considered both from the cost saving and the possible breadth of validation tests on the models they can provide.

Only model builders themselves can comprehend the excitement of developing a successful modelling theory where none had previously existed. Perhaps one of the best ways to explain this to a layman is to say that, for example, in an experiment somewhere in the middle of a hot Australian desert, the sight of rows of cubes of different weights and sizes which have been blown back so that they remain in straight rows, can give rise to the exhilaration of achievement not unlike that experienced by the first men to climb the Matterhorn. Such then are the enthusiasms and rewards of the model builder.

One final comment. Do not be too discouraged if the model fails to perform adequately the first time it is tested. Like Rome, few really successful models were built in a day, and careful thought and application will usually allow a satisfactory solution to be obtained in the end.

Appendix 1 Allocation using a 'transport' algorithm

A number of different allocation techniques are available, and the best method tends to depend on the problem. Here is a very simple example to illustrate the kind of approach which can be used.

Example. During a battle 3,3,4 and 5 aircraft are urgently required to attack four targets at locations A, B, C and D respectively. The aircraft can be drawn from 3 airfields F_1, F_2 and F_3. As a consequence of enemy actions only 2, 6 and 7 aircraft are available at fields F_1, F_2 and F_3 respectively. Find the allocation to give the fastest possible attack on the targets. The estimated times in minutes to load the weapons and reach the targets are as given in Figure A1.1.

Let the number of aircraft from F_1 to A be x_{11}
Let the number of aircraft from F_1 to B be x_{12}
Let the number of aircraft from F_1 to C be x_{13}
Let the number of aircraft from F_1 to D be x_{14}
Let the number of aircraft from F_2 to A be x_{21}
Let the number of aircraft from F_2 to B be x_{22}

and so on.

		Targets			
		A	B	C	D
Airfields	F_1	13	11	15	20
	F_2	17	14	12	13
	F_3	18	18	15	12

Figure A1.1 Matrix showing flying times to four different targets in aircraft allocation problem

Then,

$$x_{11} + x_{12} + x_{13} + x_{14} = 2$$

$$x_{21} + x_{22} + x_{23} + x_{24} = 6$$

$$x_{31} + x_{32} + x_{33} + x_{34} = 7$$

And,

$$x_{11} + x_{21} + x_{31} = 3$$

$$x_{12} + x_{22} + x_{32} = 3$$

$$x_{13} + x_{23} + x_{33} = 4$$

$$x_{14} + x_{24} + x_{34} = 5$$

It is required to minimise:

$$13x_{11}+11x_{12}+15x_{13}+20x_{14}+17x_{21}+14x_{22}+12x_{23}+13x_{24}+18x_{31}+ 18x_{32}+15x_{33}+12x_{34}$$

It is possible to solve this kind of problem by linear programming methods. However, let us consider some simple allocation process which, superficially at least, does not involve full-scale linear programming. First rewrite the matrix given at Figure A1.1 as shown in Figure A1.2

with the times in brackets in the bottom corner of each cell. Let us also write the x values in the top left-hand corner of each cell.

			Targets			
			A	B	C	D
			3	3	4	5
Airfields	F_1	2	x_{11} (13)	x_{12} **2** (11)	x_{13} (15)	x_{14} (20)
	F_2	6	x_{21} (17)	x_{22} (14)	x_{23} **4** (12)	x_{24} (13)
	F_3	7	x_{31} (18)	x_{32} (18)	x_{33} (15)	x_{34} **5** (12)

Figure A1.2 First steps in the allocation algorithm

			Targets			
			A	B	C	D
			3	3	4	5
Airfields	F_1	2	(13)	**2** (11)	(15)	(20)
	F_2	6	**1** (17)	**1** (14)	**4** (12)	(13)
	F_3	7	**2** (18)	(18)	(15)	**5** (12)

Figure A1.3

Proceed as follows:

1 Take cell with shortest time (11) and allocate 2.

2 As this completes line take minimum (12) in next line and allocate the 4 (this leaves 2 spare).
3 The next minimum in row 3 is also (12). Allocate as many as possible (5) (leaves 2 spare and completes figure A1.2).
4 Next minimum is x_{22} but only 1 required to complete column.
5 Next minimum is x_{21} (17) but only 1 available.
6 Fill in remainder at x_{31} to complete the allocation (Figure A1.3).

Is this solution optimal?

The total time is 2x11+1x17+1x14+4x12+2x18+5x12

$$= 197 \text{ minutes.}$$

Test optimality by filling in some of the empty cells. Thus say we add 1 to x_{11} we must take 1 from x_{12}, add one to x_{22} and take one from x_{21}. If t_{11} is the time from F_1 to target A_1; t_{12} the time from F_1 to target B; etc., then

$$t_{11} - t_{12} + t_{22} - t_{21} = 13-11+14-17 = -1$$

so the total time is reduced to 196 minutes.

On the other hand if we add 1 to x_{14}, the total time is increased thus:

$$t_{14} - t_{34} + t_{31} - t_{21} + t_{22} - t_{12} = 12.$$

A simpler method consists of the notion that for each used route the travel time is made up of a despatch and a receipt time. If an arbitrary value is assigned to one of the despatch points, then all other despatch and receipt costs can then be calculated. (This is because there are $m+n$ equations to satisfy giving $m+n-1$ independent equations. This gives $m+n-1$ variables in the solution in general, although a zero value needs special attention.) If u_i are taken as the despatch costs and v_j the receipt costs then setting u_i=0 gives the matrix shown in Figure A1.4.

			Target				
			A	B	C	D	
			3	3	4	5	u_i
Airfield	F_1	2		11			$u_1 = 0$
	F_2	6	17	14	12		u_2
	F_3	7	18			12	u_3
		v_j	v_1	$v_2 = 11$	v_3	v_4	

Figure A1.4

Then $v_2=11$ since $u_1 + v_2=11$.

			Targets				
			A	B	C	D	
			3	3	4	5	u_i
Airfield	F_1	2		11			0
	F_2	6	17	14	12		
	F_3	7	18			12	4
		v_j	14	11			

Figure A1.5

Also, from Figure A1.5 it can be seen that $u_2+v_2= u_2+11=14$, so $u_2=3$ and so on for all u_i and v_j:

$$u_2+v_1 = 3+v_1 = 17 \text{ so } v_1 = 14$$

$$u_3+v_1 = u_3+14 = 18 \text{ so } u_3 = 4$$

$$u_2+v_3 = 3+v_3 = 12 \text{ so } v_3 = 9$$

$$u_3+v_4 = 4+v_4 = 12 \text{ so } v_4 = 8$$

The despatch and receipts times, commonly known as 'shadow' costs or times, can now be used to assess the unoccupied cells of the matrix. First we calculate the despatch plus receipt time for each unoccupied cell, as in Figure A1.6.

			Target				
			A	B	C	D	
			3	3	4	5	u_j
Airfield	F_1	2	14	–	9	8	0
	F_2	6	–	–	–	11	3
	F_3	7	–	15	13	–	4
		v_j	14	11	9	8	

Figure A1.6

If we compare these times with the actual times in the unoccupied cells, we can see if any of the unoccupied cells indicate a potential saving over these shadow times. The cell F_1 to A has a time of 13 which is less than the shadow time of 14 and hence will produce a saving if used. We now look at the old allocation (Figure A1.7).

	3	3	4	5
2		2		
6	1	1	4	
7	2			5

The Old Allocation

Figure A1.7

We reallocate adding and subtracting 1 in an anti-clockwise manner (Figure A1.8).

	3	3	4	5
2	1	1		
6		2	4	
7	2			5

This gives this matrix

Figure A1.8

We now test this allocation for optimality by calculating shadow costs again as before (Figure A1.9).

Let $u_1 = 0$

$$u_1 + v_1 = 0 + v_1 = 13 \text{ so } v_1 = 13$$

$$u_1 + v_2 = 0 + v_2 = 11 \text{ so } v_2 = 11$$

$$u_2 + v_2 = u_2 + 11 = 14 \text{ so } u_2 = 3$$

$$u_3 + v_1 = u_3 + 13 = 18 \text{ so } u_3 = 5$$

$$u_2 + v_3 = 3 + v_3 = 12 \text{ so } v_3 = 9$$

$$u_3 + v_4 = 5 + v_4 = 12 \text{ so } v_4 = 7$$

			Target				
			A	B	C	D	
			3	3	4	5	u_i
Airfields	F_1	2	13	11			0
	F_2	6	17	14	12		3
	F_3	7	18			12	
		v_j	13	11	9	7	

Figure A1.9

Calculating total shadow costs for each unoccupied cell gives the matrix as in Figure A1.10.

			Target				
			A	B	C	D	
			3	3	4	5	u_i
Airfields	F_1	2	–	–	9	7	0
	F_2	6	–	–	–	10	3
	F_3	7	–	16	14	–	5
		v_j	13	11	9	7	

Figure A1.10

In every case the shadow costs or times in the empty cells are less than the actual times so there are no potential savings from reallocating this solution. This solution then is optimal (Figure A1.8) with a minimum total time of 196 minutes.

Appendix 2 Linear programming

A graphical approach to simple LP problems

Example: Suppose that in blending three grades of wine, a rather corrupt importer decides that he can 'stretch' a very good wine by adding one or two types of very inferior wine. One of the inferior wines has a reasonable 'bouquet' and the other one has quite a good 'body'. He is, however, worried about the level of uncombined tannic acid, which can impart a harsh taste to the wine, and the sulphuric acid content, which can give a bitter flavour. Experience has shown him that if he can keep the tannic and sulphuric acid at certain levels, the good quality wine will tend to dominate the others and he will be able to sell the resulting produce at a fairly high price.

The cost of the three wines to the importer is as follows:

1st wine : 'Vin du Plonk Premier'
costs £1.50 per bottle (Type A)

2nd wine : 'Vin du Plonk Deuxième'
costs £1.00 per bottle (Type B)

3rd wine : 'Le Roi de Vin'
costs £3.00 per bottle (Type C).

The acid content is as shown in Figure A2.1.

Type of Wine	Tannic Acid per cent	Sulphuric Acid per cent
A	0.4	0.2
B	0.3	0.35
C	0.2	0.25

Figure A2.1 The acid content of the three types of wine

In blending the wines the importer believes that the tannic acid should not exceed 0.3 per cent and the sulphuric acid should also not exceed a similar figure.

Let the quantity of wine of type A used in blending be represented by the symbol A. Similarly for B and C.

The restrictions or 'constraints' as they are called can be written as follows:

$$A + B + C = 1 \quad \ldots\ldots\ldots\ldots \quad (1)$$

This is known as the balance equation and means that the total quantity for one unit, say one bottle, must add up to one, which is pretty obvious anyway. It is, however, one of the most often accidentally omitted equations, but since we are concerned with the exact logic of mathematics even the obvious must be stated and nothing is obvious to the computer other than the stated facts.

Since in any one unit, or bottle, the quantity of type A is A, the quantity of type B is an amount B and of type C and amount C, the total tannic acid in one unit is equal to $.4A + .3B + .2C$. Moreover the importer has specified that the tannic acid shall not exceed 0.3 per cent.

Thus we can state the following constraint for the quantity of tannic acid permissible:

$$.4A + .3B + .2C \leqslant .3$$

Similarly the sulphuric acid constraint is:

$$.2A + .35B + .25C \leqslant .3$$

It is also necessary that:

$$A \geqslant O \; ; \; B \geqslant 0 \text{ and } C \geqslant 0 \; \ldots\ldots\ldots\ldots \qquad (2)$$

From these equations and from the balance equations C can be expressed in terms of A and B thus:

$$C = 1 - A - B$$

Substituting in the tannic acid and sulphuric acid constraints and slightly simplifying leads to:

$$2A + B \leqslant 1 \quad \ldots\ldots\ldots\ldots \qquad (3)$$

and

$$-A + 2B \leqslant 1 \quad \ldots\ldots\ldots\ldots \qquad (4)$$

The constraints numbered 1, 2, 3 and 4 above explicitly state the problem, but give no indication of how to obtain the required result at the least cost.

Referring to the cost of the wine per bottle, the cost of one bottle of the blended wine will be as follows:

$$1.50A + 1.00B + 3.00C$$

and this cost function must be minimised.

If the signs are reversed and the expression is maximised instead of minimised, the result will be the same as minimising the original cost function, i.e.

Maximise $- 1.50A - 1.00B - 3.00C$ or, using (1),

Maximise $1.5A + 2B - 3$.

The expression is known as the *objective function*.

The constraints can be illustrated in the form of a graph as shown in Figure A2.2. For example, if the constraint $2A + B \leqslant 1$ is plotted on a graph having axes A and B and a line is drawn joining $A = \frac{1}{2}$ on the A axis and $B = 1$ on the B axis, then the constraint is satisfied for all values of A and B which lie toward the origin in the direction shown by the arrows.

Since we also know that A and B cannot be less than zero, the values of A and B which satisfy all the constraints must lie within the shaded zone. It will also

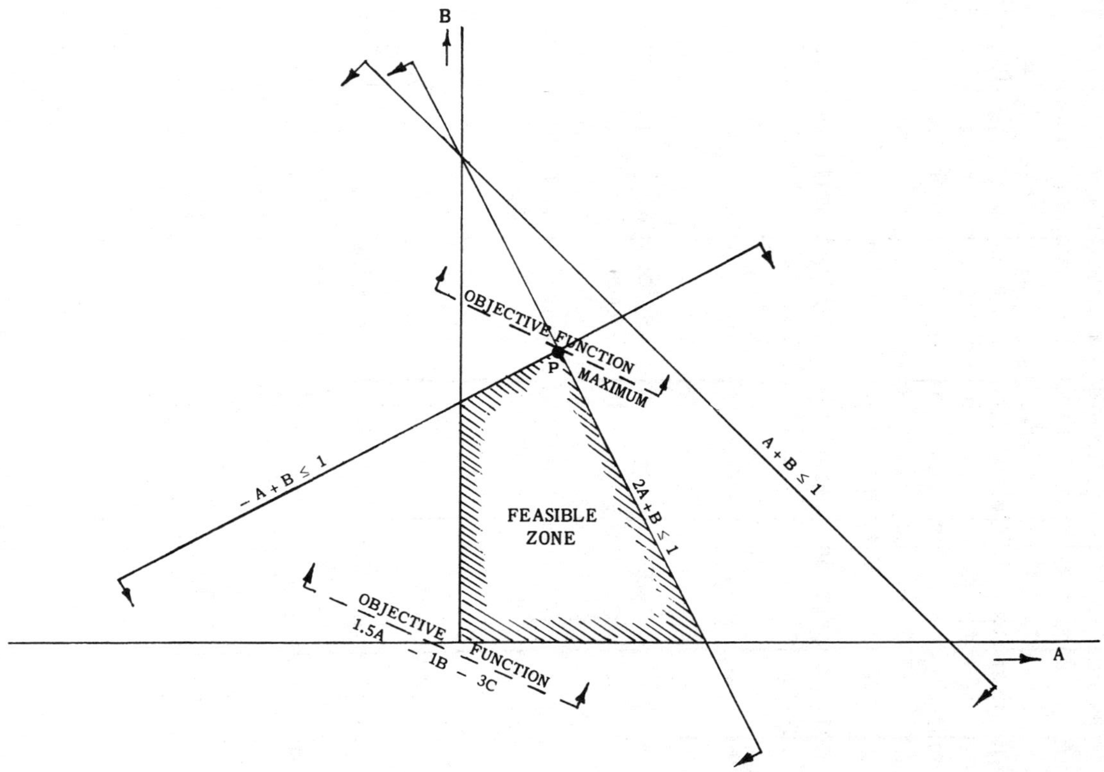

Figure A2.2 Illustrating the feasible solution zone for the linear programming example and how the objective function can be moved upwards to obtain a maximum value

be observed that the constraint $A + B \leqslant 1$ has no effect on the shape of the shaded area and so for this reason it is called a redundant constraint.

Now draw in the slope of the objective function to be maximised so that it passes near the origin. Imagine this line being moved across the shaded area until it is pushed as far out as possible from the origin but still just touches the zone which satisfies the constraints. It will be found that this occurs at the point marked P which has the value A = .2 and B = .6. From the balance equation it is obvious that for these values C = .2. These values are the optimum values which satisfy the constraints (i.e. the restrictions imposed on the levels of tannic and sulphuric acid) and do so at the least cost to the importer.

Thus, in order to make one bottle of wine to the required standard, the importer will need:

.2 bottles of type A costing £1.50 per bottle
.6 bottles of type B costing £1.00 per bottle
and
.2 bottles of type C costing £3.00 per bottle.

It will therefore cost him £1.50 to produce a bottle of what he will doubtlessly describe as 'Specially Blended Roi de Vin', which he may sell for perhaps £7.00 per bottle. All of which just goes to show that whilst the saying 'honesty is the best policy' may be true, a little operational research can sometimes pay good dividends!

Computer solutions

This graphical method has been described because it is a useful way of visualising the problem. Clearly, although a three-dimensional graph could be constructed, there is little point in using this technique as other methods are available. In fact, the normal algorithm using 'hand' calculating procedures is known as the simplex method, and a revised version of this method is generally employed when a digital computer is used.

In this method, the problem is arranged in tableau or matrix form and a number of special variables called 'slack' and 'artificial' are introduced to change the inequality signs to equal signs so that it is possible to manipulate the expressions more easily. The slack variables can be negative or positive. For example, in the problem just given, $2A + B \leqslant 1$ is in the positive quadrant, and so any compensatory slack variable S added to make the inequality into an equation will be positive, i.e. $2A + B + S = 1$.

If the constraint operated in the reverse direction (e.g. $2A + B \geqslant 1$), the solution would be in the negative quadrant and the slack variable S would have to carry a minus sign, i.e. $2A + B - S = 1$. However, in order to manipulate the system to find a solution, it is necessary to introduce another variable, called an artificial variable, which has a negative value so that the slack variable can be made positive. These artificial variables must be removed from the final solution and, if they cannot be removed, no basic feasible solution to the problem exists.

As mentioned earlier, the 'revised simplex' or, as it is sometimes called, the 'inverse matrix' method, is usually employed with computers rather than the normal simplex method which is generally used for 'hand' calculations. The amount of information storage required for the revised simplex is often less than with the normal simplex method in that it is only necessary to know a transformation matrix and certain row and column data. In addition, a number of unnecessary calculations are avoided so that, generally, there is a saving in time as well as in computer storage space. There is little point in going through these methods in detail since they are admirably described in most textbooks [17-19] dealing with LP for the main objective here is to briefly compare different methods of model building and to give some indications of the methods of problem formulation.

A more complex example

Example: A chemical plant produces three products A, B and C which require various quantities of two chemicals X and Y. The chemicals pass through different plants to be processed before they are converted to the final products. For instance, the output from Plant No. 1 may be sold directly as Product A or may be diverted to Plant No. 4 to form Product B. Alternatively, it could go to Plant No. 2 and hence to Plant No. 3 where it can either be sold directly as Product C, or be combined with the output from Plant No. 1 to form Product B. If the capacity of Plant No. 1 is say 200 tons, Plant No. 2, 150 tons, Plant

No. 3, 100 tons and Plant No. 4, 80 tons; the cost of chemical X is £10 a ton and chemical Y is £5 a ton; the process cost of the plants are £1, £1, £3, £4 per ton for Plants 1, 2, 3 and 4 respectively; and Products A, B and C sell for £50, £40 and £30 per ton respectively; how should the plant be operated in order to obtain the maximum profit?

This can be shown diagrammatically as in Figure A2.3.

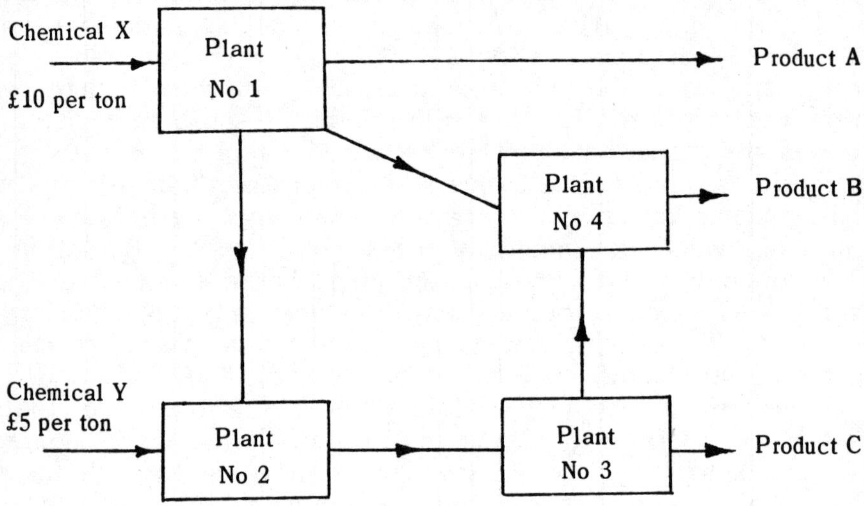

Figure A2.3 The arrangement of chemical plants in a manufacturing company

Let the amount of chemical X input into Plant No. 1 be x_1 and the amount from Plant No. 1 to Plant No. 2 be x_2 and so on, as shown in Figure A2.4.

Hence, we can state the following:

The input to Plant No. 1 must be equal to the output, so

$$x_1 = x_2 + x_3 + x_4$$

Similarly

$$x_6 = x_2 + x_5$$

and

$$x_3 + x_7 = x_8$$

$$x_6 = x_7 + x_9$$

the plant restrictions are:

$$x_2 + x_3 + x_4 \leqslant 200$$

$$x_6 \leqslant 150$$

$$x_7 + x_9 \leqslant 100$$

$$x_8 \leqslant 80$$

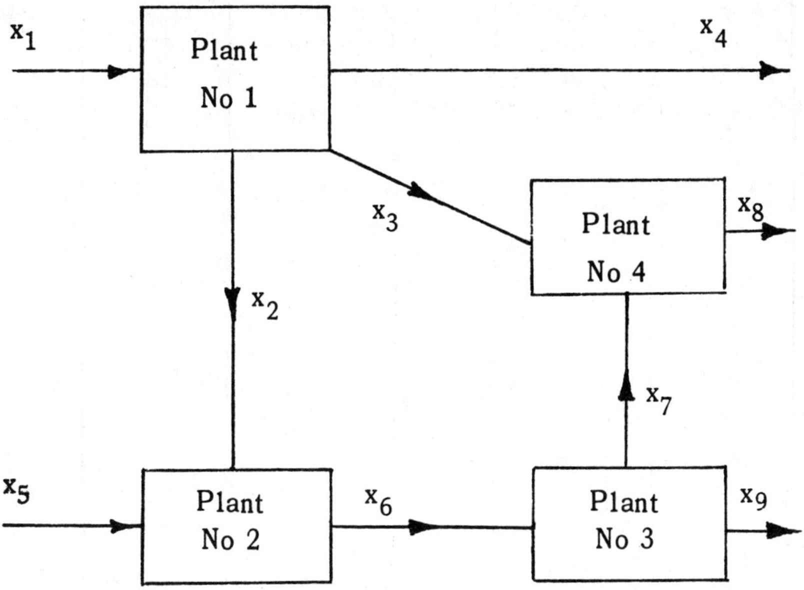

Figure A2.4 The *x* values represent material flows through various parts of the company

Necessarily also all x_r, r = 1,2, ... 9 are $\leqslant$ 0.

The purchasing costs are:

$$10x_1 + 5x_5$$

The process costs are:

$$1x_1 + 2x_6 + 3x_6 + 4x_8$$

and the return on sales is:

$$50x_4 + 40x_8 + 30x_9$$

Since the profit = sales return - purchase costs, the profit is given by:

$$50x_4 + 40x_8 + 30x_9 - 11x_1 - 5x_5 - 5x_6 - 4x_7$$

This simplifies to:

$$25x_3 + 39x_4 + 11x_5 + 6x_6 + 36x_7$$

This is the objective function which is to be maximised.

Hence the problem has now been completely formulated as a linear program. A very large linear program could be formulated in exactly the same way, but could involve hundreds, and possibly even thousands, of variables and constraints.

Appendix 3 Dynamic programming

Dynamic programming provides a method of obtaining the solution to problems by means of an optimisation iterative process. It concerns states, stages, decisions, paths and transformation operators. The concepts of these items can be briefly described as follows:

State

The state describes the condition of the system, for example the condition of an instrument, the number available for use, etc.

Stage

The system undergoes a sequence of transitions from one state to another. The index ordering the direction of these transitions is called a stage. This is usually a time, for example, day n or day $n+1$, etc.

Decisions

Decisions are made at each stage of the process. A policy prescribes precisely which action shall be taken at a given stage when the state at that stage is known. For example, how many instruments to have in reserve on each $k_n(i)$ on a particular trial.

Transformation operator

If the state at stage *n is* i_n and decision $k_n(i_n)$ is taken, a transformed state results at stage n-1 which is denoted by:

$$i_{n-1} = T_n(k_n, i_n)$$

There are four kinds of transformation operator:

Deterministic	Level 0
Stochastic	Level 1
Adaptive	Level 2
Residual	Level 3

Path

A path is an ordered sequence of decisions. For example:

$$i_n, k_n; i_{n-1}, k_{n-1} \; ..... \; i_1, k_1; i_0 \text{ etc.}$$

Dynamic programming depends upon, and uses, the 'principle of optimality'. An 'optimal policy' has the property that whatever the initial state and initial decisions are, the remaining decisions must constitute an optimal policy with regard to the state resulting from the first decision.

This principle of optimality is essential in processes which involve a sequence of decisions which evolve with time and must recognise the current state of the world and its probable development.

Until recently most attempts to study these effects have been very crude and the various mathematical attempts to rationalise equipment replacement problems and, simultaneously, to allow for technological development, have not been satisfactory. The growth of operational research mathematics, linked with the wide use of computers, has enabled techniques such as dynamic programming, which can involve considerable computation, to be effectively employed. Even so, although DP may be the logical model to use it may still not be practical as a consequence of the magnitude of the computational task.

Elaborating further on the principle of optimality, which characterises the whole dynamic programming approach, it can be expressed more simply as follows: When making a decision the only factors which should influence this decision are those which are embodied in the present condition of the system (past events are only relevant in that they are reflected in the current state) and whatever past decisions have been made, decisions must be optimal for the remaining future activities.

Thus, if a factory purchased a piece of equipment six months ago, the fact that the cost has not been recovered has little to do with whether or not it should be replaced immediately. What really matters are the profit–time curves for available equipment and the probability of still more profitable equipment being available in the future.

Generally, in dynamic programming, we are concerned with achieving a number of steps with the least cost, where there is a different cost in moving from any one step (or as it is usually called, 'stage') to any other.

Consider a set of points (i) where $i = 1,2,3 \ldots j,k,N$ and a cost matrix (c_{ij}) which defines the cost of moving from one point to another in the matrix. Thus c_{ij} is the cost of moving directly from i *to* j in one move and $0 \leqslant c_{ij} \leqslant \infty$. The task is to find the minimal cost of moving from any point i *to* N and the optimum sequence of moves given that there can be as many as required.

The plan is, in fact, to find some decision rule which tells us to which point we should take the next step if we are now at the point i. This optimal decision is therefore some function $S(i)$ such that if we stand at point i the policy $S(i)$ prescribes that we shall move to a point j which is on the least cost route, the cost of such a move being c_{ij}. When the same policy, or decision rule, is applied to the new starting point j this brings us to a new point k on the least cost route with a new cost c_{jk}.

Let us denote the cost of the policy which is optimal, or least cost, in getting from i to N as $f(i)$. If we are at point j we have given rise to a cost in getting there from point i *of* c_{ij}. However, when we get to j our objective continues to be to move towards N in a least cost manner so that the cost of getting from i *to* j must be added to the cost of getting from j *to* N in an optimal fashion, i.e.

$$c_{ij} + f(j)$$

Thus the optimal route for going from *i to N* is equal to the minimum cost route to *j* plus the optimal route from *j to N*, i.e.

$$f(i) = \min_j \left(c_{ij} + f(j) \right)$$

This is known as the 'forward' DP formulation since if we work 'backwards' from the last point, or required state, say 1 to some final point *i*, the functional equation becomes:

$$f(i) = \min_j \left(f(j) + c_{ji} \right)$$

In this case $f(i)$ is the least cost of getting from 1 to *i*.

Bellman's method (81) is normally used to solve these equations.

Suppose $f_n(i) = c_{iN}$ we can define a sequence of functions:

$$f_n(i) = \min_j \left(c_{ij} + f_{n-1}(j) \right)$$

where $f_n(i)$ is the minimal cost of getting from *i* to *N* using *n* not more than *n* steps.

In other words for $n=2$ and $n=3$ the following will hold:

$$f_2(i) = \min_j \left(c_{ij} + f_1(j) \right)$$

$$f_3(i) = \min_j \left(c_{ij} + f_2(j) \right)$$

It is worthwhile considering one or two simple practical examples in order to illustrate the technique.

Example: An army commander is advancing through a series of map positions marked 1 to 5. His aim is to take position 5 and the enemy are dug in at positions 2,3,4 and 5. He does not necessarily have to take every position, but he may pay a high cost in casualties if he does things in the wrong order. One or two options, such as moving directly from his present position at 1 to position 5, are deemed to be extremely damaging as a consequence of flank attacks (see Figure A3.1). It is the task of the commander to advance from position 1 to position 5 at the least cost to his forces in terms of casualties. The estimated cost in casualties of going from one position to another is given in the matrix in Figure A3.2.

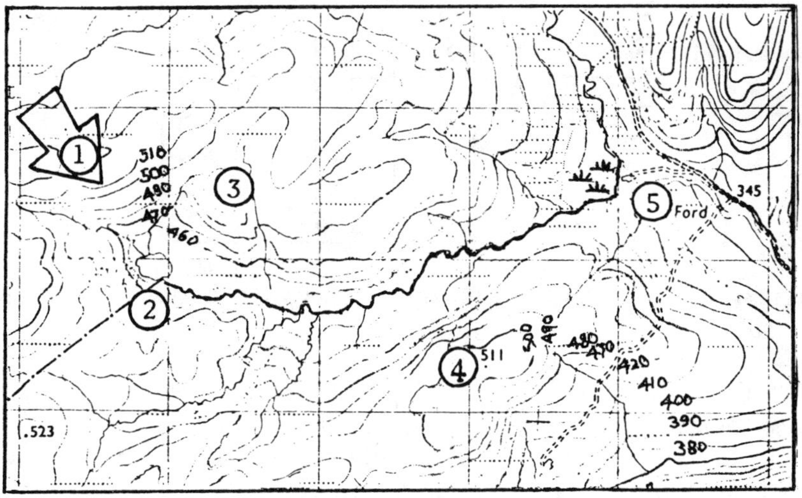

Figure A3.1 Terrain map showing attack and defence positions

		POSITION				
		1	2	3	4	5
POSITION	1	0	20	5	100	150
	2	0	0	9	8	18
	3	0	11	0	40	45
	4	0	10	20	0	6
	5	0	12	40	50	0

Figure A3.2 Casualty matrix (all values have been divided by 100)

Of course for such a simple example the listing of the total possibilities gives the quickest answer.

Let $f(i)$ be the cost of advancing from position 1 to position 5 under an optimal minimum loss policy.

If we carry out attack i to j, we must proceed optimally thereafter, so the loss of cost to do this, as explained earlier, is:

$$c(i,j) + f(j)$$

and j can be chosen to minimise the cost in terms of casualties, so that:

$$f(i) = \min_j \left(c(i,j) + f(j) \right)$$

and $j=s(i)$ which will give the decision rule.

Defining $f_n(i) = c(i,5)$ in n stages, so that

$$f_n(i) = \min_j \left(c(i,j) + f_{n-1}(j) \right)$$

If $c(i,j) \geqslant 0$, it can be shown [81] that the sequence $f_n(i)$ converges in, at most, four steps to a function $f(i)$ satisfying:

$$f(i) = \min_j \left[c(i,j) + f(j) \right]$$

(If converging did not happen looping would occur.)

Hence, from the table:

$$f_1(1) = 150 \qquad f_1(2) = 18 \qquad f_1(3) = 45$$

$$f_1(4) = 6 \text{ and } f_1(5) = 0$$

The next stage is to calculate the f_2 values, thus:

$$f_2(1) = \min_j \left[c(1,j) + f_1(j) \right] \tag{1}$$

$$j = \begin{matrix} 1 \\ 2 \\ 3 \\ 4 \\ 5 \end{matrix} \begin{bmatrix} c(1,1) + f_1(1) & = 0 + 150 = 150 \\ c(1,2) + f_1(2) & = 20 + 18 = (38) \\ c(1,3) + f_1(3) & = 5 + 45 = 50 \\ c(1,4) + f_1(4) & = 100 + 6 = 106 \\ c(1,5) + f_1(5) & = 150 + 0 = 150 \end{bmatrix} \begin{matrix} \\ \leftarrow \text{Min} \\ \\ \\ \\ \end{matrix}$$

$$f_2(2) = \min_j \left[c(2,j) + f_1(j) \right] \tag{2}$$

$$j = \begin{matrix} 1 \\ 2 \\ 3 \\ 4 \\ 5 \end{matrix} \begin{bmatrix} c(2,1) + f_1(1) & = 0 + 150 = 150 \\ c(2,2) + f_1(2) & = 0 + 18 = 18 \\ c(2,3) + f_1(3) & = 9 + 45 = 54 \\ c(2,4) + f_1(4) & = 8 + 6 = (14) \\ c(2,5) + f_1(5) & = 18 + 0 = 18 \end{bmatrix} \begin{matrix} \\ \\ \\ \leftarrow \text{Min} \\ \\ \end{matrix}$$

$$f_2(3) = \min_j \left(c(3,j) + f_1(j) \right) \tag{3}$$

$$j = \begin{array}{c} 1 \\ 2 \\ 3 \\ 4 \\ 5 \end{array} \left[\begin{array}{lll} c(3,1) + f_1(1) & = 0 + 150 & = 150 \\ c(3,2) + f_1(2) & = 11 + 18 & = (29) \\ c(3,3) + f_1(3) & = 0 + 45 & = 45 \\ c(3,4) + f_1(4) & = 40 + 6 & = 46 \\ c(3,5) + f_1(5) & = 45 + 0 & = 45 \end{array} \right] \begin{array}{l} \\ \leftarrow \text{Min} \\ \\ \\ \\ \end{array}$$

$$f_2(4) = \min_j \left(c(4,j) + f_1(j) \right) \tag{4}$$

$$j = \begin{array}{c} 1 \\ 2 \\ 3 \\ 4 \\ 5 \end{array} \left[\begin{array}{lll} c(4,1) + f_1(1) & = 0 + 150 & = 150 \\ c(4,2) + f_1(2) & = 10 + 18 & = 28 \\ c(4,3) + f_1(3) & = 20 + 45 & = 65 \\ c(4,4) + f_1(4) & = 0 + 6 & = (6) \\ c(4,5) + f_1(5) & = 6 + 0 & = (6) \end{array} \right] \begin{array}{l} \\ \\ \\ \leftarrow \text{Min} \\ \\ \end{array}$$

$$f_2(5) = \min_j \left(c(5,j) + f_1(j) \right) \tag{5}$$

$$j = \begin{array}{c} 1 \\ 2 \\ 3 \\ 4 \\ 5 \end{array} \left[\begin{array}{lll} c(5,1) + f_1(1) & = 0 + 150 & = 150 \\ c(5,2) + f_1(2) & = 12 + 18 & = 30 \\ c(5,3) + f_1(3) & = 40 + 45 & = 85 \\ c(5,4) + f_1(4) & = 50 + 6 & = 56 \\ c(5,5) + f_1(5) & = 0 + 0 & = (0) \end{array} \right] \begin{array}{l} \\ \\ \\ \\ \leftarrow \text{Min} \end{array}$$

$f_2(1)=38 \quad f_2(2)=14 \quad f_2(3)=29 \quad f_2(4)=6 \quad f_2(5)=0$

$$f_3(1) = \min_j \left(c(1,j) + f_2(j) \right) \tag{6}$$

$$j = \begin{array}{c} 1 \\ 2 \\ 3 \\ 4 \\ 5 \end{array} \begin{bmatrix} c(1,1) + f_2(1) & = 0 + 38 & = 38 \\ c(1,2) + f_2(2) & = 20 + 14 & = (34) \\ c(1,3) + f_2(3) & = 5 + 29 & = (34) \\ c(1,4) + f_2(4) & = 100 + 6 & = 106 \\ c(1,5) + f_2(5) & = 150 + 0 & = 150 \end{bmatrix} \leftarrow \text{Min}$$

$$f_3(2) = \min_j \left(c(2,j) + f_2(j) \right) \tag{7}$$

$$j = \begin{array}{c} 1 \\ 2 \\ 3 \\ 4 \\ 5 \end{array} \begin{bmatrix} c(2,1) + f_2(1) & = 0 + 38 & = 38 \\ c(2,2) + f_2(2) & = 0 + 14 & = (14) \\ c(2,3) + f_2(3) & = 9 + 29 & = 38 \\ c(2,3) + f_2(4) & = 8 + 6 & = (14) \\ c(2,5) + f_2(5) & = 18 + 0 & = 18 \end{bmatrix} \leftarrow \text{Min}$$

$$f_3(3) = \min_j \left(c(3,j) + f_2(j) \right) \tag{8}$$

$$j = \begin{array}{c} 1 \\ 2 \\ 3 \\ 4 \\ 5 \end{array} \begin{bmatrix} c(3,1) + f_2(1) & = 0 + 38 & = 38 \\ c(3,2) + f_2(2) & = 11 + 14 & = (25) \\ c(3,3) + f_2(3) & = 0 + 29 & = 29 \\ c(3,4) + f_2(4) & = 40 + 6 & = 46 \\ c(3,5) + f_2(5) & = 45 + 0 & = 45 \end{bmatrix} \leftarrow \textit{Min}$$

$$f_3(4) = \min_j \left(c(4,j) + f_2(j) \right) \tag{9}$$

$$j = \begin{array}{c} 1 \\ 2 \\ 3 \\ 4 \\ 5 \end{array} \left[\begin{array}{llll} c(4,1) + f_2(1) & = 0 + 38 & = 38 & \\ c(4,2) + f_2(2) & = 10 + 14 & = 24 & \\ c(4,3) + f_2(3) & = 20 + 29 & = 49 & \\ c(4,4) + f_2(4) & = 0 + 6 & = (6) & \rbrace \leftarrow \text{Min} \\ c(4,5) + f_2(5) & = 6 + 0 & = (6) & \end{array} \right]$$

$$f_3(5) = \min_j \left(c(5,j) + f_2(j) \right) \tag{10}$$

$$j = \begin{array}{c} 1 \\ 2 \\ 3 \\ 4 \\ 5 \end{array} \left[\begin{array}{lll} c(5,1) + f_2(1) & = 0 + 38 & = 38 \\ c(5,2) + f_2(2) & = 12 + 14 & = 26 \\ c(5,3) + f_2(3) & = 40 + 29 & = 69 \\ c(5,4) + f_2(4) & = 50 + 6 & = 56 \\ c(5,5) + f_2(5) & = 0 + 0 & = (0) \leftarrow \text{Min} \end{array} \right]$$

$$f_4(1) = \min_j \left(c(1,j) + f_3(j) \right) \tag{11}$$

$$j = \begin{array}{c} 1 \\ 2 \\ 3 \\ 4 \\ 5 \end{array} \left[\begin{array}{lll} c(1,1) + f_3(1) & = 0 + 34 & = 34 \\ c(1,2) + f_3(2) & = 20 + 14 & = 34 \\ c(1,3) + f_3(3) & = 5 + 25 & = (30) \leftarrow \text{Min} \\ c(1,4) + f_3(4) & = 100 + 6 & = 106 \\ c(1,5) + f_3(5) & = 150 + 0 & = 150 \end{array} \right]$$

$$f_4(2) = \min_j \left(c(2,j) + f_3(j) \right) \tag{12}$$

$$j = \begin{array}{c} 1 \\ 2 \\ 3 \\ 4 \\ 5 \end{array} \left[\begin{array}{lll} c(2,1) + f_3(1) & = 0 + 34 & = 34 \\ c(2,2) + f_3(2) & = 0 + 14 & = (14) \\ c(2,3) + f_3(3) & = 9 + 25 & = 34 \\ c(2,4) + f_3(4) & = 8 + 6 & = (14) \\ c(2,5) + f_3(5) & = 18 + 0 & = 18 \end{array} \right] \leftarrow \text{Min}$$

$$f_4(3) = \min_j \left(c(3,j) + f_3(j) \right) \tag{13}$$

$$j = \begin{array}{c} 1 \\ 2 \\ 3 \\ 4 \\ 5 \end{array} \left[\begin{array}{lll} c(3,1) + f_3(1) & = 0 + 34 & = 34 \\ c(3,2) + f_3(2) & = 11 + 14 & = (25) \\ c(3,3) + f_3(3) & = 0 + 25 & = (25) \\ c(3,4) + f_3(4) & = 40 + 6 & = 46 \\ c(3,5) + f_3(5) & = 45 + 0 & = 45 \end{array} \right] \leftarrow \text{Min}$$

$$f_4(4) = \min_j \left(c(4,j) + f_3(j) \right) \tag{14}$$

$$j = \begin{array}{c} 1 \\ 2 \\ 3 \\ 4 \\ 5 \end{array} \left[\begin{array}{lll} c(4,1) + f_3(1) & = 0 + 34 & = 34 \\ c(4,2) + f_3(2) & = 10 + 14 & = 24 \\ c(4,3) + f_3(3) & = 20 + 25 & = 45 \\ c(4,4) + f_3(4) & = 0 + 6 & = (6) \\ c(4,5) + f_3(5) & = 6 + 0 & = (6) \end{array} \right] \leftarrow \text{Min}$$

$$f_5(5) = \min_j \left[c(5,j) + f_3(j) \right] \tag{15}$$

$$j = \begin{matrix} 1 \\ 2 \\ 3 \\ 4 \\ 5 \end{matrix} \begin{bmatrix} c(5,1) + f_3(1) & = 0 + 34 & = 34 \\ c(5,2) + f_3(2) & = 12 + 14 & = 26 \\ c(5,3) + f_3(3) & = 40 + 25 & = 65 \\ c(5,4) + f_3(4) & = 50 + 6 & = 56 \\ c(5,5) + f_3(5) & = 0 + 0 & = (0) \end{bmatrix} \leftarrow \text{Min}$$

It is now possible to construct the table shown in Figure A3.3.

i	1	2	3	4	5
f_1 (i)	150	18	45	6*	0
s_1 (i)	5	5	5	(5)	5
f_2 (i)	38	14*	29	6	0
s_2 (i)	2	(4)	2	(4, 5)	5
f_3 (i)	34	14	25*	6	0
s_3 (i)	(2, 3)	(2, 4)	(2)	(4, 5)	5
f_4 (i)	30*	14	25	6	0
s_4 (i)	(3)	(2, 4)	(2, 3)	(4, 5)	5

Figure A3.3 Table showing method for constructing least cost route

Select minimum in top row but not including cost of going from 5 to 5.

Select value in next row which is least cost above the minimum value in row 1 and not in same column. In this case 14 is the value next highest in the $f_2(i)$ row with value greater than 6. The decision is therefore that the route lies on 5,4 (but in practice in reversed order).

The next least cost on the $f_3(i)$ row is 25 so point 2 is adjacent on the route to the 4.

The next least cost on the $f_4(i)$ row is 30 so we go to point 3 before arriving at point 1.

Thus, the route the commander should take (the correct way round) is 1→3→2→4→5. The least cost value, 30, is given above the last point on the bottom row.

The foregoing method is very lengthy so let us see how quickly the final table can be derived using a simple algorithm.

Take the last column in the matrix and add the value in the first row to column 1 of the original matrix. Then add the second value in the last column to column 2 of the original matrix, and so on until a new matrix is formed.

$f_1(1)=150$	$f_1(2)=18$	$f_1(3)=45$	$f_1(4)=6$	$f_1(5)=0$
$i=1$	$i=2$	$i=3$	$i=4$	$i=5$

150	(38)	50	106	150
150	18	54	(14)	18
150	(29)	45	46	45
150	28	65	(6)	(6)
150	30	85	56	0

$f_2(1)=38$	$f_2(2)=14$	$f_2(3)=29$	$f_2(4)=6$	$f_2(5)=0$
$i=2$	$i=4$	$i=2$	$i=4,5$	$i=5$

Now add $f_2(1)$ to column 1 of the original matrix and $f_2(2)$ to column 2 and so on to form a new matrix, thus:

38	(34)	(34)	106	150
38	(14)	38	(14)	18
38	(25)	29	46	45
38	24	49	(6)	(6)
38	26	69	56	(0)

$f_3(1)=34$	$f_3(2)=14$	$f_3(3)=25$	$f_3(4)=6$	$f_3(5)=0$
$i=2,3$	$i=2,4$	$i=2$	$i=4,5$	$i=5$

A new matrix can be formed from the row minima and the original matrix, thus:

34	34	(30)	106	150
34	(14)	34	(14)	18
34	(25)	(25)	46	45
34	24	45	(6)	(6)
34	26	65	56	(0)

$f_4(1)=30$	$f_4(2)=14$	$f_4(3)=25$	$f_4(4)=6$	$f_4(5)=0$
$i=3$	$i=2,4$	$i=2,3$	$i=4,5$	$i=5$

This enables the final decision rule to be entered into the matrix in Figure A3.3.

The defence force problem

Another problem which can be treated as a dynamic program is the composition of defence forces for a given level of expenditure. In some respects this is similar to the problem of loading aircraft although the similarity may not be immediately obvious.

The problem is to:

$$\text{maximise} \quad \sum_{i=1}^{N} g_i(x_i) \quad \text{subject to} \quad \sum_{i=1}^{N} \alpha_i x_i = C \qquad x_i \geqslant 0$$

One way of solving this is to use Langrangian multipliers and:

$$\text{maximise} \quad \sum_{i=1}^{N} \{g_i(x_i) + \lambda\alpha_i x_i\} - \lambda_C$$

For a given value of λ it is possible to differentiate with respect to x_i and solve the equations:

$$g_i(x) = -\lambda_i\alpha_i \qquad i = 1 \; ..... \; n$$

and then to find:

$$\sum_{i=1}^{n} \alpha_i x_i$$

amending λ until the expression equates to the constant C.

The difficulties are that $g_i(x_i)$ may not be capable of being differentiated everywhere; that there may be boundary solutions (x_i=0); and x_i may be required to be integer.

The dynamic programming approach is to produce an optimal return by dividing a resource S amongst the first n items. The return from the nth item is $g_n(x_n)$ leaving $S - \alpha_n x_n$ to be allocated amongst the first n-1 products.

Thus

$$f_n(S) = \max_{0 \leq x_n \leq \frac{S}{\alpha_n}} \left[g_n(x_n) + f_{n-1}(S-\alpha_n x_n) \right]$$

Stochastic dynamic programming

The deterministic problems so far considered, where in the function $f_n(i_n)$ the i_{n-1} is known in advance, are sometimes referred to as 'level 0' dynamic programming. 'Level 1' dynamic programming is concerned with stochastic processes where the event i has a known probability distribution.

Dynamic programming may involve stochastic processes. For example, consider the situation where, when an instrument fails during a trial, causing a large quantity of material and manpower to be wasted.

Let us suppose that the cost of repeating the trial is c_t, that the cost of repairing an instrument which fails during a trial is c_r and that the cost of purchasing a new instrument system is c_p.

If $f_n(m)$ is the least expected cost of carrying out n more trials when m trials have already been undertaken, and p_m is the probability of failure on the next trial when m trials have already been undertaken, then:

$$f_{n}(m) = \min \begin{bmatrix} f_n(0) + c_p \\ (1-p_m)f_{n-1}(m+1) + p_m\ (c_t+c_r) + f_n(0) \end{bmatrix}$$

Also:

$$f_n(0) = f_{n-1}(1) + \frac{p_0(c_t+c_r)}{1-p_0}$$

And:

$$f_1(0) = \frac{p_0(c_t+c_r)}{1-p_0}$$

Let us suppose that the following values of p_m have been calculated from past data:

$$p_0 = 0, \quad p_1 = .1, \quad p_2 = .2, \quad p_3 = .4,$$

$$p_4 = .6, \quad p_5 = 1.0$$

Also suppose the cost of repeating a trial c_t is £2,500, the cost of purchasing a new instrument system c_p is £1,000 and the cost of repairing an instrument which fails during a trial c_r is £500.

The optimal policy is determined as follows: Clearly the cost of carrying out 0 trials no matter how many have been undertaken is zero. So

$$f_0(0) = f_0(1) = f_0(2) = f_0(3) = \ldots\ldots f_0(m) = 0$$

The policy is:

$$f_n(m) = \min \begin{bmatrix} f_n(0) + c_p & \text{Replace} \\ (1-p_m)f_{n-1}(m+1)+p_m\left((c_t+c_r)+f_n(0)\right) & \text{Keep} \end{bmatrix}$$

For $n=1$ and $m=0$

$$f_1(0) = \min \begin{matrix} f_1(0) + 1000 & \text{Replace} \\ f_0(1) & \text{Keep} \leftarrow \end{matrix} = 0$$

$m=1$

$$f_1(1) = \min \begin{bmatrix} f_1(0) + 1000 \\ .9f_0(2) + .1\left((3000+f_1(0)\right) \end{bmatrix} \begin{matrix} \text{Replace} \\ \text{Keep} \leftarrow \end{matrix} = 300$$

$m=2$

$$f_1(2) = \min \begin{bmatrix} f_1(0) + 1000 \\ .8f_0(3) + .2\left((3000+f_1(0)\right) \end{bmatrix} \begin{matrix} \text{Replace} \\ \text{Keep} \leftarrow \end{matrix} = 600$$

$m=3$

$$f_1(3) = \min \begin{bmatrix} f_1(0) + 1000 \\ .6f_0(3) + .4\left((3000+f_1(0)\right) \end{bmatrix} \begin{matrix} \text{Replace} \leftarrow \\ \text{Keep} \end{matrix} = 1000$$

Similarly, $f_1(4)$, $f_1(5)$ ..... etc.

Now for $n=2$ and $m=0$

$$f_2(0) = \min \begin{bmatrix} f_2(0) + 1000 \\ f_1(1) \end{bmatrix} \begin{matrix} \text{Replace} \\ \text{Keep} \leftarrow \end{matrix} = 300$$

$m=1$

$$f_2(1) = \min \begin{bmatrix} f_2(0) + 1000 \\ .9f_1(2) + .1\left((3000+f_2(0)\right) \end{bmatrix} \begin{matrix} \text{Replace} \\ \text{Keep} \leftarrow \end{matrix} = 1140$$

$m=2$

$$f_2(2) = \min \begin{bmatrix} f_2(0) + 1000 & \text{Replace} \leftarrow \\ .8f_1(3) + .2\big((3000+f_2(0)\big) & \text{Keep} \end{bmatrix} = 1300$$

And so on for various values of n and m.

The taxi-driver problem

Example: A typical well known example of stochastic dynamic programming is known as 'the taxi-driver problem'. Briefly, this problem can be stated as follows: A taxi-driver operates in several areas and has several choices, to cruise or to drive to a taxi rank. Depending on which area he is in, and which policy he decides to adopt, so the probability of having to drive to another area, and the value of the reward from the fare itself, also varies. A matrix or table of the possibility of a fare requiring him to drive to another area when in the various areas is given, as is the return he will receive for this action (Figure A3.4).

	Probability of Staying in Area I	Return	Probability of Going to Area II	Return
Cruising in Area I	½	4	½	6
At ranks in Area I	⅓	6	⅔	9
Cruising in Area II	½	3	½	7
At ranks in Area II	⅓	3	⅔	6

Figure A3.4

For example, when cruising in Area I the probability of a fare requesting that the driver stay in Area I is say $\frac{1}{2}$ and the fare reward is then 4 units. Similarly, when cruising in Area I, the probability of a fare requesting he go to Area II is $\frac{1}{2}$ with a reward of 6 units and so on.

An iterative method of solving this problem is as follows:

Expected return if cruising in Area I = $(\frac{1}{2}x4) + (\frac{1}{2}x6) = 7$
Expected return if at rank in Area I = $(\frac{1}{3}x6) + (\frac{2}{3}x9) = 8*$
Expected return if cruising in Area II = $(\frac{1}{2}x3) + (\frac{1}{2}x7) = 5*$
Expected return if at rank in Area II = $(\frac{1}{3}x3) + (\frac{2}{3}x6) = 5.$

Thus if the driver is in Area I he should go to the cab rank and if in Area II he should cruise. (The asterisk denotes optimum policy.)

Now suppose that the taxi-driver feels that he has sufficient time for two fares. What policy should he adopt?

Expected return if cruising in Area I = $\frac{1}{2}(4+8) + \frac{1}{2}(6x5) = 11\frac{1}{2}$
Expected return if at rank in Area II = $\frac{1}{3}(6+8) + \frac{2}{3}(9+5) = 14*$

Similarly,

Expected return if cruising in Area II = $\frac{1}{2}(3x8) + (7x5) = 11\frac{1}{2}*$
Expected return if at rank in Area I = $\frac{1}{3}(3+8) + \frac{2}{3}(6x5) = 11$

So his policy remains the same for two fares and in the same way further iterations give the optimal policy if he expects more than two fares. In fact a further iteration will show that with three or more fares the optimal policy when in Area II changes, and that he should drive to a cab rank. Thus if he expects to have three fares or more he should go to a cab rank irrespective of which of the areas he is in.

It is also possible to transform the military attack problem described earlier into a stochastic dynamic program. A variety of possibilities exist. For example, the problem could consider the expected return if attacking from Point 1 against the expected return from just defending Point 1 (since the enemy may withdraw from Point 2 anyway). There is also a similar situation at Point 2. The problem can also be converted to read 'probable losses if attacking from Point 1' etc. The direction and shape of an enemy retreat is to some extent stochastic so there may be scope for using dynamic programming to determine the optimum direction of attack based on figures obtained from military judgement.

Some people look with disfavour on dynamic programming and regard it as no more than a vaguely scientific number-crunching procedure. In some respects this is true but it is worth noting that it can prove very useful at times.

Typical dynamic programs for instrument replacement

PROGRAM 1

```
@1 RUN PEE006,JM2,S20,50
@NXT FOR JM2,JM2
  DIMENSION F(99,99)P(99),A(99),B(99),X(3),Y(3),O(99)
C OPTIMAL INSTRUMENT REPLACEMENT.
C DYNAMIC PROGRAMMING APPROACH.

C INITIALISE ARRAYS.
      DO 1 1=1,99
      A(1)=0
      DO 2 J=1,99
    2 F(I,J)=0.
    1 P(I)=0.

C READ DATA.
      READ(5,1000)N,CP,(P(K),K=1,N),(C(K),K=1,N)
 1000 FORMAT(I2,F10.2/(F8.4))
C CALCULATE FUNCTIONAL VALUES.
      DO 3 1=2,N
      F(I,1)=P(1)*C(1)+F(I-1,2)
      F1=F(I,1)+CP
      DO 4 J=2,N
      F2=P(J)*C(J)+F(I-1,J+1)
      IF(F2-F1)11,11,12
   11 F(I,J)=F2
      GO TO 4
   12 F(I,J)=F1
    4 CONTINUE
    3 CONTINUE
C COMPUTE COST PER YEAR FOR ALL N.
      DO 5 I=2,N
      A(I-1)=(F(I,1)+CP)/(I-1)
    5 B(I-1)=A(I-1)
      DO 7 J+1,3
C FIND MINIMUM COSTS.
      CALL LMIN(A,AMIN,N,K)
      X(J)=AMIN
      Y(J)=K
      IF(J-1)41,41,30
   41 L=K-2
      L1=K-1
      L2=K+1
      L3=K+2
```

```
    30 DO 8 I=1,N
       IF(A)I)-X(J))15,16,15
    16 A(I)=999999
    15 GO TO 8
     8 CONTINUE
     7 CONTINUE
       M=N+1
       IF(L2-M)31,32,33
    31 WRITE(6,1115)
       GO TO 40
    32 WRITE(6,1113)(B(I),I=L1,L2)
       GO TO 40
    33 WRITE(6,1114)(B(I),I=L,L3)

 1115 FORMAT(///,1X,
     145HOPTIMUM POLICY IS TO KEEP AS LONG AS POSSIBLE)
 1113 FORMAT(///,1X,
     146HTHREE VALUES IN NEIGHBOURHOOD OF OPTIMUM ARE:-
     1,/,3F10.3)
 1114 FORMAT(///,1X,
     145HFIVE VALUES IN NEIGHBOURHOOD OF OPTIMUM ARE:-
     1,/,5F10.3)
C PRINT THREE OPTIMUM CASES.
   40 WRITE(6,1111)
 1111 FORMAT(///,1X,22HOPTIMAL POLICIES ARE:-)
      WRITE(6,1112)(Y(J),X(J)),J=1,3)
 1112 FORMAT(///,1X,13HREPLACE EVERY,F3.0,
     136HYEARS AT AN AVERAGE COST PER YEAR OF
     ,F11.4,6H PNDS)
      END
@NXT FOR JM2,JM3
      SUBROUTINE LYMIN(A,AMIN,N,K)
      DIMENSION A(99)
      AMIN=A(1)
      M=N-1
      DO 6 I=1,M
      IF(AMIN-A(1))13,13,14
   14 AMIN=A(I)
      K=I
   13 GO TO 6
    6 CONTINUE
      RETURN
      END
@ XQT JM2
 6
      150.0
```

0.0
0.32
0.52
0.64
0.76
0.864
0.0
40.0
56.0
64.0
75.0
90.0

PROGRAM 2

```
@1 RUN PEE006,JM2,S20,50
@NXT FOR JM2,JM2
      DIMENSION F(99,99),P(99),A(99),B(99),X(3),Y(3)
C OPTIMAL INSTRUMENT REPLACEMENT.
C DYNAMIC PROGRAMMING APPROACH.

C INITIALISE ARRAYS.
      DO 1 1=1,99
      A(I)=0
      DO 2 J=1,99
    2 F(I,J)=0.
    1 P(I)=0

C READ DATA.
      READ(5,1000)N,CP,CR,CT,(P(K),K=1,N)
 1000 FORMAT(I2,3F10.2/(F8.6))
      CF=CT+CR
C CALCULATE FUNCTIONAL VALUES.
      DO 3 I=2,N
      F(I,1)=F(I-1,2)+P(1)*CF/(1-P(1))
      F1=F(I,1)+CP
      DO 4 J=2,N
      F2=(1-P(J))*F(I-1,J+1)+P(J)*(CF+F(I,1))
      IF(F2-F1)11,11,12
   11 F(I,J)=F2
      GO TO 4
   12 F(I,J)=F1
    4 CONTINUE
    3 CONTINUE
C COMPUTE COST PER TRIAL FOR ALL N.
      DO 5 1=2,N
      A(I-1)=(F(1,1)+CP)/I-1)
    5 B(I-1)=A(I-1)
      DO 7 J=1,3
C FIND MINIMUM COSTS.
      CALL LMIN(A,AMIN,N,K)
      X(J)=AMIN
      Y(J)=K
      IF(J-1)41,41,30
   41 L=K-2
      L1=K-1
      L2=K+1
```

```
      L3=K+2
   30 DO 8 I=1,N
      IF(A(I)-X(J))15,16,15
   16 A(I)=999999
   15 GO TO 8
    8 CONTINUE
    7 CONTINUE
      M=N+1
      IF(L2-M)31,32,33
   31 WRITE(6,1115)
      GO TO 40
   32 WRITE(6,1113)(B(I),I=L1,L2)
      GO TO 40
   33 WRITE(6,1114)(B(I),I=L,L3)
 1115 FORMAT (///,1X,
     145HOPTIMUM POLICY IS TO KEEP AS LONG AS POSSIBLE)
 1113 FORMAT(///,1X,
     146HTHREE VALUES IN NEIGHBOURHOOD OF OPTIMUM ARE:-
     1,/,3F10.3)
 1114 FORMAT(///,1X,
     145HFIVE VALUES IN NEIGHBOURHOOD OF OPTIMUM ARE:-
     1,/,5F10.3)
C PRINT THREE OPTIMUM CASES.
   40 WRITE(6,1111)
 1111 FORMAT(///,1X,22HOPTIMAL POLICIES ARE:-)
      WRITE(6,1112)((Y(J),X(J)),J=1,3)
 1112 FORMAT (///,1X,13HREPLACE EVERY,F3.0,
     138HTRIALS AT AN AVERAGE COST PER
     1TRIAL OF,F11,4,6H PNDS)
      END
@NXT FOR JM2,JM3
      SUBROUTINE LMIN(A,AMIN,N,K)
      DIMENSION A(99)
      AMIN=A(1)
      M=N-1
      DO 6 I=1,M
      IF(AMIN-A(I))13,13,14
   14 AMIN=A(I)
      K=I
   13 GO TO 6
    6 CONTINUE
      RETURN
      END
```

```
@ XQT JM2
 5  2340.      120.      180.
 .0
0.3
0.6
0.9
1.0
```

Appendix 4 Branch and bound models

The method known as 'branch and bound' provides a useful way of dealing with many combinatorial problems. It enables a general purpose strategy to be obtained with curtailed enumeration and, as its name implies, the approach consists of two fundamental procedures. *Branching* is the part of the process of partitioning a large problem into two or more sub-problems, whilst *bounding* involves calculating a lower bound on the optimal solution of a given sub-problem.

The branching procedure replaces an original problem by a set of new problems that are mutually exclusive and exhaustive sub-problems of the original; partially solved versions of the original; and smaller problems than the original. Moreover, the sub-problems can themselves be partitioned in a similar manner.

In the bounding procedure a lower bound is calculated for the solution to each sub-problem which has been generated in the branching process. Suppose that at some intermediate stage a complete solution has been obtained that has an associated performance measure P. Suppose also that a sub-problem encountered in the branching process has an associated lower bound b, where $b > P$. In this case the sub-problem need not be considered any further in the search for an optimum. Since no matter how the remainder of the sub-problem is solved, the resulting solution can never have a value better than *P*. When such a sub-problem is found, its branch is said to be *fathomed*. The enumeration process is therefore curtailed by not branching any further from fathomed branches, and the entire set of feasible solutions is enumerated implicitly with respect to the optimum solution.

The solution used in comparisons that allow branches to be fathomed is called a ***trial solution.*** It may be obtained at the very outset by applying a heuristic procedure; or alternatively it can be obtained directly from tree search, by obtaining a single solution from one branch to tree bottom as quickly as possible.

Let us consider a very much simplified defence cost/effectiveness problem.

Example: Suppose that a force can be made up from three different types of military equipment. Let us suppose that the first equipment, which we will call Type I, costs £64,000, whilst the second, Type II, costs £59,000 and the third, Type III, costs £30,000. From careful analysis supported by field studies it is estimated that the 'effectiveness' of the Type I equipment is 64 units, the Type II is 50 units and the Type III is 29 units. The total cost we are prepared to pay for equipment for the force is £216,000. How should we make up the force in order that it shall have the maximum effectiveness?

It is first necessary to decide which type of equipment has the maximum effectiveness per unit cost.

$$\text{Type I effectiveness per unit cost} = \frac{64}{64{,}000} = 1 \times 10^{-3}$$

$$\text{Type II effectiveness per unit cost} = \frac{50}{59{,}000} = .847 \times 10^{-3}$$

$$\text{Type III effectiveness per unit cost} = \frac{29}{30{,}000} = .967 \times 10^{-3}$$

Now consider the types in descending order, taking the most effective per unit cost first, i.e. Type I, then Type III and lastly Type II.

The maximum effectiveness it would be possible to obtain if fractions of a unit were allowed can now be calculated. In this case it would be:

$$216{,}000 \times 1 \times 10^{-3} = 216$$

The next step is to investigate the highest value if 0,1,2,3 ... *m* equipments of Type I are deployed with the cost of $m \leqslant$ £216,000, and assuming unrestricted divisibility of the second most cost-effective type (Figure A4.1).

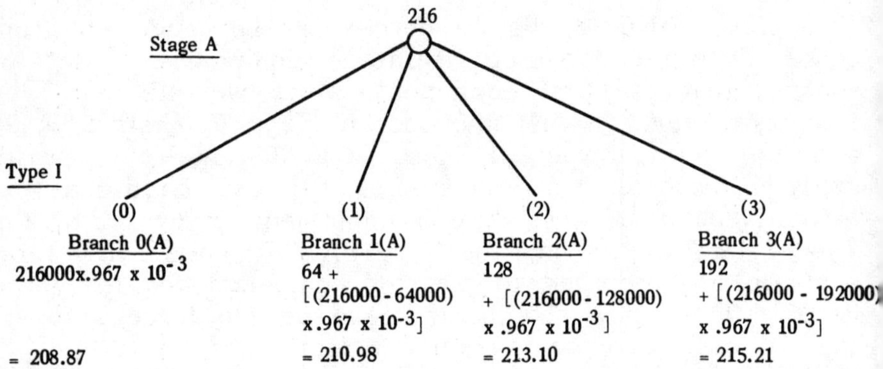

Figure A4.1

The highest of these values is 215.21 and if 3 equipments of Type I are obtained no further item can be purchased. The total effectiveness obtained in this way is therefore 3 x 64 = 192.

The upper bounds of the other possibilities are higher, so take the next highest value 213.10 for Branch 2 and consider the various possibilities for Type III (Figure A4.2).

The largest values actually obtainable with integers for Branch 1(B) and Branch 2(B) and for Type III are:

$$128 + 29 = 157 \qquad \text{and} \qquad 128 + 58 = 186$$

This is because no further complete equipments of Type II can be added. These are both less than Branch 3(A) which will yield at least 192 effectiveness units.

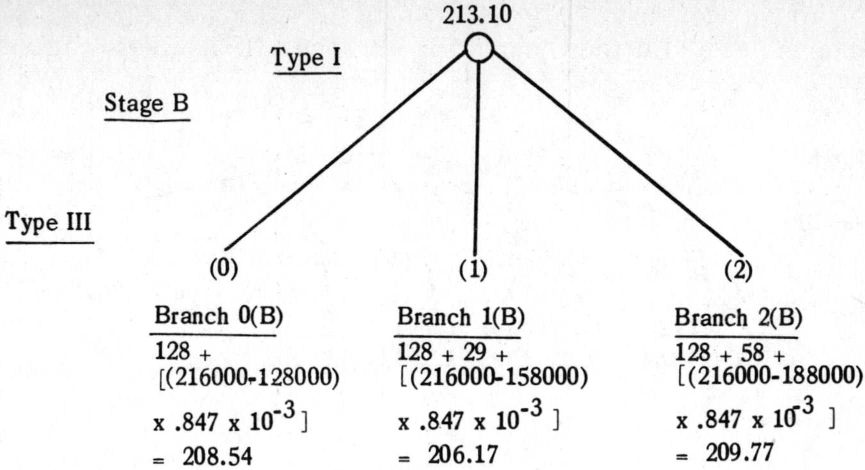

Figure A4.2

Continuing where we have left off at Branch 0(A), Branch 1(A) and Branch 0(B) we have Figure A4.3. The highest is 210.98, so we continue with Branch 1(B) (Figure A4.4).

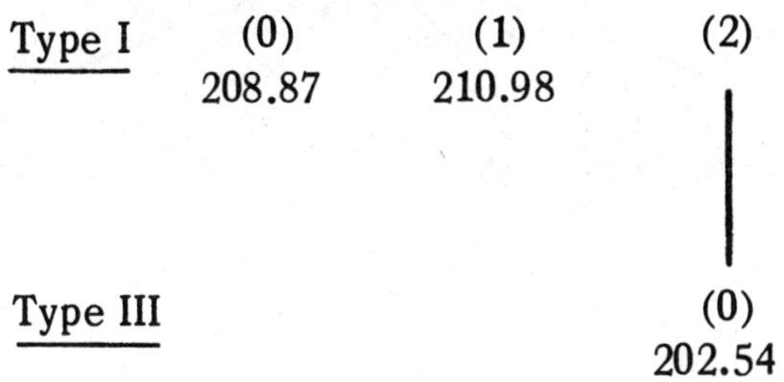

Figure A4.3

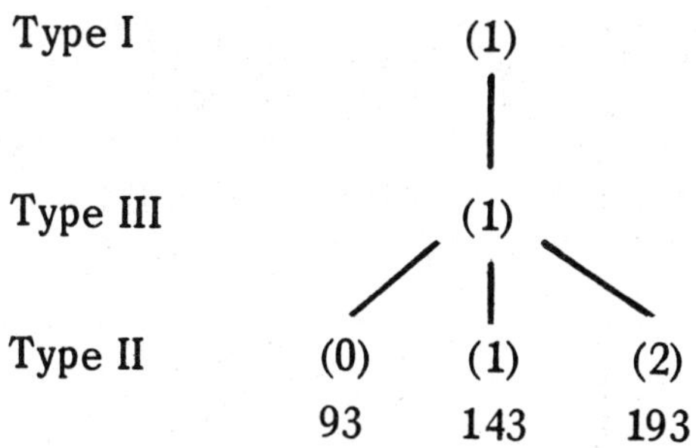

Figure A4.4

Thus we have a new higher value of 193 now attainable. Further exploring Branch 1(A) gives Figure A4.5.

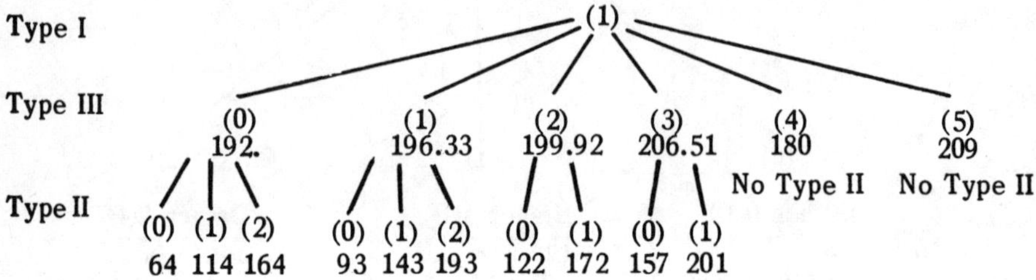

Figure A4.5

Thus the best solution is 1 Type I and 5 Type III which gives an effectiveness of 209 units. The next best is 1 Type I, 1 Type II and 3 Type III with an effectiveness value of 201.

Finally, investigating Branch 0(A) gives Figure A4.6.

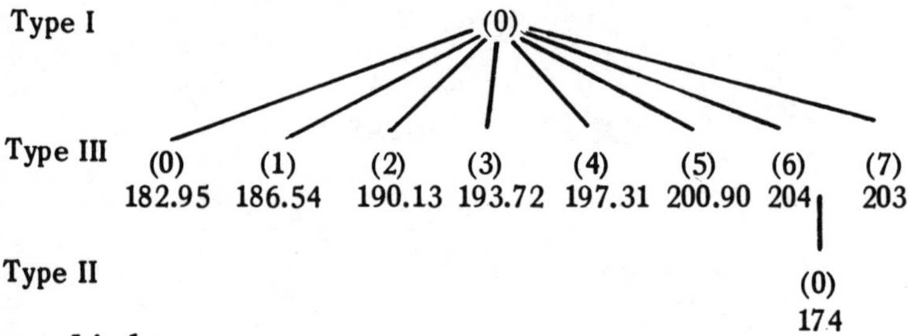

Figure A4.6

Thus, although we can obtain 201 effectiveness units with 1 of Type I, 1 of Type II and 3 of Type III, and an effectiveness of 203 with 7 of Type III, the highest effectiveness of 209 is obtained with 1 of Type I and 5 of Type III.

We could have imposed various other constraints with this type of model. For example, we could have stated that the number of Type III units must not exceed three. In this case we would have obtained the third best solution to the unconstrained problem.

Zero-one branch and bound models

It is also worth outlining briefly a particular form of branch and bound programming known as zero-one branch and bound programming. In the example given here, a number of rather different constraints will be imposed in order to show how these can be handled.

Example: Consider a national defence problem where we are examining the possible purchase of a number of different weapon systems, S_1, S_2, S_3, S_4 and S_5 for inclusion in the long-term estimates. The relative effectiveness of these systems is set out below:

$$S_1 = 50 \text{ effectiveness units}$$

$$S_2 = 250 \text{ effectiveness units}$$

$$S_3 = 100 \text{ effectiveness units}$$

$$S_4 = 50 \text{ effectiveness units}$$

$$S_5 = 100 \text{ effectiveness units}$$

The cost constraints are such that we can afford only two of S_1, S_2 and S_3. Similarly, on grounds of requirements and cost we can choose one only of S_2 or S_5. Also if S_2 is deployed then from technical and military considerations S_3 and S_4 cannot be used together. On the other hand if S_1 is deployed then we must also deploy S_4.

Briefly the constraints are therefore:

If S_1 then also S_4

Choose only two of S_1, S_2 and S_3

S_2 or S_5

If S_2 then also S_3 and S_4.

Let us write:

$x_1' = 1$ if S_1 chosen or 0 if not present

$x_2' = 1$ if S_2 chosen or 0 if not present

$x_3' = 1$ if S_3 chosen or 0 if not present

$x_4' = 1$ if S_4 chosen or 0 if not present

$x_5' = 1$ if S_5 chosen or 0 if not present

So the constraints are:

$x_1' - x_4' \leqslant 0$	(If S_1 then also S_4)
$x_1' + x_2' + x_3' \leqslant 2$	(Cost limitation of only two of S_1, S_2 and S_3)
$x_2' + x_5' \leqslant 1$	(Choose either S_2 or S_5 for cost and technical considerations)
$x_2' - x_3' + x_4' \leqslant 1$	(If S_2 and S_4 are present then S_3 is not possible)
$x_2' + x_3' - x_4' \leqslant 1$	(If S_2 and S_3 are present then S_4 is not possible)

Maximise $y_0 = 50x_1' + 250x_2' + 100x_3' + 50x_4' + 100x_5'$

We wish to reformulate the problem so that an increase in each variable from 0 to 1 reduces the function. Introduce new variables such that $x_i = 1 - x_i'$

Thus the constraints now become:

$$x_4 - x_1 \leqslant 0$$

$$-x_1 - x_2 - x_3 \leqslant -1$$

$$-x_2 - x_5 \leqslant -1$$

$$-x_2 + x_3 - x_4 \leqslant 0$$

$$-x_2 - x_3 + x_4 \leqslant 0$$

The maximum effectiveness it would be possible to obtain if every x value were feasible is:

$$50 + 250 + 100 + 50 + 100 = 550$$

Thus we can maximise the transformed function which is:

$$550 - 50x_1 - 250x_2 - 100x_3 - 50x_4 - 100x_5$$

Introduce slack variables y_j (all y's ≥ 0):

$$y_1 = x_1 - x_4$$

$$y_2 = -1 + x_1 + x_2 + x_3$$

$$y_3 = -1 + x_2 + x_5$$

$$y_4 = x_2 - x_3 + x_4$$

$$y_5 = x_2 + x_3 - x_4$$

If all $x_i = 0$, y_2 and y_3 would be negative and so we make one of the variables equal to 1. We will choose (arbitrarily) in this case the variable x_1. Thus if $x_1 = 1$ and all other x_i are zero we get:

$$y_1 = 1 - x_4$$

$$y_2 = x_2 + x_3$$

$$y_3 = -1 + x_2 + x_5$$

$$y_4 = x_2 - x_3 + x_4$$

$$y_5 = x_2 + x_3 - x_4$$

and

$$y_o = 500 - 250x_2 - 100x_3 - 50x_4 - 100x_5$$

$$x_3 = 1 \quad \text{or} \quad x_4 = 1 \quad \text{and} \quad x_i = 0 \quad \text{for all other } i$$

This i gives two negative y_j's when substituted in the new constraint equations.

$x_2 = 1$ or $x_5 = 1$ both do not give any y_j negative but y_o is larger for $x_5 = 1$ than for $x_2 = 1$, so choose x_5.

$$x_1 = \quad (1)$$

$$x_4 = (1)$$

The problem is now:

$$y_1 = 1 - x_4$$

$$y_2 = x_2 + x_3$$

$$y_3 = x_2$$

$$y_4 = x_2 - x_3 + x_4$$

$$y_5 = x_2 + x_3 - x_4$$

$$y_o = 400 - 250x_2 - 100x_3 - 50x_4$$

x_2 is now best so:

$x_1 =$ (1)

/

$x_5 =$ (1)

/

$x_2 =$ (1)

So:

$$y_1 = 1 - x_4$$

$$y_2 = 1 + x_3$$

$$y_3 = 1$$

$$y_4 = 1 - x_3 + x_4$$

$$y_5 = 1 + x_3 - x_4$$

$$y_o = 150 - 100x_3 - 50x_4$$

Now use $x_3 = 1$

$x_1 =$ (1)

/

$x_5 =$ (1)

/

$x_2 =$ (1)

/

$x_3 =$ (1)

and

$$y_1 = 1 - x_4$$

$$y_2 = 2$$

$$y_3 = 1$$

$$y_4 = x_4$$

$$y_5 = 2 - x_4$$

$$y_o = 50 - 50x_4$$

The remaining variable x_4 can be 0 or 1 and 0 gives the higher value:

$$y_o = 50 - 50x_4$$

i.e. $\quad x_4 = 0 \qquad y_o = 50$

The question is now whether a higher value than $y_o = 1$ is possible so go back up the tree to $x_3 = 1$.

Try $x_3 = 0$ in the previous equations.

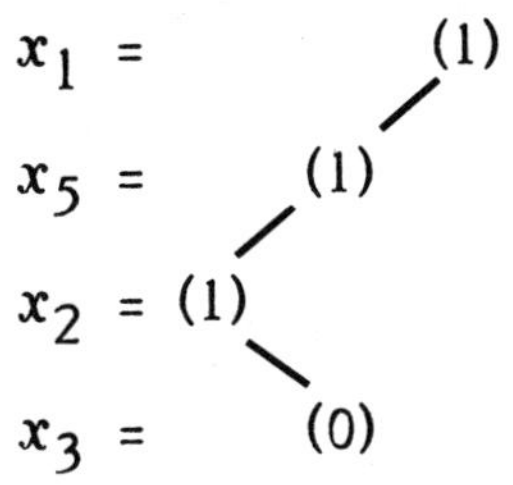

So

$$y_1 = 1 - x_4$$

$$y_2 = 1$$

$$y_3 = 1$$

$$y_4 = 1 + x_4$$

$$y_5 = 1 - x_4$$

$$y_o = 150 - 50x_4$$

And as x_4 could be 0 or 1 we find for $x_4 = 0$ y_o = 150 and this is the higher value.

So now go back a further step

$x_1 =$ (1)

$x_5 =$ (1)

$x_2 =$ (0)

These equations become:

$$y_1 = 1 - x_4$$

$$y_2 = x_3$$

$$y_3 = 0$$

$$y_4 = -x_3 + x_4$$

$$y_5 = x_3 - x_4$$

$$y_o = 400 - 100x_3 - 50x_4$$

Notice from y_4 and y_5 that x_3 and x_4 must either both be 0 or both be 1. Now y_o is largest if both are 0. In which case $y_o = 400$.

Going back further:

$x_1 =$ (1)

$x_5 =$ (0)

$$y_1 = 1 - x_4$$

$$y_2 = x_2 + x_3$$

$$y_3 = -1 + x_2$$

$$y_4 = x_2 - x_3 + x_4$$

$$y_5 = x_2 + x_3 - x_4$$

$$y_o = 500 - 250x_2 - 100x_3 - 50x_4$$

The equation y_3 involves a forced choice in that x_2 must be 1 so the new problem is:

$$y_1 = 1 - x_4$$

$$y_2 = 1 + x_3$$

$$y_3 = 0$$

$$y_4 = 1 - x_3 + x_4$$

$$y_5 = 1 + x_3 - x_4$$

$$y_o = 250 - 100x_3 - 50x_4$$

This case can be ignored because y_o cannot be made larger than 250 and we have already obtained $y_o \geqslant 400$ earlier.

Return to $x_1 = 0$, so $y_1 = -x_4$ and by another forced choice $x_4 = 0$

$x_1 =$ (0)

$x_2 =$ (0)

In this case we have:

$$y_1 = 0$$

$$y_2 = -1 - x_2 + x_3$$

$$y_3 = -1 + x_2 + x_5$$

$$y_4 = x_2 - x_3$$

$$y_5 = x_2 + x_3$$

$$y_o = 550 - 50x_2 - 100x_3 - 50x_5$$

If all the other variables are zero, $x_2 = 1$ is the only case which does not give a negative value for y_j and gives an objective function $y_o = 300 - 100x_3 - 50x_5$. This value is less than that already obtained.

We should therefore select weapon systems S_2, S_3 and S_4 for which we will obtain an effectiveness return of 400 units.

It will be recalled that quite early in the solution to this problem we chose, quite arbitrarily, to make x_1 equal to 1. A useful decision rule, normally having a better efficiency than arbitrary choice, is to select from those variables still to be considered one which, if made equal to 1 on its own, will produce the smallest number of negative y_j.

General remarks

This approach to integer programs (the Land–Doig algorithm) [30] does not enlarge the problem, as for example is the case with the Gomory algorithm. [31] In fact the problem is considered as a series of new problems and a smaller number of variables. The computational procedures are therefore generally more efficient. (See also the Balas algorithm.[82]) A good overall exposition is given in Chapter 4 of Garfinkel and Nemhauser. [20]

An advantage often claimed for the branch and bound technique is that it is frequently possible to obtain a 'good' integer solution which is satisfactory and feasible in very much less time than that required to obtain the optimum. Moreover, in most cases this is likely to be accompanied by some knowledge of how far the solution is from the true optimum. This can be a considerable advantage in very large problems.

Appendix 5 Notes on an example of a simulation model for a computer

The following notes are largely extracted from Chapter IX of a thesis on the 'Limits to Growth in Business Organisations', (32) undertaken in the Department of Computational Methods in the Statistics Faculty at the London School of Economics.

The reason for including this example is simply to show the way in which it was gradually built up through a series of models of increasing complexity with the results of each added complexity seeming to have a reasonably qualitative, if not quantitative, explanation.

It was intended, in the simulation, to build a model of a company on a computer in order to study the effects of changing various features, such as retained capital, dividend policy, taxation, etc. In the course of the study some eight different models were constructed, and at the end they were combined into a single general model which could be used to produce the same effects as all the individual models.

The first model was fairly simple and had only four variables. This model allowed the company to either make a fixed profit G or a loss L. The probability Q of obtaining a loss could be changed, but remained constant throughout the run. The working capital was denoted by C, and if the value of this variable fell below zero the company was deemed to be bankrupt. It was assumed that all profits which raised the capital in excess of a fixed percentage of the total capital were paid out as dividends, so that C could never rise above a certain fixed level.

The company was then allowed to operate for 100 years and the number of times it became bankrupt was noted. A further 99 runs were then carried out in order to obtain a good statistical sample. The process was a continuous birth–death process so that when one company was liquidated a new one was formed. Usually the value of C could not fall to zero in the first year or two by virtue of the total losses needed to reduce the capital, and consequently the company was, in effect, prevented from bankruptcy during this period.

A similar situation applied during the first few years in the lives of all the 100 companies, and it was considered that this created an unreal starting up effect in the model. It was therefore decided to run each company for 110 years and to discard the first 10 years of operation. In this way the simulation was initialised into a fairly typical 'real world' situation where some companies might go bankrupt during the first year in which they were being observed.

The program output was arranged so that the probability of the company becoming bankrupt, the percentage of companies failing in any one year, the expected life of the company, the total number of companies (out of 100) failing in 100 years, and the total number of companies (out of 100) surviving 100 years, were listed.

The foregoing model is extremely simple and has many limitations. For example, many companies are involved in more than one kind of major work (or project), and each major activity may have a large number of individual sales lines.

Whether a number of lines can be regarded as spreading the risk in quite the same manner as diversification of activities, requires careful assessment in each case. A new invention can greatly affect an activity and all lines. An instance of this was the invention of the gramophone, which affected the total sales of pianolas irrespective of model, mark or make. In the same way television affected the majority of cinemas, irrespective of their size or quality.

The second program to be developed (Program B) was concerned with determining the effect of spreading the risk over a number of projects, or activities. Initially, the same value of C was used, and the sum of the projects' gain or loss was similar to the single project values.

A third model (Program C) was developed to allow for fluctuations in trade. A four-year cycle was selected and the probability of profit and loss was decreased and increased in a cyclical manner.

The extent to which this kind of changing probability is realistic is questionable. In adverse trading conditions money is usually restricted, whilst in good times it is more readily available. It therefore seemed more sensible to increase G and L by fixed amounts in good times, and to decrease them in times of poor trading. The cycle adopted therefore followed one year of normal trading, one good year, one normal year, and one bad year, and so on.

Accordingly, a program was developed along these lines for a simple model in which all capital over a certain set amount was paid out in dividends, and where a single project, or activity, was involved.

It was considered worthwhile extending the effect of trade fluctuations to the multi-project model and this formed 'Program D' of the series of simulation models. Once again it was found that spreading the risk rapidly diminished the chance of a complete disaster for the company.

It was thought that tax and dividend policies might have a considerable effect upon the rate of company liquidations. Taxation can act as a kind of cushion for a company over a short period when it incurs serious losses, since in this event it is able to claim against taxes paid in earlier years. In this model it was possible to claim against taxes paid during the previous three years.

Dividend policy tends to fluctuate and is usually based on several rules rather than a simple one of, say, paying out all capital above that required to maintain the business at its present level. Quite apart from the requirement for expansion and growth, a public company usually seeks to satisfy Stock Exchange brokers, trustee fund rules and regulations, and so on. This means that the earlier programs, being based on relatively simple assumptions, are, to a certain extent, only generally indicative of the real world situation.

A single-project model, which included taxation and what is hoped was a typical general dividend policy, was therefore developed. This program was described as

'Program E'. It was found that allowing a company to recall its back taxes over a three-year period when it encountered difficulties increased its chances of survival approximately twofold.

This model was developed further to include the multi-project situation to give 'Program F'. It was run with various back tax recalls of 0,1,2 and 3 years so that the model could be checked against the earlier results.

The model was then developed to include the effects of trade fluctuations in addition to the tax and dividend algorithms. This particular version was designated 'Program G'. It was run in various ways to check the results which had been obtained with the previous models and then uniquely to obtain the data which only this model could provide. It was found that for the single-project model when allowing trade fluctuations the three-year back tax recall reduced the rate of company failures by between threefold and sevenfold. This clearly illustrated the importance of dividend policies and tax laws on companies. The effect with the multi-project model was even more pronounced.

The model was then extended to enable the effects of corporation tax changes to be studied more readily, in that variable values for corporation tax could be examined without changing the values in the program itself. This resulted in 'Program H' which was again checked as far as possible against its predecessors. This program yielded some interesting data. Changes in the variable RI (rate of interest) in this model provided a dramatic predictor of the British industrial difficulties of the early 1980s when a high interest rate policy was followed.

Finally, the whole program was 'tidied up' so that all features were incorporated into one model and the variables input to the model, and the output of the model, could be controlled by program 'flags'. This program, known as 'Program I' was again checked against all previous models. A measure of the complexity and scope of the final model can be judged from the comment cards explaining the program and for this reason they are included in this Appendix, although for reasons of brevity the entire program, which was written in FORTRAN, is not. At every stage as the program was developed the results were examined, qualitatively, and tested as far as possible.

PROGRAM I : FINAL GENERAL COMPANY MODEL

NOTES: THIS MODEL INCLUDES A NUMBER OF ADDITIONAL VARIABLES, SOME OF WHICH MAY BE USED TO ESTIMATE THE PARAMETERS IN MODEL H. A NUMBER OF RATIOS ARE CALCULATED IN ORDER TO ASSIST IN ASSESSING THE VULNERABILITY OF THE COMPANY TO FAILURE, AND AS BEFORE THE MODEL CAN INCLUDE A NUMBER OF PROJECTS, DETERMINISTIC ECONOMIC CYCLES, TAX VARIATIONS, AND CHANGES IN THE GAINS LOSSES, WORKING CAPITAL, ETC. THE MODEL CAN BE CHANGED BY USING THE DIFFERENT FLAGS.

SYMBOLS:

TX = TAX
C = CAPITAL
GG1 = GAIN FOR PROJECT 1
LL1 = LOSS FOR PROJECT 1
QQ1 = PROBABILITY OF LOSS ON PROJECT 1
ZZ1 = GAIN FLUCTUATION ON PROJECT 1
TT1 = LOSS FLUCTUATION ON PROJECT 1
A = FRACTION OF PROFITS PAID OUT
GG2 = GAIN FOR PROJECT 2
LL2 = LOSS FOR ... ETC.
TP = TOTAL PROFITS
TA = TOTAL ASSETS
CR = INTEREST RECEIVED ON DEPOSITS
RI = INTEREST RATE PAID
FR = PROPORTION OF CAPITAL ASSETS WHICH CAN BE REALISED
VS1 = VALUE OF SALES FOR PROJECT 1
CB1 = COST OF BORROWING TO SUPPORT PROJECT 1
CL1 = COST OF LABOUR FOR PROJECT 1
CM1 = COST OF MATERIAL FOR PROJECT 1
OH1 = PORTION OF OVERHEADS ATTRIBUTABLE TO PROJECT 1
EP1 = EXPECTED PROFIT ON PROJECT 1
VS2 = VALUE OF SALES FOR PROJECT 2
CB2 = COST OF BORROWING ... ETC.
TEP = TOTAL EXPECTED PROFITS
BFP = BEST FEASIBLE PROFITS
MFL = MAXIMUM FEASIBLE LOSS

PP = PROFIT PERCENTAGE
PSR = PROFIT SENSITIVITY RATIO
VL = LIFE VULNERABILITY

ALL VALUES EXCEPT A, TA, TP(1), BS(1), CB(1), CL(1), CM(1), OH(1), PSR AND VL ARE EXPRESSED AS PERCENTAGES.

PROGRAM CONTROL BY FLAGS : IFLAG, JFLAG, KFLAG, LFLAG, NFLAG AND M1, M2, M3, M4, M5, M6.

FLAGS:

IFLAG CONTROLS LENGTH OF PRINT FORMATS AND PROGRAM TERMINATION.
ABS. VALUE OF IFLAG=1 FOR SHORT FORMAT,
ABS. VALUE OF IFLAG≠1 FOR LONG FORMAT
IFLAG POSITIVE INDICATES LAST SET OF DATA.

JFLAG CONTROLS PRINTING OF COMPANY HISTORIES.
JFLAG=0 FOR PRINTOUT SUPPRESSION,
JFLAG≠0 FOR PRINTOUT INCLUSION.

KFLAG CONTROLS PRINTING OF DIVIDEND ARRAY.
KFLAG=0 FOR PRINTOUT SUPPRESSION,
KFLAG≠0 FOR PRINTOUT INCLUSION.

LFLAG CONTROLS INCLUSION OF TAX AND DIVIDEND ALGORITHMS.
LFLAG=0 FOR SUPPRESSION OF ALGORITHMS,
LFLAG≠0 FOR INCLUSION OF ALGORITHMS.

NFLAG CONTROLS NUMBER OF YEARS OVER WHICH TAX CAN BE RECLAIMED.
NFLAG=0 FOR 0 YEARS BACK TAX,
NFLAG=1 FOR 1 YEARS BACK TAX,
NFLAG=2 FOR 2 YEARS BACK TAX,
NFLAG=3 FOR 3 YEARS BACK TAX.

M1 CONTROLS CALCULATION AND PRINTING TOTAL EXPECTED PROFITS.
M1=0 FOR PRINTOUT SUPPRESSION,
M1=1 FOR PRINTOUT INCLUSION.

M2 CONTROLS CALCULATION AND PRINTING OF FEASIBLE PROFITS AND LOSSES.
M2=0 FOR PRINTOUT SUPPRESSION,
M2=1 FOR PRINTOUT INCLUSION.

M3	CONTROLS CALCULATION AND PRINTING OF TOTAL PROFITS. M3=0 FOR PRINTOUT SUPPRESSION, M3=1 FOR PRINTOUT INCLUSION.
M4	CONTROLS CALCULATION AND PRINTING OF PROFITS PERCENTAGE. M4=0 FOR PRINTOUT SUPPRESSION, M4=1 FOR PRINTOUT INCLUSION.
M5	CONTROLS CALCULATION AND PRINTING OF PROFIT SENSITIVITY RATIO. M5=0 FOR PRINTOUT SUPPRESSION, M5=1 FOR PRINTOUT INCLUSION.
M6	CONTROLS CALCULATION AND PRINTING OF LIFE VULNERABILITY RATIO. M6=0 FOR PRINTOUT SUPPRESSION, M6=1 FOR PRINTOUT INCLUSION.

Appendix 6 Some comments on simulation languages

A reasonable selection of simulation languages is now available and, unlike the situation in the late 1960s and early 1970s, many languages can be used on different types of machine. It has long been the aim to develop a universal simulation language but there are some inherent difficulties in that the 'best' simulation language for one task is not necessarily ideal for another. In part these difficulties often involve decisions as to whether or not to base the simulation on 'events' or 'activities'.

In the early simulation language, SIMSCRIPT, a system is defined in terms of 'entities' and 'attributes'. The latter are used to describe the properties possessed by the entities. This language also uses 'sets' which are groups of entities, and both entities and sets can be changed dynamically.

The main differences between the early languages are the dominant entities in the language. For example, we must consider two types of entities, those which are active and those which are passive. Languages in which the passive entities are dominant are said to be 'material-based' languages and those in which the active entities feature are described as 'machine-based'. The reason is that material exemplifies a passive system whilst machines which act on the material are regarded as active entities. Hence a language in which active entities are dominant is known as a machine-based language.

The system being studied consists of a sequence of processes which, in the simulation, are modelled by the processes being executed *by* entities or *upon* entities, depending on whether the system is 'machine'- or 'material'-oriented.

Another early simulation language was GPSS. This is formulated in such a way that the simulation models can be constructed directly from block diagrams. Transactions take place using temporary elements so they can have different parameters and priority levels.

In many languages a list is used to decide the time the next event will occur. The time advance may be achieved by scanning the times associated with all entities and by selecting only those greater than the current clock time. The user may also be responsible for adding and deleting times from the scanned list. Alternatively, the language may consist of statements which alter the times and imply adding the time to the list. In the last case, for example, if an entity is associated with a process its time is the forecast of the end of the process. In this case it is said to be committed. (It is regarded as being 'available' otherwise.) If the list of entities is long, it is usually possible to use dynamic listing techniques.

Every process has associated conditions for its commitment and rules which signify its effect on the system. These are described as 'activities'. In certain instances the completion of one process implies the commencement of another, so that with this concept of 'bound' activities one activity may directly call up the next one.

Programs may be organised in three ways: those based on 'conditional' activities (C activities); 'bound'-based activities (B activities); and those based on a combination of both types of activity.

B activity programs are often more difficult to plan and write than C-based programs but the latter are generally slower running. They are also difficult to change. Sometimes a combination of the two methods may provide a good compromise solution. Many of the newer languages tend to have more flexibility. Thus in GPSS and SIMSCRIPT activity organisation is primarily in the B form and in ECSL in the C form, but in some other languages such as ESP and GSP it is in both the B and C form.

Regarding the question as to whether machine-based or material-based languages are ideal for any particular simulation, we must ask if the characteristics of the 'machines' are to be the most complex and dominant, or are the main characteristics likely to be 'material'? One interesting advantage may be claimed for machine-based systems in that the number of machines in a system is usually fixed, but the number of batches of material can vary. Hence, in a material-based system dynamic storage allocation can be used and new items created as others are destroyed.

Important features of language facilities are 'find' and 'scan' loops. 'Find' loops test entities in turn until a successful candidate is found, whilst 'scan' loops check all entities for successful candidates from which a selection decision is then necessary. Many languages also have 'first', 'last', minimum, maximum and random choice facilities. Languages based on ALGOL usually have good Boolean facilities. Some languages such as GPSS, GSP, ECSL and SIMSCRIPT allow tests but do not have Boolean facilities.

Important aspects of simulation facilities are the sampling procedures. In order to model 'real life' situations it is necessary to introduce a stochastic element and this entails sampling from a distribution. Most languages use a so-called random number generator of the Lehmer multiplication congruence type. This random number is then used to sample some type of distribution, such as a cumulative or a Poisson distribution, etc.

The objective of simulation studies is to obtain distribution data of the parts of the system such as queues, delays, idle time, etc. Practically all the languages have histogram-forming facilities and most of them will calculate the mean and standard deviation in addition. Most of the modern languages have good reporting features, and the computer will present a report to a specified format filling in the missing information where appropriate.

The ability to monitor simulation and to change the initial conditions quickly is quite important. Many languages permit the elimination of an initialisation phase of the program.

Simulation languages and their peripheral software have improved appreciably during the past decade. Most of the modern languages have good facilities for queue handling and selecting stochastic variates. The data from the simulation can be tested against various distributions and histograms plotted. The data collection and output are simple and effective, and the output can be in graphical form. Furthermore, simulation languages are no longer confined to large and expensive machines. Some are now available on microcomputers.

To sum up briefly: the choice of simulation languages is nowadays quite a wide one, but some care is still needed to select the most suitable language for the task. The question also remains whether, in the case of small organisations, it is advisable to write programs using any dedicated simulation languages at all. The use of simulation languages means that a specialist programmer with knowledge of the language must continually be available to carry out any necessary modifications to the program. However, in the case of large organisations which repeatedly develop certain types of simulation, special languages and additional software can be very time-saving both in writing programs and in the actual running times on the computer.

Appendix 7 The system dynamics approach

Professor Platt of the University of Chicago, writing in the *New York Times*, hailed Jay Forrester's book on industrial dynamics as 'a book which might be comparable in ultimate importance to Galileo, or Malthus, or Rousseau or Mill'. Many will agree that in the light of experience he was somewhat overstating the case although Forrester's book remains as a valued classical exposition of the subject. (83)

The concept behind system dynamics is the analagous behaviour of many real world situations to machines with servo-mechanisms whose response can be studied using control theory. It is suggested that information feedback control loops exist in business, and that orders and inventory levels lead to manufacturing decisions to fulfil orders, restock inventories, and these in turn lead to changing manufacturing decisions. In performing these processes temporary over- and under-correction can lead to wild oscillatory swings in some of the variables, and where the information feedback is inadequate, the situation may rapidly get out of control.

The first step in system dynamics is very much the same as in most other approaches to model building and involves a clear identification of the problem to be studied and a careful assessment of the variables which are most likely to be of major effect.

For example, Forrester states that the process most central to many industrial companies is that of production and distribution. One of the principal tasks of a company is to match the production rate to the rate of consumer sales. From experience we know that factory production rates (e.g. boom and slump conditions) often fluctuate much more than consumer purchases. Distribution systems usually contain cascaded (i.e. a series of) inventories and ordering procedures which tend to amplify the relatively small disturbances that occur at the retail level.

System dynamics attempts to model this kind of system using feedback and control theory and to answer questions with regard to how management policies can assist in damping out these oscillations and at the same time remain reasonably responsive to fairly rapid changes in retail sales. Figure A7.1 shows a typical industrial situation. The encircled numbers indicate the kind of time delays Forrester suggests are typical of many manufacturing processes. The broken line shows the passage of information requesting that goods be supplied or that routine supplies be reduced, whilst the solid lines denote the actual movement of manufactured goods.

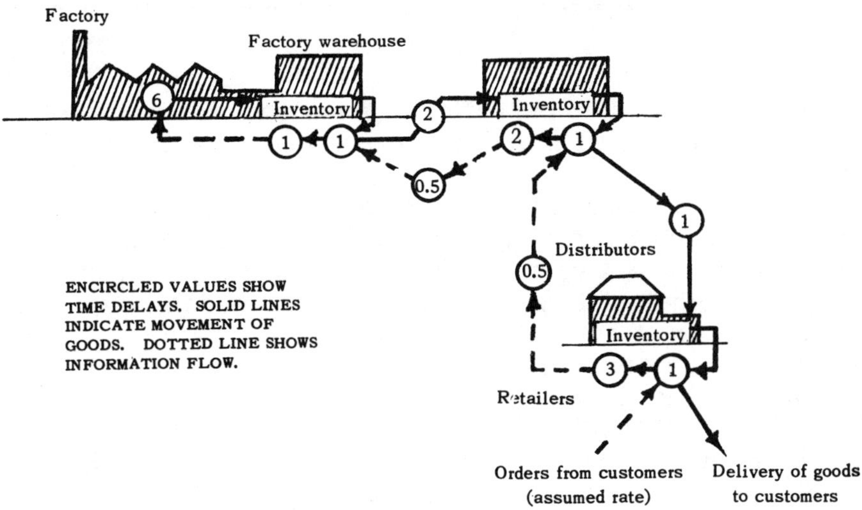

Figure A7.1 Forrester's model of an industrial dynamic situation

It is possible to carry out a simple simulation with this model by following through, step by step, the orders for goods, the time to manufacture these and the times for new instructions and orders for goods from the retailer to reach the manufacturer. The result obtained for the system shown in Figure A7.1 is given in Figure A7.2. This shows the oscillations which occur in building up to a stable retail level.

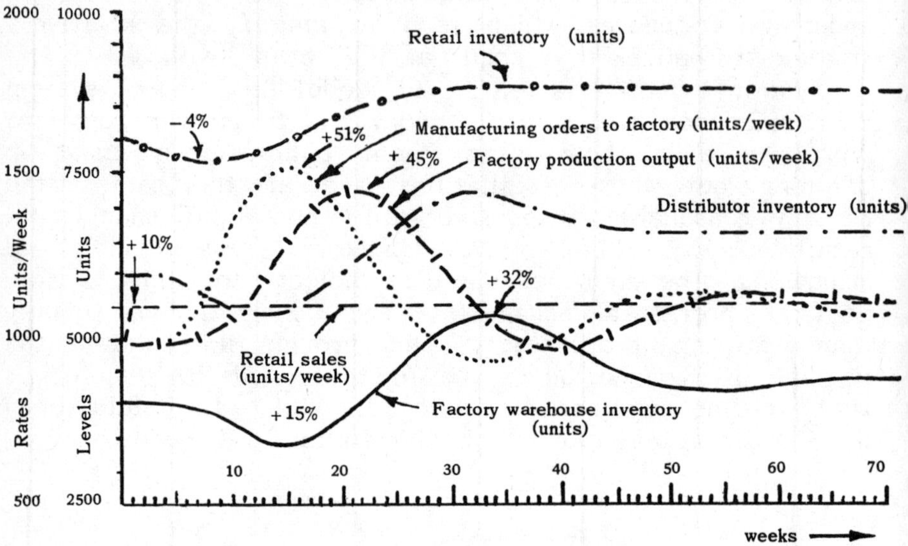

Figure A7.2 Oscillatory effects demonstrated by Forrester as a consequence of a sudden 10 per cent rise in demand

The more complex the model, the more difficult it becomes to conduct a dynamic simulation by hand. Once the model has been constructed and the equations expressing the different parts of the model have been formulated in terms of the variables and rate of change equations, a computer can be used to obtain a rapid solution. The levels of the different variables can be plotted automatically and the effects of system changes investigated. One programming language which was developed for this purpose is known as DYNAMO (Forrester at MIT). Another language uses a vaguely similar title DYSMAP (Coyle at Bradford University). A further language called MICRODYN (Parker at Lancaster University) is available for the Apple II microcomputer at a cost of approximately £500, so the technique is no longer exclusive to large machines.

It is usual, when commencing to solve a problem by system dynamics methods, to construct a conceptual diagram (or some slight variant thereof) for the entire system. Forrester in his book on industrial dynamics [83] recommended flow diagrams in which the different kinds of element have different symbols similar to those used by control engineers. He suggested six different types of flow; but normally two types, namely information and materials, are sufficient.

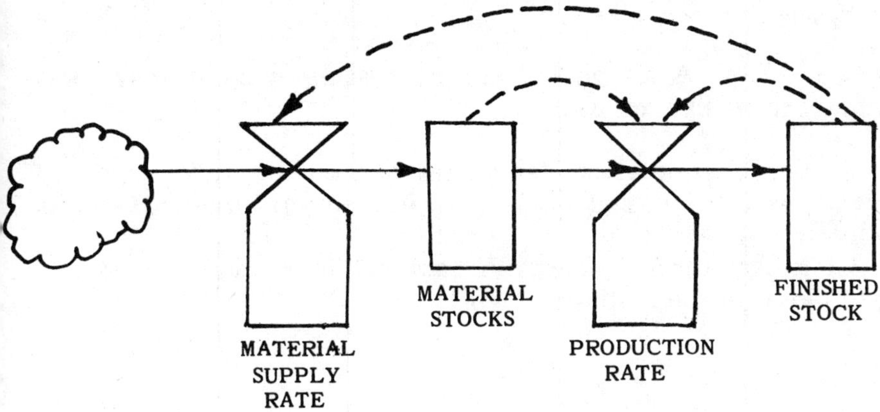

Figure A7.3 Typical system dynamics symbols

Figure A7.3 shows some of the symbols used in system dynamics. In the diagram the symbol shaped like a cloud represents what is described as the source, or start, of the problem. The rectangles represent 'levels' and one can consider these as being analogous to reservoirs. The other symbols in the diagrams, which are vaguely reminiscent of milk churns, are supposed to represent valves and indicate 'rates'. They do, in fact, control the rates of flow, whilst the dotted lines show how information affects the model. It is doubtful whether the elaborate symbols for rates shown in Figure A7.3 are necessary, and the symbols in Figure A7.4 are easier to construct and probably also less confusing.

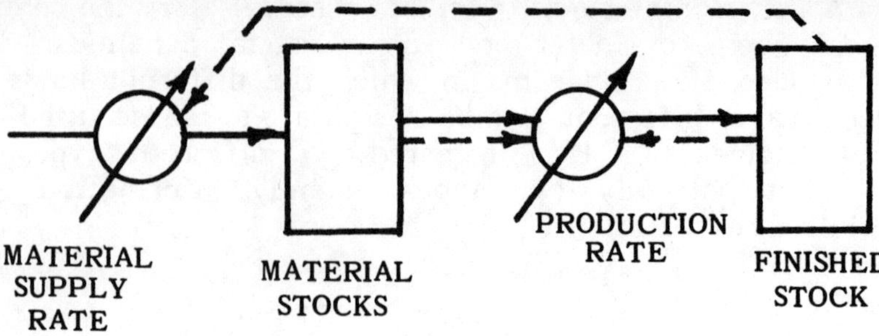

Figure A7.4 A simplified set of system dynamics symbols which are easier to use

Dynamic simulation languages increment time in fixed steps, usually called DT. The general method is as follows:

Consider three successive points J, K and L, spaced apart in time such that:

$$K = J + DT$$

and

$$L = K + DT$$

The conventional system dynamics model contains a number of 'rate equations' and 'level equations'. Thus a level equation which is used to calculate the new levels after a time advance DT is of the form:

LEVEL.K = LEVEL.J + DT*(IN.JK - OUT.JK)

where

LEVEL.K is the value of the level at time K
LEVEL.J is the value of the level at time J
IN.JK is the input rate constant over the interval J to K

and

OUT.JK is the output rate constant over the interval J to K.

A 'rate' in the rate equation is, in much the same way, expressed as a function of levels, other rates, or constants. Thus, we might have:

R1.KL = CONST*R2.KL

where R1.KL and R2.KL are two different rates of flow and CONST is some constant. R2.KL is evaluated before R1.KL.

The rate equations represent a translation of decisions and the rate at which certain actions are taking place in the real world.

The simulation proceeds in the following stages:

1 Increment time to K.
2 Evaluate level equations at time K.
3 Evaluate the rate equations (these rates being held constant over the time interval KL).
4 'Rates' and 'levels' are relabelled so that K becomes J, and L becomes K.
5 The time is incremented by DT and the time L is renamed time K.
6 Continue in the same manner until the simulation is completed.

System dynamics can provide a useful way of looking at various kinds of problem. It is likely that its greatest value lies in considering the initial approach to a problem, and in developing a rapid simple understanding of a situation. The timings of induced oscillatory effects can often prove to be fairly accurate using even quite simple system dynamics models. However, the precise quantification of the magnitude of the peaks and troughs of the variables usually necessitates resorting to more complex models in which the delightful simplicity of the system dynamics approach tends to get lost. Nevertheless, model builders should remain alert to the advantages of this technique for certain kinds of problem.

Appendix 8 System dynamics using 'spreadsheet' software

Occasionally we are aware of the mathematical relationships between the variables and merely wish to study the effects of simple changes in the values of the variables on the objective function, or outcome of the model. There is an interesting way in which models for this purpose can be constructed with the help of conceptual diagrams. It is, in fact, normally necessary to reconstruct the diagram in a new form so that it can ultimately be used in conjunction with certain standard software packages.

After the relationship between the variables, and the precise sequence of their actions, has been established, it is possible to enter them in a mathematical form into a matrix in an action sequence. In order to do this one has to visualise a large empty matrix as being sequenced in some way (usually with regard to time) in either an horizontal or vertical direction, so that the mathematical relationships may be entered in their correct sequence.

Once the problem has been reformulated in this manner it is possible to transfer it to a microcomputer which uses an electronic worksheet package such as 'Visicalc'. This kind of package was primarily designed for accounting and budgeting activities and, as a consequence, the originators do not seem to be aware of its full potential.

A Visicalc worksheet consists of 63 columns and 254 rows. Each cell can contain an individual formula precisely describing how the value in the cell is calculated. This formula is displayed on the visual display unit by moving the cursor on to the particular cell under review. As the cursor is moved to the next cell, the value in the old cell is calculated and entered into the matrix. Formulae can be replicated in rows and columns relative to other values, and the whole matrix calculated automatically.

Using Visicalc for military models

Let us consider an example of how this process can be used for battle model development. The Lanchester Square Law battle equations when laid out in the form of a conceptual framework diagram are extremely simplistic, as can be seen from Figure A8.1. It only takes a few minutes to build what is known as a one-by-one battle (one category of weapon on each side) using this method. The matrix can be laid out to time sequence in either direction, depending on the number of time sequences or weapon categories required. A one-by-one battle would obviously be laid out with the different surviving forces, casualties and lethality coefficients separately in columns. This would give some 254 possible states for the variables and far more than are necessary to reasonably represent the non-linear curves given by a set of equations. (It should be noted that linear equations do not have to be used anyway. They just help to keep the model simple in this case.)

In this particular example the matrix was time-sequenced across each row and then downwards. It was then a simple matter to make changes and to demonstrate the effect of one side having twice the size of force and the other side having twice the weapon power. The well known Lanchester effect is then obtained and the numerically superior side wins despite the better weapon effectiveness of the other side. It is immediately possible to go on to show that weapons four times as good are needed to balance out the effect of doubling the numbers of one of the combatants.

Very much more complicated battle models can be quickly produced using this method and it is possible to investigate the effects of more than a hundred different categories of weapon on either side, i.e. a 'one hundred by one hundred' battle.

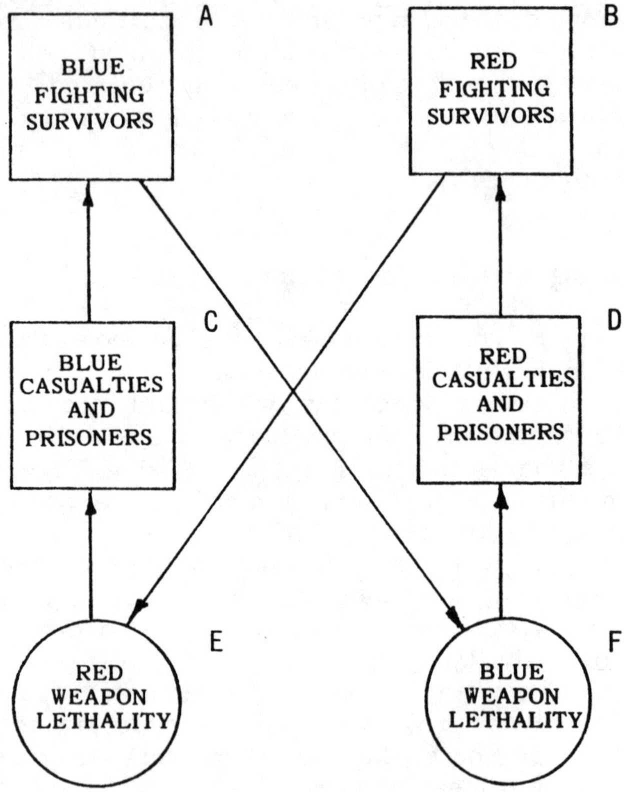

<u>Note</u>:

Boxes denote variables

Circles denote coefficients

A2 = A1 − C1 B2 = B1 − D1

C2 = B1 x E1 D2 = A1 x F1

(where the numbers indicate time relationships)

Figure A8.1 A model of the simple 'one by one' battle

Visicalc does not have a random number generator and so the model cannot be made stochastic without additional effort. However, a simple mathematical random number generator can be included in the matrix and this can be arranged to change the attrition coefficients after every run. It is possible to log the outcome of all the runs quite quickly or send them to file. As a consequence one can carry out and log, say, 20 runs in as many minutes for a moderate sized model. This will normally provide a stochastic sample of adequate size for most purposes.

In this way the whole process of model building becomes a relatively simple activity. Naturally, all the difficulties normally encountered in building models are not entirely removed. The task of selecting the important variables and their relationships is not obviated. Nor are any of the problems which can sometimes be experienced when other variables, outside those currently incorporated in the model system, suddenly assume importance. However, by carefully using the process just outlined, together with the Visicalc software, the model can be developed logically within a prescribed framework which ensures that every variable is considered individually as an entity without any serious programming language distractions, or worries as to precisely where it fits into the program, or added considerations as to whether the additional program will in some way invalidate previous programming, possibly by creating looping errors.

A wide variety of military models can be developed using this technique. For example, a simple reconnaissance model can be produced in which a number of reconnaissance aircraft are flown against different areas, there being certain probabilities of useful detection and certain probabilities of aircraft being shot down. These probabilities can be changed with time in order to allow for learning processes.

The Visicalc page can be arranged so that the commander of the reconnaissance force is progressively able to view the left-hand portion of the page. The opposing commander must move his forces through various areas deploying his counter-aircraft support as he wishes. This can form the centre section of the Visicalc page and can be viewed separately so that the information of each commander is discreetly stored. The ground commander must plan his movements the requisite time ahead on this portion of the page. The value of the information available in each area may be modified by data acquired as a consequence of successful reconnaissance flights.

The right-hand part of the Visicalc page can simply be used as a work area to enable the program to run effectively. It can contain calculations to determine the cumulative probability of an aircraft being shot down over one of the reconnaissance zones and the cumulative probability of finding a target therein. The model may be deterministic, but it can be made stochastic by the inclusion of random number streams in the matrix.

System dynamics for Everyman

An investigation was made into the feasibility of studying more dynamic problems using standard 'worksheet' software packages. One of the objectives was to see whether they could provide a reasonable alternative to the use of special simulation languages such as DYNAMO, DYSMAP and MICRODYN.

In many respects the models already described can be said to be 'dynamic', but the questions remained as to just how far a classical system dynamics solution could be emulated. Perhaps one of the most effective of Forrester's system dynamics models is his model of an industrial situation. (83) It was therefore decided to use this model for a comparative study rather than his urban dynamic model or his model of world dynamic behaviour.

The classical problem set out by Forrester involved a factory producing items which passed through a distributor to a retailer, there also being various storage points in the system. Figure A7.1 in Appendix 7 shows the kind of arrangement he proposed as being typical of production and distribution networks. It is usual to talk about 'rates' and 'levels'. 'Rates' are the equations which govern the flows, and the 'levels' are stock levels or levels attained by the variables under consideration.

Forrester applied step and ramp functions to the retail demand levels and showed that this could cause serious oscillations in the distributor's supplies, in factory production, and in other variables. Figure A7.2 in Appendix 7 shows the oscillatory effects produced in the system by a sudden 'stepped' rise in the demand.

The first step in constructing the model was to sketch out a rough chain of events in conceptual diagram form starting with the initial receipt of an order for a product. These were then laid out in approximate sequence across the top of the Visicalc page, and the time delays between each event were estimated. The numbers entered into the columns beneath represent either a stock level or a rate of flow at some point in the system.

As explained earlier, each cell in the Visicalc matrix can contain a small program which enables the value at that point to be calculated. Since the steps in this model are small, the interactions between the variables are reasonably simple. This means that the programs can be

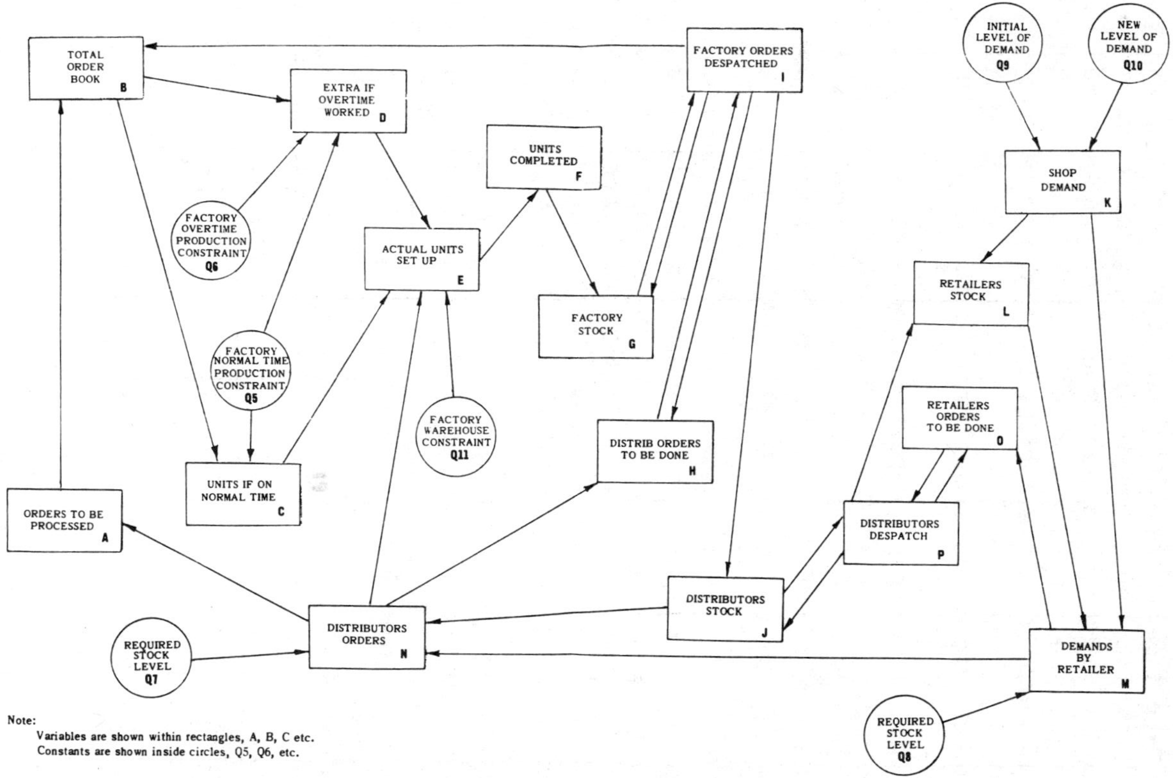

Note:
Variables are shown within rectangles, A, B, C etc.
Constants are shown inside circles, Q5, Q6, etc.

Figure A8.2 Interactions in the industrial model

accommodated within the cells and that virtually no further columns need to be added exclusively for calculation purposes.

It therefore follows that the decision rules which are applied to each column correspond to the decisions which have to be taken by a departmental manager or by a section head within the department. Hence, each column must be treated separately and a mathematical expression, or program, must be devised which corresponds to the decision rules of the department being considered. Thus, for example, in the case of a distributor deciding what orders to place on a manufacturer it may be necessary for him to consider such things as his current stock level, the demands he is receiving from retailers and his desired stock level.

Considering each department or section separately can cause the system to grow appreciably, so that what started as a fairly simple conceptual diagram ultimately becomes rather more complex, as can be seen from Figure A8.2. Nevertheless, this corresponds more closely to the real world situation with fixed, or steadily changing, flow rates.

After all, what happens in practice is that policies are decided after a finite time interval and decisions are based on all the affective input variables at that time. Thus we tend to think of placing orders at, say, weekly or monthly intervals. Some would argue that this kind of discrete model may be more realistic than a system dynamic model where continuous flow rates have to be estimated from observation and prediction. It is also rather more meaningful, for example, to say to a manager of a purchasing department 'What causes you to decide to place a new order?' rather than to say 'What is your anticipated dy/dt and why?'.

Perhaps, and this may be even more important, it simplifies the task of model building. With this technique elaborate courses or training are not necessary and anyone with a rudimentary knowledge of programming can learn to build their own dynamic models within the space of a few hours.

Figure A8.3 gives an example of the output appearing on the visual display unit. The time periods sequence down the page and it will be observed that stable conditions are obtained fairly quickly. After seven time periods of complete stability the shop demand is changed by creating a 16 per cent step function increase. The

VARIABLE	A	B	C	D	E	F	G	H	I	J	K	L	M	N	O	P
TIME STEP	ORDERS TO BE PROCESSED	TOTAL ORDER BOOK	UNITS IF ON NORMAL TIME	EXTRA IF OVERTIME WORKED	ACTUAL UNITS SET UP	UNITS COMPLETED	FACTORY STOCK	DISTRIB ORDERS TO BE DONE	FACTORY ORDERS DES-PATCHED	DIST-RIBUTORS STOCK	SHOP DEMAND	RETAILERS STOCK	DEMANDS BY RETAILER	DIST-RIBUTORS ORDERS	RETAILERS ORDERS TO BE DONE	DIST-RIBUTORS DESPATCH
STARTING CONDITIONS																
14	150	150	150	0	150	150	300	150	150	300	150	150	150	150	150	150
15	150	150	150	0	150	150	300	150	150	300	150	150	150	150	150	150
16	150	150	150	0	150	150	300	150	150	300	150	150	150	150	150	150
17	150	150	150	0	150	150	300	150	150	300	150	150	150	150	150	150
18	150	150	150	0	150	150	300	150	150	300	150	150	150	150	150	150
19	150	150	150	0	150	150	300	150	150	300	150	150	150	150	150	150
20	150	150	150	0	150	150	300	150	150	300	174	0	174	174	174	174
21	174	150	150	0	150	150	300	174	174	300	174	0	174	174	174	174
22	174	150	150	0	150	150	276	174	174	300	174	0	174	174	174	174
23	174	150	150	0	150	150	252	174	174	300	174	0	174	174	174	174
24	174	150	150	0	150	150	228	174	174	300	174	0	174	174	174	174
25	174	150	150	0	150	150	204	174	174	300	174	0	174	174	174	174
26	174	150	150	0	150	150	180	174	174	300	174	0	174	174	174	174
27	174	150	150	0	150	150	156	174	156	282	174	0	174	204	174	174
28	204	168	150	0	150	150	150	222	150	258	174	0	174	204	174	174
29	204	222	150	0	150	150	150	276	150	234	174	0	174	204	174	174
30	204	276	150	50	200	150	150	330	150	210	174	0	174	204	174	174
31	204	330	150	50	200	150	150	384	150	186	174	0	174	204	174	174
32	204	384	150	50	200	150	150	438	150	162	174	0	174	204	174	162
33	204	438	150	50	200	150	150	492	150	150	174	0	174	204	186	150
34	204	492	150	50	200	200	200	546	200	200	174	0	174	204	210	200
35	204	496	150	50	200	200	200	550	200	200	174	0	174	204	184	184
36	204	500	150	50	200	200	200	554	200	216	174	0	174	204	174	174
37	204	504	150	50	200	200	200	558	200	242	174	0	174	204	174	174
38	204	508	150	50	200	200	200	562	200	268	174	0	174	204	174	174
39	204	512	150	50	200	200	200	566	200	294	174	0	174	204	174	174
40	204	516	150	50	200	200	206	570	200	320	174	0	174	144	174	174
41	144	520	150	50	200	200	200	514	200	346	174	0	174	144	174	174
42	144	464	150	50	200	200	200	458	200	372	174	0	174	144	174	174
43	144	408	150	50	200	200	200	402	200	398	174	0	174	144	174	174
44	144	352	150	50	200	200	200	346	200	424	174	0	174	144	174	174
45	144	296	150	50	200	200	200	290	200	450	174	0	174	144	174	174
46	144	240	150	50	200	200	200	234	200	476	174	0	174	144	174	174
47	144	184	150	50	200	200	200	178	178	480	174	0	174	144	174	174
48	144	150	150	50	200	200	222	144	144	450	174	0	174	144	174	174
49	144	150	150	50	200	200	278	144	144	420	174	0	174	144	174	174
50	144	150	150	0	150	200	334	144	144	390	174	0	174	144	174	174
51	144	150	150	0	150	200	390	144	144	360	174	0	174	144	174	174
52	144	150	150	0	150	200	446	144	144	330	174	0	174	144	174	174
53	144	150	150	0	150	200	502	144	144	300	174	0	174	174	174	174
54	174	150	150	0	87	150	508	174	174	300	174	0	174	174	174	174
55	174	150	150	0	87	150	484	174	174	300	174	0	174	174	174	174
56	174	150	150	0	150	150	460	174	174	300	174	0	174	174	174	174
57	174	150	150	0	150	150	436	174	174	300	174	0	174	174	174	174
58	174	150	150	0	150	87	349	174	174	300	174	0	174	174	174	174
59	174	150	150	0	150	87	262	174	174	300	174	0	174	174	174	174
60	174	150	150	0	150	150	238	174	174	300	174	0	174	174	174	174

FACTORY CAPACITY AND STOCK ETC (CONSTANTS)

Q5 - 150 Units per week if No Overtime

Q6 - 50 Units per week extra on Overtime

Q7 - 300 Distributors Initial Stock

Q8 - 150 Retailers Stock

Q9 - 150 Units Sold Per Week

Q10 - 174 New Sales

Q11 - 500 Factory Stock Constraint regarding execution of further orders

Figure A8.3 Example of output

oscillatory effects are still evident some 41 time periods after this step function has been applied.

Naturally, not all conditions lead to this kind of oscillatory action. If large stocks are carried, or manufacturing delays are small, the effect can be quite negligible. Of course stocks cost money, and another dimension could be added by investigating the effects of stock changes on oscillatory actions and on the cost of stock-outs. Any constant or variable in the matrix can be changed and the effect immediately observed. The values can be read to 'DIF' files (a file having a suitable interchange format) and then displayed on a monitor in graphical form using 'Visiplot' or the output sent to a graph plotter.

Figure A8.4 shows the results plotted in graphical form. The movement away from the conditions of complete stability which held up to time period 20, when the step function was applied, is dramatically illustrated. The oscillatory swings compare well with those obtained by Forrester and given earlier at Figure A7.2. The linearity imposed on parts of the oscillations shown at Figure A8.4 are due to the different purchasing policies resulting from changed circumstances. It is suggested that in a real world these are very much a part of the pattern of events which take place.

It is now possible to consider different purchasing and stocking policies for the distributors and the retailers, and to assess the effects of holding larger stocks at the factory, or speeding-up production, or working longer hours. The results of such changes are displayed in a few seconds and so, as the various factors are modified, the effects can be compared in order to achieve the most favourable situation at least cost.

A fact which emerges quite quickly as a consequence of investigating the effect of changes in policy on the variables is that under certain conditions bad ordering policies by a distributor can have very serious effects on the factory producing the goods. Indeed, in this particular example it was all too easy to have the factory working flat out on full overtime in one period, whilst only a short time later it would be completely shut down until the next rush of activity came along. The moral is, of course, to choose your distributor carefully and work closely with him. Equally, the distributor should cooperate with the manufacturer and be aware of his production limitations

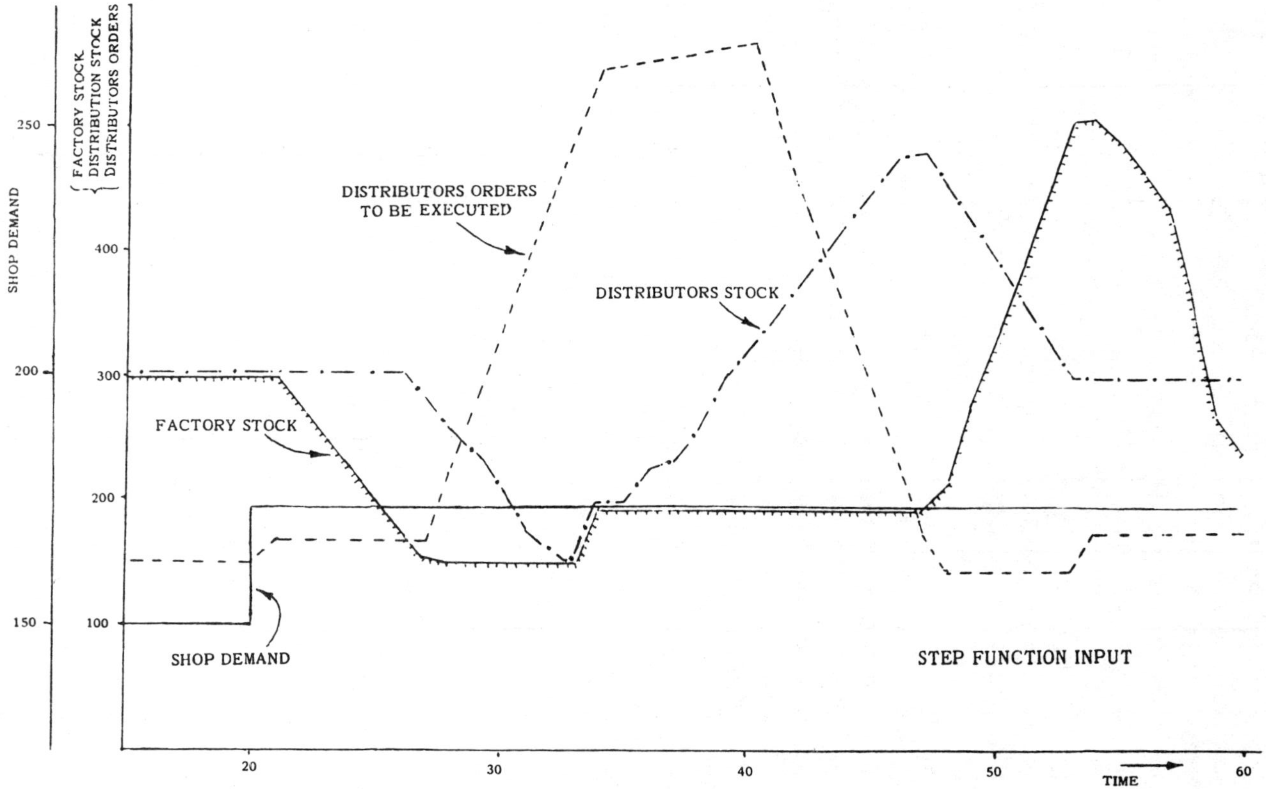

Figure A8.4 The dynamic response of a factory to a 16 per cent increase in retail demand

and normal manufacturing delay times. Together they should try to decide whether the distributor will order on the basis of his current stock position, or on his current sales. In practice, a combination of these variables may provide a good solution, but this will depend on various other factors such as time delays, etc.

It is likely that in this particular example one of the worst cases has been chosen in that a single product, single distributor chain only, has been considered. Clearly, the oscillatory effects with regard to single-product systems must exist in many instances, but a multiplicity of products can help to smooth out some of the fluctuations. Even so, a company may experience severe difficulties when the peaks or troughs of the individual beats happen to coincide, as is likely to occur occasionally.

Thus, the reformulation of conceptual diagrams in a time-sequenced mode together with the employment of 'electronic worksheet' type software packages provides a powerful means of constructing models quickly. Moreover, in the industrial problem just described, it did seem to take the analysis one stage nearer the real world and to cause the program to be formatted more directly from the statements made by the decision makers.

This method is not, of course, confined to the use of Visicalc and it seems likely that most electronic worksheet packages offering a reasonable matrix size could be adapted for this purpose. Thus Supercalc, Plannercalc, Masterplanner, Calcstar, etc. (all rather similar packages) could also be used in this way.

Appendix 9 A negotiation simulation game

The following research game was played under the auspices of the University of Southern California as part of a course in advanced international politics. With one exception, the student team members were all of American nationality although it seems probable that similar results might have been obtained with most nationalities. Some of the team members probably had a little diplomatic experience.

The scenario outlined the economic and military state of two countries in Africa. One country was relatively large and powerful, and the other small, non-industrial, but with quite large oil resources, recently discovered. These resources, unfortunately for the small country, were located in a zone long disputed by the two countries, but occupied for the last 20 years, as a consequence of a rather arbitrary UK Commission decision, by the smaller power. The area was not originally deemed to be important so this decision was not seriously contested by the larger country. The large country was now suffering from an economic recession made worse by a rapid increase in oil prices. Some years previously it had made a very big, low interest, loan to the smaller country, and a delegation from the large country had now arrived in the capital of the small country ostensibly to discuss an early repayment of this loan. The delegation also had secret instructions to try to improve its oil supply position as this commodity was proving uncertain both in supply and cost.

Participants in the game were assigned roles as members of the delegation and asked to negotiate the best possible terms for their countries.

The large country opened the negotiations by making almost less than minimal demands, since it was rather embarrassed by having to make requests to improve the oil situation under cover of the loan talks. Moreover, it was aware that it could, by a determined attack on the small country, take the oil if the talks failed, although this would clearly involve some uproar and boundary arguments at the UN.

Unfortunately, as it transpired, the small country made a very poor response to this request. It offered an earlier than promised repayment of the loan, but was extremely difficult about giving any concessions about oil. Secretly it was very scared of its own weakness and it would have been delighted to settle for a 50 per cent share of the disputed area.

The large country pressed gently for better terms and would have been willing to settle for oil at 10 per cent below the previous year's price in return for a guaranteed market. It would have been very happy indeed with a 10 per cent share of the oil zone and given the fulfilment of its wildest militaristic aims it would not have thought it fair to ask for more than a 50 per cent share.

However, very few improvements in the terms were forthcoming and so the delegates for the larger country (and I, alas, was amongst their number) set out on an incredibly Machiavellian course.

It was decided to offer further aid to the small country in return for the trivial concessions it had given, and included in this was considerable military aid. However, it was aid of a particular type, being confined to 'Tiger Moth' type trainers and very large numbers of military advisers. The economic aid also included advisers for the oil rigs and this was another key factor in the plan.

The leaders of the small country were extremely suspicious of such generous behaviour but more or less had to accept the offers since they were mindful of public opinion. Moreover, as time went on and the oil began to flow they were less and less worried.

Suddenly the partisans of an unruly tribe began attacking and burning oil rigs. These actions were secretly created by the large country who eventually also deliberately sponsored the injury of several, and the killing of one, of their own people serving as technical advisers on the oil rigs.

At that instant the whole military weight of the large country was thrown against the small country, although not ostensibly so, since it claimed to be moving into the disputed area to protect people from further injury. In this situation, with some deliberate communication confusion, and with its forces paralysed by large numbers of alien advisers in quite high places in the military network, the small country was rapidly overrun.

After suitable apologies for 'inadvertently' destroying any vital pieces of military equipment possessed by the small country, the forces of the larger country withdrew from most of the occupied zone. However, it remained in possession of 50 per cent of the oil producing area, which it felt was a fair division of the natural resources.

In this way the larger country had secured the balance of the loan repayments in the two-year preparation phase prior to the attack. Indeed, discussion on the original request for more rapid repayments or higher interest, which were required because of balance of payment problems, were deliberately channelled into talks on the repayment of the loan within two years because of the plan to occupy the oil zone at that time. The large country, by its deviousness, had also successfully prevented the small country from seeking military support from the USSR, a course which it would have taken at the outset had the negotiations appeared less friendly.

Of course this is not necessarily the end of the story. The question arises as to what the future would hold for the relations between these two countries. Perhaps if both prospered and over the years a spirit of interdependence developed, all would be well. If the small country did not prosper, then it seems only too likely that the enmity and hatred would grow. And yet it could be argued that the large country was almost 'forced' to take some action if an internal disaster was to be prevented, since it was in some senses a 'prisoner' to its own technological developments, and hence its needs had to be continually satisfied somehow.

The sad moral to the story is that, if both sides had been completely honest at the outset, no lives would have been lost, no bitterness would have been created, both countries would have been fully satisfied and there could have been a basis for a lasting friendship and cooperation.

Appendix 10 Utility and risk aversion

In decision making the relative values between a series of different choices are not always immediately obvious. It is true that in some business problems the choices can be related to company profitability and this may be accepted as a valid decision criterion. Nevertheless even in business there are many occasions when it is not always possible to relate the choice between decisions directly to profitability.

A more generally accepted concept of the value of a decision is the utility, or usefulness, of the result or outcome. Difficulties can arise when we try to compare one decision outcome with another outcome which may be very different in kind. For example, if we are given a choice of a numbers of presents such as a new car, a hi-fi radio, a pair of gloves, a handkerchief or a dozen paper clips, we could easily arrange these in rank in order of preference. Indeed, many people might be inclined to arrange them in the order given unless some special sentimentality attaches to one of the items. Having arranged these items in order of preference, it may then be possible to provide some indication as to how much more we prefer one to another. It may take 10 or 20 items similar to a hi-fi radio to be equivalent to a car and so we could award 10 or 20 points to the car and only one to the radio and similarly with the other items.

Monetary value alone is unlikely to be the sole criterion in the number of points we award to each item. If we happen to be stone deaf, or are unable to drive, a radio or a motor car may be much less important. Naturally these items would have a resale value but basically, in themselves, it is possible for them to have quite low utility to certain persons, whilst appearing highly desirable to others.

Utility is therefore a subjective measure and is unlikely to be viewed by any two people in an exactly similar manner. Moreover, since different people often value the same things very differently and do not have an identical willingness to undertake risk, conflicts can readily arise in formulating policy irrespective of whether we are concerned with commercial business or government control. In view of the subjective nature of utility, it is difficult to formulate any measure which does not include in some way a common reference scale such as money. Yet money is not in itself useful; it is only a means to usefulness. It must be remembered that in utility theory studies we are transforming scales in an attempt to compare unlike things with all the subsequent difficulties this involves.

One of the earliest persons to try to measure utility was Jeremy Bentham, who in 1789 sought to compare the utility of objects, events and states. In 1947 Von Neumann and Morgenstern made some progress in developing the mathematical theory.

In many operational research problems it is adequate to rank the possible outcomes in order of preference and then to assign numbers of 'values' to the outcomes. (Care must be taken in doing this - see Appendix 13, which deals with value judgements.) It is obviously a useful step forward to be able to order a preference for the outcomes. Such a step is said to describe a ranking which is unique up to a monotone transformation. In other words, there is still some option in the choice of the number 'valuations' but the system has been constrained into a single scale.

Cardinal utility theory is intended to provide much more information than that which the simple method just described can offer. Given that we have the required information about two items in isolation, cardinal utility theory enables us to predict something about their usefulness together, in that the utility properties are additive. It is therefore of especial value in deciding between risky alternatives and provides information about the kind of risks a decision maker is prepared to take. As a direct consequence it is therefore possible for a second person to assess whether or not a completely different risk situation would be acceptable to the decision maker without any further direct reference to him.

The general procedure in utility assessment is to consider the probability and usefulness of an outcome and to compare this with another outcome which has a different amount of usefulness and which occurs with a different probability. It will be seen that this method at least provides a semi-scientific way of comparing possible outcomes or 'prospects' as they are often called.

Example: Suppose that we possess a ticket in a lottery in which a total of 100 tickets have been sold. There is only one prize, a transistor radio set, and so we have a 1 in 100 chance of winning this prize (i.e. .01 probability). Let us call this prospect P_1. Now suppose that a friend of ours has purchased three tickets in another lottery for which 100 tickets have been sold. Again there is only one prize, which is a bottle of Scotch whisky, and we will call the prospect of a single ticket in this lottery P_2. Our friend only drinks Bourbon and so offers to change his three tickets for the whisky lottery for our one ticket in the radio lottery. Now suppose that as far as we are concerned the monetary value of the items is virtually their utility value since we are down on whisky and will have to get a new radio set as ours is failing anyway. Consequently we agree to this trade since we believe that the monetary value of three bottles of whisky is equivalent to the monetary value of the transistor radio set.

In other words, considering the two prospects P_1 and P_2 and the utility U of these prospects to ourselves we can state that:

$$U(P_1) = U(3P_2)$$

The fact that we now have three chances of winning in 100 instead of one accounts for the fact that P_1 is a three times more attractive prospect than P_2; moreover if p_1 and p_2 are the respective probabilities of these prospects we can write:

$$p_1 \; U(P_1) = p_2 \; U(P_2)$$

i.e.

$$.01 \; U(P_1) = .03 \; U(P_2)$$

This statement is very informative to others with regard to our preferences concerning these two prospects. Now just suppose that we had originally purchased several of the radio lottery tickets and another person was holding the tickets on our behalf. He would know from the foregoing equation and, without having to ask our permission, that we would be prepared to carry out a trade for tickets in the whisky lottery. Of course, most utility values change with time and if we won the whisky lottery and a similar lottery was held later we might value prospect P_2 much less, possibly only half, i.e.

$$.01\ U(P_1) = .06\ U(P_2)$$

In other words we would now want six tickets with a 1 in 100 chance of winning to trade for our single 1 in 100 radio lottery ticket, although as stated earlier in monetary terms three bottles of whisky were equivalent to the value of the radio.

Now let us suppose that in the original problem we regard the usefulness of the radio as 210 units (call them utils) and a bottle of whisky as 70 utils. The useful value of a lottery ticket can be expressed as the probability of winning multiplied by the utility of the prize. Hence in the first lottery this was:

$$.01 \times 210 = 2.1 \text{ utils}$$

and for the whisky lottery:

$$.01 \times 70 = .7 \text{ utils}$$

Hence three lottery tickets at .7 utils each compared with one at 2.1 utils was a fair deal, and we were indifferent as to which we held.

Later, having won a whisky lottery the useful value of tickets in a further whisky lottery fell by half to 35 utils so that to us the tickets in this lottery were only worth:

$$.01 \times 35 = .35 \text{ utils}$$

Naturally some value attaches to the probability itself. For example, it may well be regarded as better to be sure of £100 than to have a 50 per cent chance of £200 or nothing depending on, say, the toss of a coin. Indeed it is likely that if we were told that we could have either £200 or nothing depending on whether we guessed the coin correctly, or as an alternative we could put in a sealed bid stating the lowest value we would accept instead of tossing the coin, it is very probable that this would be less than £100, perhaps £99. If we were very short of cash, and needed £50 badly, we might put in a bid for £50 rather than risk gaining nothing as might happen when accepting the £200 or nothing option.

It is therefore possible to obtain a utility function for a person which indicates how he values money and shows his willingness to undertake risk. Let us suppose that we offer a person a certain £500 and then tell him that as an alternative we will offer him a 50–50 chance of £1,500 or nothing. He may well decide to take the £500. Now suppose that we go on increasing the stake. When it is raised from £1,500 to £2,000 he decides to take the 50–50 chance. Suppose we arbitrarily set the person's utility for £2,000 at 10 utils, then:

$$p_o \; U(P_0) = p_1 \; U(P_1) + (1-p_1) \; U(P_2)$$

where p_o is the probability of P_0

P_0 is the prospect of £500

p_1 is the probability of P_1

P_1 is the prospect of £2,000

and P_2 is the prospect of £0.

Hence 1.0 U(£500) = .5 U(£2,000) + .5 U(0)

(Note: p_o is 1 because this was offered as a certainty.)

Substituting the util value of 10 for U(£2,000) and assuming that the utility of £0 is zero, we obtain:

$$1.0 \quad U(£500) = .5 \times 10 + .5 \times 0$$

i.e.

$$U(£500) = 5 \text{ utils}$$

so that the utility of £500 is 5 utils to this particular person. We have thus established three points marked with crosses on the utility curve in Figure A10.1 although, of course, the upper and lower utility points were chosen arbitrarily.

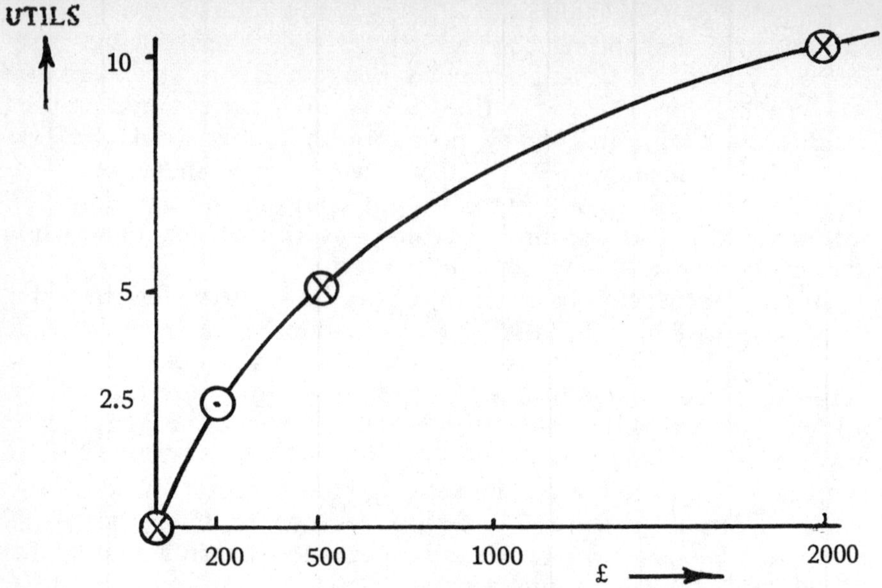

Figure A10.1 A graph expressing the value to an individual of pounds sterling in utils

Next suppose that we ask the person to consider a 50–50 chance of receiving £500 or nothing and to decide for what amount he would be willing to sell his chance. Let us suppose that he would sell the opportunity for £200. Thus:

$$1.0\ U(£200) = .5\ U(£500) + .5\ U(0)$$

Remembering that U(£500) has been found to be 5 utils and substituting this in the above equation gives:

$$1.0\ U(£200) = .5 \times 5 + .5 \times 0$$

i.e. $$U(£200) = 2.5 \text{ utils}$$

This gives therefore another point on the curve (marked on the graph with a dot in a circle). Other values can be obtained by similar methods or by changing the probability values. It should be noted that the latter can sometimes introduce anomalies as most people find it easiest to think clearly about a 50–50 gamble, but it can also provide a useful check on the consistency of the estimates.

Axioms

It is probably advisable to state the axioms of utility theory so that the assumptions which have been made can be appreciated and the limitations of the method understood. The major axioms are as follows:

1 Given two prospects P_1 and P_2, it is possible to decide which of these prospects is preferred or whether they are equal (i.e. using the symbol $>$ to mean preferred we have either $P_1 > P_2$, or $P_2 > P_1$ or $P_1 = P_2$). In other words it is possible to arrange the prospects in the order of their desirability.

2 Given that P_1 is as well regarded as P_2 and P_2 at least as well as P_3, then P_1 is regarded at least as well as P_3 (i.e. $P_1 \geqslant P_2$; $P_2 \geqslant P_3$, so $P_1 \geqslant P_3$). This is known as the transitive property of ordering.

3 Given that P_1 is preferred to P_2 and P_2 preferred to P_3, there is a mixture of P_2 and P_3 which is preferred to P_2 (i.e. $P_1 > P_2$; $P_2 > P_3$ so $\delta P_1 + \psi P_3 > P_2$ where δ and ψ denote fractions of P_1 and P_3 respectively).

4 Given that P_1 is preferred to P_2 and P_3 is a new prospect, then a given mixture of P_1 and P_3 is preferred to a mixture in the same proportions of P_2 and P_3 (i.e. $P_1 > P_2$ and there is also P_3. Then $\{\delta P_1 + \psi P_3\} > \{\delta P_2 + \psi P_3\}$).

Decisions to accept risk

We are all concerned in making some risk decisions, but they are probably most critical for politicians, military commanders and the chairmen and managing directors of companies. Let us look at the kind of utility curve for working capital for a company derived from discussions with the senior executives. This provides some indication of the company's willingness to accept risk.

Example: A utility curve for the estimated usefulness of maintaining a working capital and reserve for a small private company has been drawn up by the directors. This is shown in Figure A10.2.

The company shows a fairly steady net profit of £2,000 per month and it has £25,000 in liquid assets. The demand for the product is reasonably steady and the market fairly assured but not really capable of expansion. A diversification prospect is placed before the directors which requires an investment of £5,000 and may give a return which, when expressed in terms of present worth (discussed later), is valued at £15,000. However, after very careful consideration the directors decide that it has only a 50–50 chance of succeeding. Moreover if it fails it does so completely and the £5,000 investment is completely lost. Equally the nature of the venture is such that if it is successful the initial investment is written off but there is a certain return of £15,000, i.e. the company gains £10,000. Should the directors take the risk?

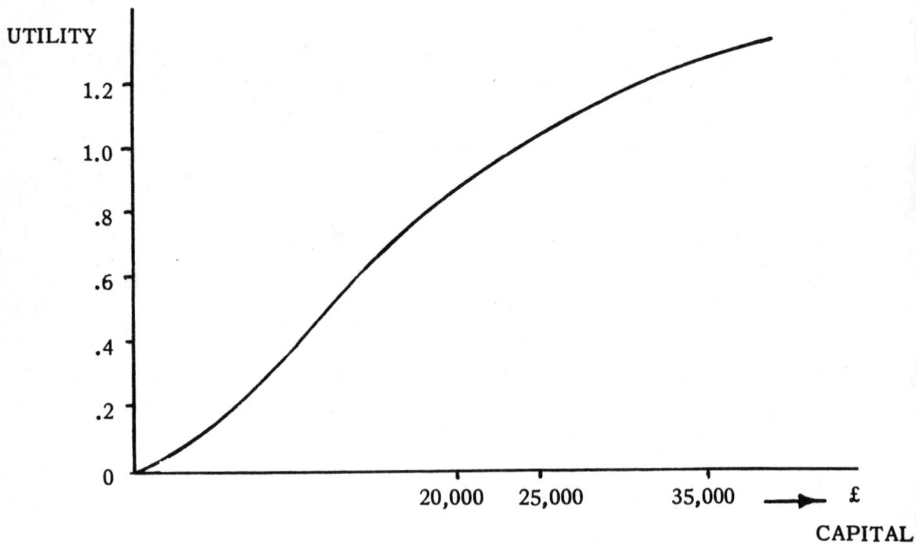

Figure A10.2 Estimated utility of capital for a small private company

Naturally the answer to this question really depends on a variety of circumstances including what other opportunities are open. For simplicity we will assume that this is the only possibility open to the directors at the time.

Basically the problem is that if the directors decide to invest in the project the company makes £10,000 with a probability of .5 and loses £5,000 with a probability of .5. In drawing up the graph in Figure A10.2 the directors arbitrarily set their present capital utility at 1.

Hence, if they take the project, there is a .5 probability of having a capital reserve of £35,000 and a .5 probability of having only £20,000. The utility values associated with these figures are 1.25 and .85. The utility of the project P can be expressed as follows:

$$U(P) = .5\ U(£35{,}000) + .5\ U(£20{,}000)$$

Expressing this in utils gives:

$$U(P) = .5 \times 1.25 + .5 \times .85$$

$$= 1.05$$

Thus, since the company has a greater expectation of capital utility, it should accept the project. Note, however, that the decision is much more marginal in this case than one might have anticipated. Indeed everything really depends on the utility values the directors originally ascribed to the capital assets.

In this example the utility curve the directors produced was concave towards the horizontal axis near the bottom. This denotes willingness to undertake risk and can readily be understood. Even if we have sufficient money to buy a bread roll, the few pennies we may have are of little use because we can hardly buy anything else to provide sustenance.

Now let us suppose that in the example the present capital were only £5,000. This sum is clearly regarded by the directors as inadequate, consequently at this stage they are apparently even more ready to make the speculative investment since (rather like the pennies and the bread rolls) they regard £5,000 as insufficient to run the business properly, i.e.

$$U(P) = .5\ U(£15{,}000) + .5\ U(0)$$

$$= .5 \times .4 + 0$$

$$= .2 \text{ utils.}$$

A present working and reserve capital of £5,000 represents .08 utils so there is a strong incentive for the company to take the risk in this case.

The foregoing method is an extension of the rather more abstract problem of when to take and when not to take bets. It will be seen the treatment is exactly similar and in a sense most business decisions involve a certain amount of risk in that for practical purposes business policy and decision making can be regarded as a series of 'bets' with generally favourable odds.

Mention should be made of the fact that the group utility function considered in the example is not necessarily easy to synthesise from the utility functions of the individuals. This is simply due to the fact that risk aversions are basically different and that groups tend to be more willing to take risks than individuals. Hence, some care should be exercised in attempting this procedure.

The question arises with regard to the extent to which utility and risk aversion are important in deciding defence policy. Should utility values be given to the successful outcome of various battle scenarios? Who should decide these so-called utility values? Within any particular service framework the individual chiefs of staff are trying to consider the utility of each proposed new weapon purchase. Implicit within this decision is some form of conscious or unconscious weighting factor they may attach to the probability of having to fight a particular scenario and its relative importance. It is what analysts like Raiffa call a 'messy' problem. The Defence Operations Analysis Establishment provides immensely valuable input into this decision process, but the whole question of precisely how to structure the defence problem is an exceedingly difficult one.

Appendix 11 A complex decision tree example

Complex decision tree analysis is concerned with the resolution of complicated probability situations and may also include such things as the analysis of the effects of conducting experiments to obtain more certain knowledge of a situation. A fairly simple example of this is given below.

Example: A large nation is planning a campaign to create a revolution in a small country which is following a policy hostile to the large one. The large nation estimates that the campaign will cost £50 million, and has decided that it is essential to mount the campaign either now or in the next three years. A great deal will depend on the support a sponsored guerilla movement can obtain from the local population, and this will affect the cost of the operation to the large nation. The popularity of the ruling regime of the small country tends to fluctuate every few years depending on the prevailing economic conditions. The economy of the small country is largely based on a single crop and it is therefore difficult to decide precisely when to mount the operation.

If the large nation decides not to mount the operation for three years, and support is currently low for the guerillas due to good crops, the large nation will save £3 million. On the other hand if support is currently high and it fails to mount the operation it will lose £2 million as a consequence of damage to its interests in the interim period.

If the large nation decides to mount the operation and there is much support for the revolution it will gain £10 million, but if there is no support it will lose £4 million.

Field intelligence is sparse but the information available suggests that there is a 40 per cent chance of immediate support for the guerillas and a 60 per cent chance of little support.

DI 1001 (one of many intelligence departments) are not very happy about the assessment and believe that a detailed intelligence survey, which would cost about £900,000, would improve the reliability of the estimate. However, an analysis of data from past operations shows that this kind of survey gives true information 90 per cent of the time when there is little guerilla support, whereas when there is good support the surveys are correct 6 times out of 10.

The large nation feels that it cannot tolerate the small country's behaviour and must mount the operation within three years. The question which faces the director of DI 1001, who must make a recommendation to the Prime Minister, is should the operation be mounted now or in three years' time?

The reader may feel that this is a very complicated problem, and indeed as it stands it is moderately complex. Let us now lay out the problem in the form of a decision tree so that we can see the various alternatives a little more clearly.

In the first place the large nation must decide whether or not to have an intelligence field survey undertaken in order to improve the state of knowledge. Thus there are two alternatives which can be expressed diagrammatically as in Figure A11.1.

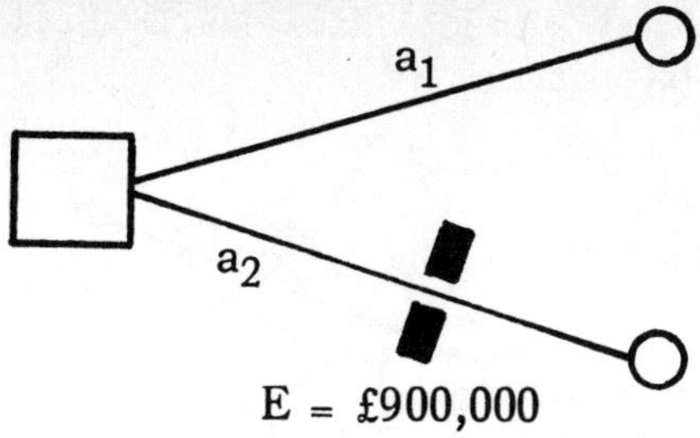

Figure A11.1 The action to have an intelligence survey undertaken can be regarded as a 'toll' if path a_2 is followed

Thus starting at *A* and taking action a_1 means that we do not have the survey undertaken. Alternatively, if action a_2 is taken we carry out experimentation E (the survey), in order to improve the chances of making a satisfactory decision. Whether or not this work should be undertaken depends on how much the experiment will improve the expectation of making a correct judgement and on the value of the improvement.

The point *E* on the decision tree can, in effect, be regarded as a toll-gate where a charge must be paid for any further progress along that branch.

However, let us first investigate the situation where action a_1 is taken and the survey is not carried out. In this case the large nation can decide either to carry out the operation immediately or to defer its plans for the present.

Let us call these actions a_3 and a_4 respectively. In each case there is a .6 probability of poor support and a .4 probability of good support.

Hence, the tree in Figure A11.2 can be constructed. It will be seen from this figure that if it is decided to take action a_3 two different events may occur with different probabilities. Support may be poor initially and the probability of this is .6 . In this event, having taken the decision to carry out the operation, the large nation will lose £40 million. On the other hand, support may increase rapidly, with a probability equal to .4, and if this happens when the large nation has mounted the operation it will gain £10 million. Should the large nation decide to defer the operation (action a_4) and support does not expand rapidly, it will gain £3 million from interest on capital, and this will occur with a probability equal to .6 .

However, this action could also lead to a loss of £2 million with a probability of .4 .

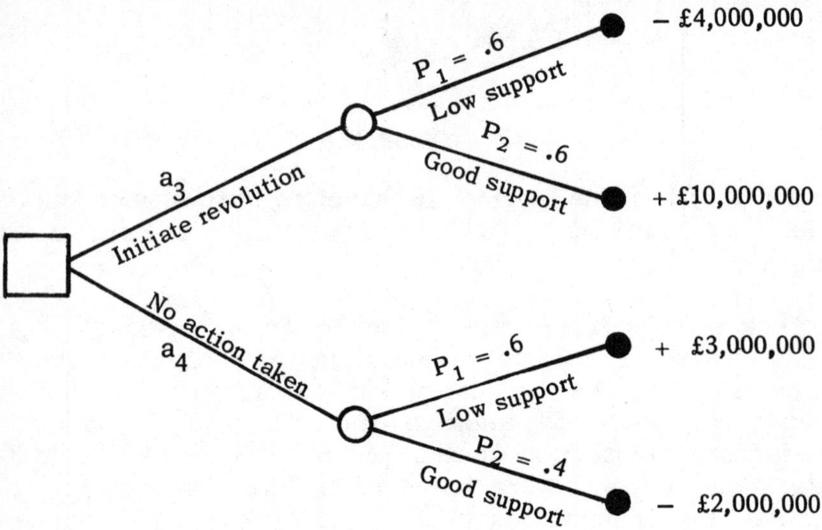

Figure A11.2 The introduction of 'good support' and 'low support' into the decision tree

Using the expected monetary value (EMV) as a decision rule (the result if the decision could be taken many times), the following values can be placed on the different actions:

Action a_3

(£10,000,000 x .4) - (£4,000,000 x .6) = £1,600,000

Action a_4

(£3,000,000 x .6) - (£2,000,000 x .4) = £1,000,000

Thus, on the basis of expected monetary value, the large nation might be expected to choose action a_3.

Now suppose that it is decided to have the intelligence survey undertaken, let us consider whether this will afford any improvement.

Let p_3 be the conditional probability of making a correct decision given poor initial support for the guerilla movement, i.e.

$$p_3 = p\begin{pmatrix}\text{Poor} & \text{Intelligence survey says} \\ \text{support} & \text{poor initial support.}\end{pmatrix}$$

Similarly,

$$p_4 = p\begin{pmatrix}\text{Poor} & \text{Intelligence survey says} \\ \text{support} & \text{good initial support.}\end{pmatrix}$$

$$p_5 = p\begin{pmatrix}\text{Good} & \text{Intelligence survey says} \\ \text{support} & \text{good immediate support.}\end{pmatrix}$$

$$p_6 = p\begin{pmatrix}\text{Good} & \text{Intelligence survey says} \\ \text{support} & \text{poor immediate support.}\end{pmatrix}$$

The probability values are:

$$p_3 = .9; \quad p_4 = .1; \quad p_5 = .6; \quad p_6 = .4$$

It is therefore possible to construct a decision tree to express these events (Figure A11.3).

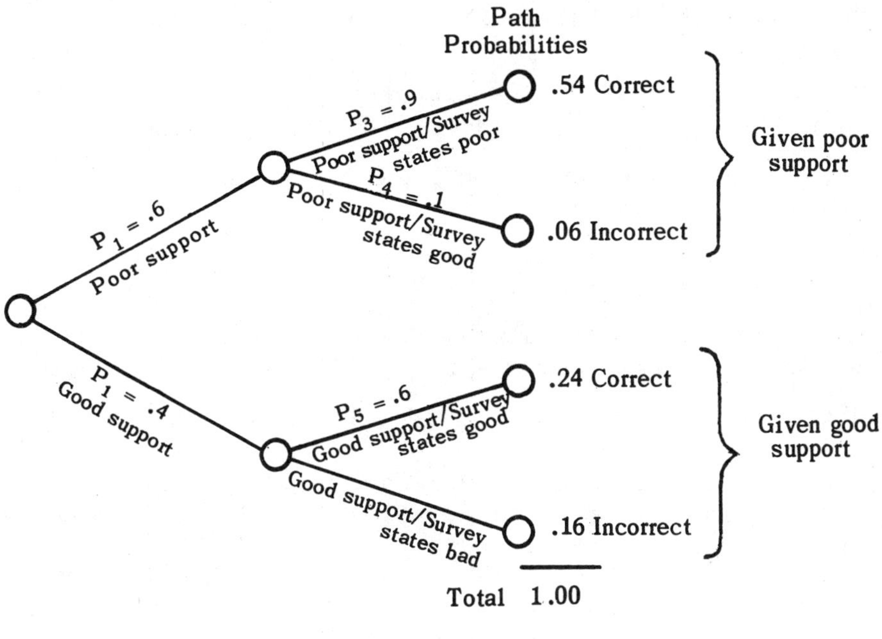

Figure A11.3 Path probabilities

The path probabilities are calculated by multiplying p_1 by p_3 (.6 x .9 = .54), and p_1 by p_3 etc. These probabilities indicate the chance of following any particular branch.

This information requires further treatment to arrange it in a form suitable for straightforward analysis.

Let us consider the probability of the intelligence survey giving a correct or an incorrect answer and develop a tree from this information. The probability of a correct report is .54 + .24 = .78, and the probability of an incorrect report is .06 + .16 = .22 .

It will also be seen that:

There is a .7 probability the survey indicates low support	.54 + 16 = .7
There is a .3 probability the survey indicates much support	.24 + .06 = .3
	Total 1.0

Having developed the first two branches of the new tree it is still necessary to determine the probabilities of the final paths to the outcomes.
Thus:

If the survey says low support the probability of being correct is	$\frac{.54}{.54+.16}$	= .77
If the survey says low support the probability of being incorrect is		= .23
If the survey says much support the probability of being correct is	$\frac{.24}{.24+.06}$	= .80
If the survey says much support the probability of being incorrect is		= .20

Hence, it is possible to construct the tree shown in Figure A11.4.

The importance of conditional probability in decision tree analysis will now be appreciated, since it is essential to assign probabilities to each path and expected values to every node. The whole decision tree for the entire problem can now be constructed and if expected monetary values are used as a decision criterion an optimal solution can be obtained. The complete decision tree is given in Figure A11.5.

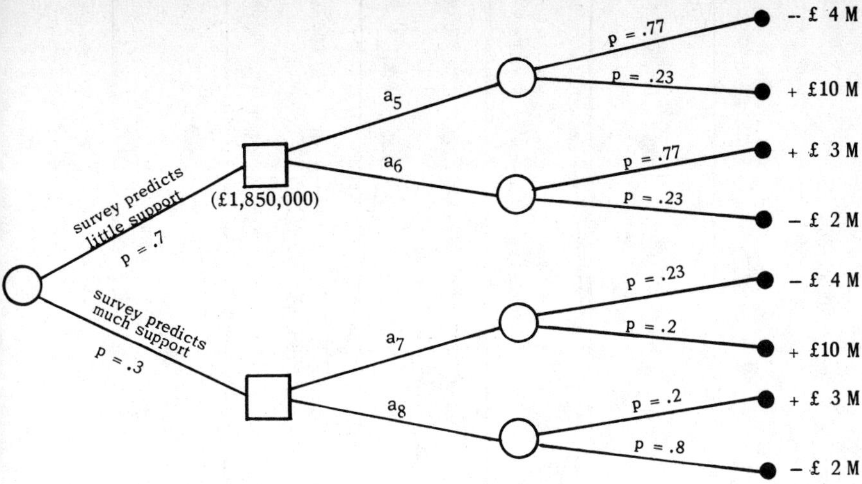

Figure A11.4 Including probabilities of the final paths to the outcomes

The expected monetary values (i.e. the value of the actions) are shown in brackets under the nodes. The value of actions a_5, a_6, a_7 and a_8 are therefore as follows:

Action a_5

(£10,000,000 x .23) - (£,000,000 x .77) = - £780,000

Action a_6

(£3,000,000 x .77) - (£20,000,000 x .23) = £1,850,000

Action a_7

(£10,000,000 x .80) - (£4,000,000 x .20) = £7,200,000

Action a_8

(£3,000,000 x .20) - (£2,000,000 x .80) = - £1,000,000

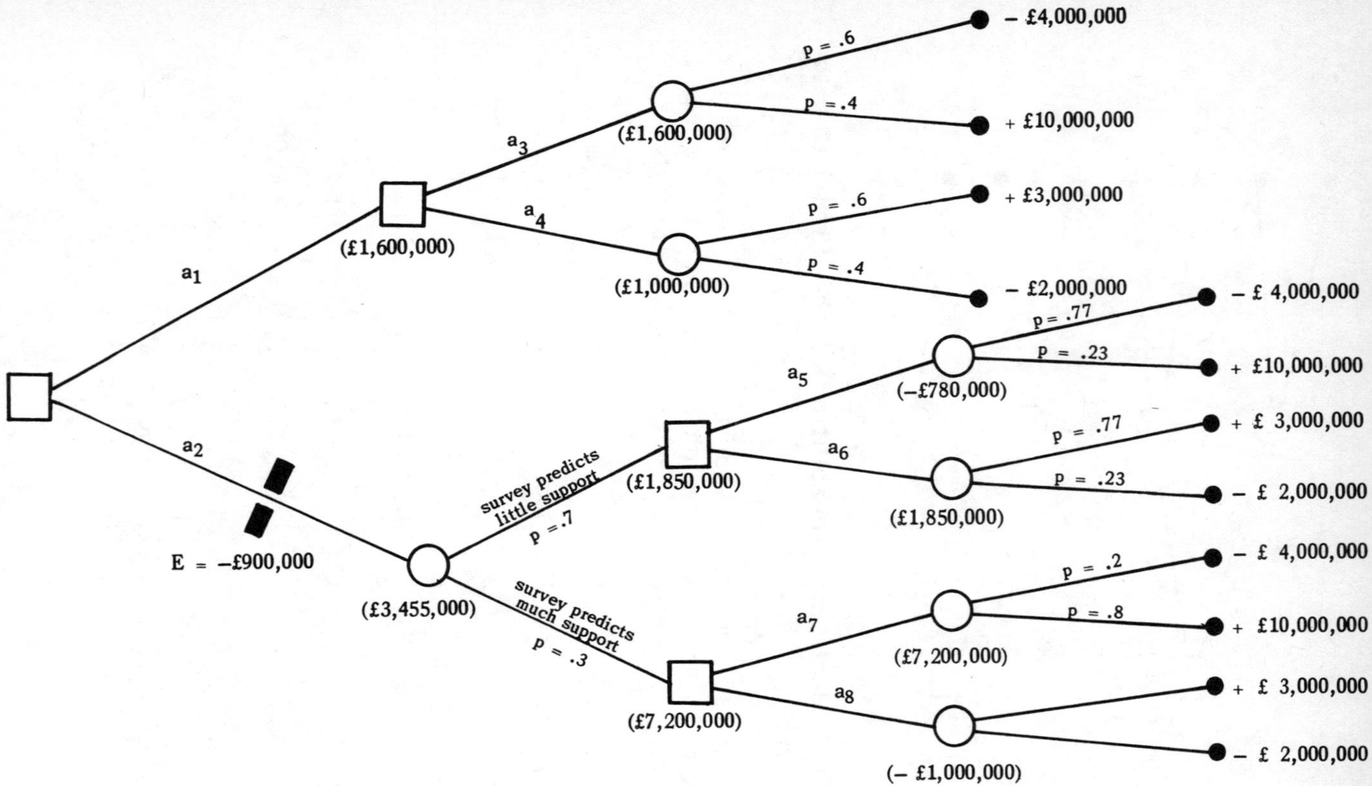

Figure A11.5 The complete decision tree

It is therefore possible to work backwards across the diagram attaching values to the square decision nodes. For example, in choosing between action a_5 valued at -£780,000 and action a_6, valued at +£185,000, the large nation would certainly select a_6, so we attach the maximum value to this node. Similarly a value of +£7,200,000 can be attached to the decision node (a_7a_8).

The node just prior to these decision nodes is a chance node and 70 per cent of the time the path to the (a_5a_6) node, valued at £1,850,000, will be followed, and only 30 per cent of the time will the survey predict much support. Hence, the value which can be attached to the chance node is:

$$(£1,850,000 \times .70) + (£7,200,000 \times .30) = £3,455,000$$

Working back still further to the first decision node (a_1a_2), it is necessary to pay the intelligence survey cost of £900,000 and so the value of the first decision is £(3,455,000-900,000), i.e. £2,555,000. There is therefore an appreciable advantage in having the expensive intelligence survey carried out and basing the decision on the outcome of this survey.

Although this particular example may have appeared fairly involved, very many practical cases are even more complicated. Indeed it is sometimes necessary to undertake what virtually amounts to a preliminary decision analysis in order to get the data into a satisfactory form so that a final, and it is hoped more accurate, analysis can be carried out.

It will be appreciated that the expected monetary values, as used in this example, are not always a satisfactory criterion in decision tree analysis. Quite frequently utility provides a better measure. The data in the example could obviously equally well have been translated into utils and the problem solved in terms of expected utility values (EUV) rather than EMV.

The subjective assessments which are required in utility theory do not have the same precision as expected monetary values, but they are often more meaningful. Hence, in evaluating the outcomes care should be taken that in every respect the true merit (or value) of the outcomes is used as a decision criterion. This may involve assessing the relationships between certain and uncertain monetary return, or between different kinds of objects.

Even when employing expected monetary values as decision criteria care should be taken to account for time differences in the various parts of the decision tree, and the use of techniques which relate money to a common time scale, such as for example 'present worth', may be advisable.

Appendix 12 Strategy in decision analysis

Many problems are concerned with uncertainty, and the actions which an individual is likely to take will depend on the instinctive estimates which are made in various circumstances. Here is an example of different strategies in an uncertain situation.

Example: An army is advancing steadily across an ocean strewn with islands. It is capturing one island after another from the enemy, but at some cost. It is possible to deploy up to 50,000 men in an assault on an island, but the commander-in-chief is anxious to use only the minimum number of men in an attack as he has considerable commitments elsewhere. However, if he fails to send sufficient troops against the enemy and the island is not taken quickly, it appears that the defenders' morale increases appreciably. This means that for every original soldier who would have been needed two more will be required as a consequence of the delay.

There is a variation in the defensibility of the islands, and the ease with which an operation can be undertaken depends to some extent on the weather conditions.

An operational analyst has studied the data from the previous assaults on the other islands and estimates the probabilities of the manpower required for a successful assault as follows:

$p(0)$, the probability that no men are required (island virtually undefended)	= .05
$p(1)$, the probability that 10,000 men are required	= .10
$p(2)$, the probability that 20,000 men are required	= .40
$p(3)$, the probability that 30,000 men are required	= .30
$p(4)$, the probability that 40,000 men are required	= .10
$p(5)$, the probability that 50,000 men are required	= .05

Although 50,000 men will virtually make certain that the island is captured without undue delay, what is the correct number of troops which should be deployed in an attack? It should also be noted that in view of the light aircraft and helicopters attached to every unit of 10,000 men, approximately 500 men can be redeployed quickly, so that only 9,500 men remain committed and are lost to other sectors as an effective force for a noticeable period. How should the commander operate?

Let us begin by constructing a pay-off matrix and listing the various possible strategies. We will call the strategies of attacking with 10,000, 20,000, 30,000, 40,000 or 50,000 men A1, A2, A3, A4, A5 respectively, and A0 as the strategy of using virtually no troops. Thus the commander can take various actions, A0, A1, A2, etc. and nature can require various quantities of troops for a successful attack depending on the terrain and weather conditions, etc. Let B1 represent nature's requirement for 10,000 men, and B2 the need for 20,000 men, etc.

It is now possible to construct the matrix shown at Figure A12.1. The figures in the matrix are calculated as follows:

AO/BO Attack without a major force, virtually no troops required, no cost.

AO/B1 Attack without a major force, 10,000 would have been sufficient, now 20,000 men needed.

Also other values in the AO row are calculated in a similar manner.

A1/BO Attack with 10,000 troops, none required, recover 500 troops quickly for another area.
A1/B1 Attack with 10,000 troops, 10,000 required.
A1/B2 Attack with 10,000 troops, 20,000 would have been sufficient, now need extra 20,000 (total 30,000).

Similarly for the rest of the row. The other rows are also calculated in the same way.

It will be seen from an inspection of the matrix that if we considered nature actively to be playing against us, it would be advisable to choose the minimax strategy and to select the strategy giving us the maximum of the row minimum. This choice would lead us to attack with 50,000 men.

Nature is not, in fact, playing against the attackers, and the probability of her moves is known and is shown at the bottom of the matrix. Consequently, it is possible to calculate the pay-off that may be expected from any strategy the commander decides to employ.

For example, the expected pay-off for strategy AO is obtained as follows:

$$-(0x.05)-(20x.1)-(40x.4)-(60x.3)-(80x.1)-(100x.05) = -49$$

The expected values of strategies A1, A2, A3, etc. have been calculated in the same way. It will be seen that in this case the maximum strategy (denoted by an asterisk) is to attack with 20,000 men. This is known as the Bayesian strategy.

If the probability of the occurrence of the various events had not been known, it would have been possible to attach equal probabilities to all of them. In order to determine the effect of equal probability it is simply necessary to add all the pay-offs in the row and divide by the number of possibilities. Such a strategy is often referred to as the strategy of Laplace, and suggests attacking with 30,000 men. Clearly the Bayesian and Laplacian strategies are identical when the *p(j)* values are all exactly the same, rather than the stochastic distribution given in the bottom row of the matrix.

It is also possible to consider using Savage's minimax regret principle, and to develop a new matrix giving the cost of mistakes. This is as shown in Figure A12.2.

Payoff in Thousands		Nature (Number of Troops Actually Required)						Row Minimum	Expected Payoff	Laplace
		B0	B1	B2	B3	B4	B5			
	A0	0	– 20	– 40	– 60	– 80	– 100	– 100	– 49.00	– 30.00
Troops	A1	– 95	– 10	– 30	– 50	– 70	– 90	– 90	– 39.97	– 25.95
Initially	A2	– 19	– 19.5	– 20	– 40	– 60	– 80	– 80	– 32.90*	– 23.85
Deployed	A3	– 28.5	– 29	– 29.5	– 30	– 50	– 70	– 70	– 33.62	– 13.75*
by C-in-C	A4	– 38	– 38.5	– 39	– 39.5	– 40	– 60	– 60	– 40.20	– 25.50
	A5	– 47.5	– 48	– 48.5	– 49	– 49.5	– 50	– 50	– 48.62	– 29.25
Probability	p(j)	.05	.10	.40	.30	.10	.05			

Figure A12.1 Pay-off matrix for various levels of attack

	B0	B1	B2	B3	B4	B5	Max regret
A0	0	10	20	30	40	50	50
A1	9.5	0	10	20	30	40	40
A2	19	9.5	0	10	20	30	30
A3	28.5	19	9.5	0	10	20	28.5*
A4	38	28	19	9.5	0	10	38
A5	47.5	19	28.5	19	9.5	0	47.5

(All values in thousands of troops)

Figure A12.2 A minimax matrix giving regrets by showing the costs of mistakes

The first row of the matrix is obtained by considering what would happen if the commander used no major force (or in fact virtually none) or 10,000, 20,000 or 30,000 men, etc. were required. Thus if no troops were needed and virtually none were deployed there would be no regrets, but if 10,000 should have been sent, and none were supplied, the commander would need to despatch 20,000 to carry out the attack. In terms of manpower needed, therefore, the commander would have 20,000 minus 10,000, or 10,000 regrets.

Similarly if the commander had ordered an attack with 10,000 men and none were required, he would have deployed 10,000 men needlessly, and would only be able to redeploy 500 of these troops quickly. In this case his regret would be 10,000 minus 500, or 9,500.

It will be seen that the minimum of the maximum regret values occurs when the attack is made with 30,000 men, and so on Savage's minimax regret criterion the commander should attack with 30,000 men.

In practice it is also necessary to consider a number of other factors such as the effect on morale of a delay or a temporary setback and the 'value' elsewhere of the men who might be tied up in an unnecessarily large attack.

Moreover, different operations are likely to vary in their value to the grand plan, or strategy of the war, so that the value of manpower to the commander is likely to depend on the amount available at any given time. This may have the effect of causing large changes in the matrix at certain stages, so that having carried out the mathematical analysis of the problem the solution should be examined carefully in order to ensure that a sensible practical decision is made.

Although a military problem has been used as an example in this case there are many instances where this kind of analysis can be useful in industry. Policy decisions on manufacturing, installing new plant, and carrying expensive spares are all examples of strategic decision analysis.

Appendix 13 Value judgements in decision theory

A further example of decision theory is provided in this appendix. It also illustrates the importance of subjective valuations and judgements in this kind of analysis. A military example has been used, but there are many instances where similar timing decisions have to be made in normal commerce.

Example: An army commander is proposing to mount a paratroop attack involving some 800 men to secure a bridge across a river surrounded by a very difficult piece of terrain. His main army is slowly fighting its way to the bridge, but he knows that, when the enemy are driven back, unless the paratroops can intervene, the bridge will be demolished and his army will suffer a great many casualties trying to cross the river.

As each day goes by the value of taking and holding the bridge diminishes as more enemy troops and vehicles are pulled back across it. Moreover the cost of taking the bridge by a paratroop attack becomes correspondingly higher as time passes.

However, this is not all that the army commander has to contend with, since it is now the beginning of the monsoon season. He can now expect rain at any time. This will either delay the progress of the attack, or even bring it to a complete standstill, as his armoured vehicles sink in the mud on the dirt roads and in the fields.

Information on the monsoon arrival times, and amount of rainfall in the first few days, is very well documented and the meteorological officer estimates the probability of no rain in the next few days as .2, the probability of some rain which will slow the attack and increase casualties as .5, and the probability of heavy rain which will completely stop the attack as .3 .

The army commander can launch the paratroop attack on any of three days ahead of the time he estimates that his main spearhead will reach the river (given no rain). The paratroops need time to construct reasonable defences and to establish their position. The most effective defence positions will take some three days to complete. However, if the paratroops are not engaged in battle almost immediately after the third day and the monsoon arrives, this will halt the progress of the spearhead towards the river and the paratroops will be cut off and completely wiped out.

Then again, if the defenders turn and fall back quickly so that the paratroops do not have the full time to dig in, then the 800 paratroops may still be lost, although in overall terms of the effort used by the defenders in trying to re-establish lines of communication, etc. the attackers may score a success with regard to the relative number of casualties. If the defending commander decides to pull back on the first day after the drop, he will not be able to take the benefit of the possible arrival of the monsoon halting the main column and so starve the paratroops into submission.

The defending commander is keen to be in a good defensive position when the monsoon arrives and is rather unaware of how ineffective his opponent will be in torrential conditions. The attacking commander believes that the value to his army in terms of casualty differentials is roughly according to the first matrix given at Figure A13.1, and he has also tried to estimate the reactions of the defending commander for various types of weather conditions. These are shown in a second matrix at Figure A13.2.

What is the best policy for the attacking commander if he wishes to maximise the casualty differential? What is the maximum regret which this policy will incur?

Nature \ Attacking Action	Drop estimated 1 day ahead of column a_1	Drop estimated 2 days ahead of column a_2	Drop estimated 3 days ahead of column a_3
No Rain θ_1	4	10	30
Moderate Rain θ_2	2	5	10
Torrential Rain θ_3	1	0	− 8

(Casualties in hundreds)

Figure A13.1 Differential casualty matrix

robab-ity of θ	Nature \ Defending Action	Course C1: Will engage paratroops within 1 day of drop	Course C2: Will engage paratroops within 2 days of drop	Course C3: Will engage paratroops within 3 days of drop
.2	No Rain θ_1	.4	.3	.3
.5	Moderate Rain θ_2	.6	.2	.2
.3	Torrential Rain θ_3	.7	.2	.1

Figure A13.2 Nature and defending action probability matrix

Course C1 taken by the defending commander corresponds to action a_1 taken, and the probabilities take into account possible inexactitude in estimating the progress of the attackers as a consequence of stubborn defence. Similarly course C2 corresponds to action a_2, etc.

In this case it is possible to determine the probability of each state and defensive actions by multiplying the probabilities. This enables a new matrix (Figure A13.3) to be constructed.

The values in the matrix are obtained by multiplying the probability of θ by the probability of C, e.g.

$$0.2 \times 0.4 = 0.08 \text{ for condition } (\theta_1 C1)$$

and $$0.2 \times 0.3 = 0.06 \text{ for condition } (\theta_1 C2)$$

Nature \ Defensive Action	Probability of condition C1	Probability of condition C2	Probability of condition C3
Weather θ_1	.08	.06	.06
Weather θ_2	.30	.10	.10
Weather θ_3	.21	.06	.03

Figure A13.3 Probability of weather and defence reaction

It is then possible to determine the expected return for actions a_1, a_2 and a_3 by multiplying the probability value by the pay-off and summing for each action.

Hence, the expected pay-off for action a_1 is given by:

$$E(a_1) = (4 \times .08) + (2 \times .30) + (1 \times .21) = 1.13$$

Similarly,

$$E(a_2) = (10 \times .06) + (5 \times .10) + (0 \times .06) = 1.10$$

and

$$E(a_3) = (30 \times .06) + (10 \times .10) + (-8 \times .03) = 2.56$$

(Note: Since the pay-off matrix is in hundreds, all the above expected values should be multiplied by 100.)

Thus, on the basis of differential casualty expectation, the attacking commander should drop his paratroops when he estimates it will take three days to reach the river, since course a_3 gives the maximum expectation. However, if things go against him, he will not only lose 800 men, but will lose the opportunity to inflict a 100-man casualty differential which he would have gained if he had held his paratroops back and dropped them when he estimated his main force was only a day's fighting from the river. His regret for condition θ_3 is therefore (800+100), some 900 men, and this is a maximum. If he cannot afford to take this risk he should follow action a_1 rather than a_3, but in this case his differential casualty expectation will be seriously curtailed.

Interesting though this example may be, the whole problem hinges on the commander's estimation of his probable casualties under various rates of advance and on the probability of good, or bad, weather conditions. Past data may provide a satisfactory estimate of the chance of good weather, but the effects of terrain, enemy reinforcements, etc. on the probable rates of advance and casualties may well be the kind of value judgement which differentiates between 'good' and 'average' army commanders.

Appendix 14 Forecasting models using the Box-Jenkins approach

Definition of terms

Consider a series of data values which we will represent as a series z_1, z_2, z_3 ..... z_n etc., where z_n is the most recent value.

From this series of values certain statistical functions are calculated. These are the estimated autocovariance, represented by $c(k)$ and the estimated autocorrelation $r(k)$.

A simple mathematical trick is also used to express certain formulae more easily. This is known as the backward shift operator B. Another rather similar operator is called the 'difference' operator ∇.

For example, in the series of data values z_1, z_2, z_3, etc., let t be the value 1, 2, 3, etc., then clearly z_{3-1} is the z_2 value, and in terms of the backward shift operator B the relationship is as follows:

$$Bz_t = z_{t-1}$$

Thus using the data values 4.1, 4.15, 4.25 ... , then:

$$B(4.25) = 4.15$$

The difference operator ∇ also acts on a value and is defined as:

$$\nabla z_t = z_t - z_{t-1}$$

Or, using the previous data values:

$$\nabla\ (4.25) = 4.25 - 4.15$$

$$= 0.1$$

It should also be mentioned at this stage that the various coefficients which will be applied to the past data to derive a forecast are described as ϕ and θ.

In addition, the term a_t is used in the equations. It is often referred to as 'white noise'. This may be a little misleading, because although in theory it should refer to the random scatter of forecasts, in practice the mean of the scatter may not prove to be zero, since after the first forecast a bias may be evident.

A further shortened notation, $\dot{z}_t$ is used to denote the difference of the data value z_t from the mean value of the date series, $\bar{z}$, i.e.

$$\bar{z} = \frac{\sum_{x=1}^{x=t} z_x}{t}$$

and

$$\dot{z}_t = z_t - \bar{z}$$

The forecast one period ahead is represented as $\hat{z}_{t+1}(1)$ and for two periods as $\hat{z}_{t+2}(2)$ etc.

The general approach

The formulae are derived by considering the different values in the series and examining the differences between certain values. If the Box–Jenkins method is to be applied, then some pattern must be observable. This process of taking the first or second differences between various values is sometimes called 'the process of obtaining stationarity'. For a discrete process to be stationary, the joint probability distribution must be unaffected by shifting all the times forwards or backwards by an integer amount k, where k is known as the 'lag'.

As a means of diagnosing the type of model the autocorrelation values are found for the original values of the series, for the first differences of the values and also for the second differences.

This is carried out as follows:

1 Take the values of the series z_1, z_2, z_3, etc.
2 Sum the series and divide by the number of values to find the mean, $\bar{z}$.
3 Find the autocovariance and the autocorrelation values for the observations, and also for the first and second differences of the observations. This is achieved as follows:

d=0 no differences - original observations

Estimated autocovariance = $\hat{\gamma}$ = $c(k)$

$$c(0) = \frac{\sum_{i=1}^{n} (z_i - \bar{z})(z_i - \bar{z})}{n}$$

$$c(1) = \frac{\sum_{i=1}^{n-1} (z_i - \bar{z})(z_{i+1} - \bar{z})}{n-1}$$

$$c(2) = \frac{\sum_{i=1}^{n-2} (z_i - \bar{z})(z_{i+2} - \bar{z})}{n-2}$$

And so on until the value $c(m)$ is reached, thus:

$$c(m) = \frac{\sum_{i=1}^{n-m} (z_i - \bar{z})(z_{i+m} - \bar{z})}{n-m}$$

The estimated autocorrelation $\hat{\rho}$ = $r(k)$ and:

$$r(0) = \frac{c(0)}{c(0)}; \qquad r(1) = \frac{c(1)}{c(0)};$$

$$r(2) = \frac{c(2)}{c(0)} \; \ldots \text{ to } \ldots \; r(m) = \frac{c(m)}{c(0)}$$

d=1 first differences ∇z_t

Let

$$y_i = z_{i+1} - z_i$$

Then the mean of the first differences $\overline{y}$ is given by:

$$\overline{y} = \frac{\sum_{i=1}^{n-1} (z_{i+1} - z_i)}{n-1}$$

and the estimated autocovariance for the first differences is given by:

$$c(0) = \frac{\sum_{i=1}^{n-1} (y_i - \overline{y})(y_i - \overline{y})}{n-1}$$

$$c(1) = \frac{\sum_{i=1}^{n-2} (y_i - \overline{y})(y_{i+1} - \overline{y})}{n-2}$$

$$c(2) = \frac{\sum_{i=1}^{n-3} (y_i - \overline{y})(y_i + 2 - \overline{y})}{n-3}$$

and so on to:

$$c(m) = \frac{\sum_{i=1}^{n-m-1} (y_i - \overline{y})(y_{i+m} - \overline{y})}{n-m-1}$$

The estimated autocorrelation values for the first difference are calculated in the same way as previously, i.e.

$$r(0) = \frac{c(0)}{c(0)} \qquad r(1) = \frac{c(1)}{c(0)} \ \ldots \text{ etc.,}$$

$$\text{to } r(m) = \frac{c(m)}{c(0)}$$

d=2 second differences $\nabla^2 z_t$

Let

$$w_i = y_{i+1} - y_i$$

$$= z_{i+2} - z_{i+1} - z_{i+1} + z_i$$

So the estimated covariance of the second differences is:

$$c(m) = \frac{\sum_{i=1}^{n-m-1} (w_i - \overline{w})(w_{i+m} - \overline{w})}{n-m-2}$$

The estimated autocorrelation values are again calculated as before, but using the new c values, i.e.

$$r(0) = \frac{c(0)}{c(0)} \qquad r(1) = \frac{c(1)}{c(0)} \ \ldots \text{ etc.,}$$

$$\text{to } r(m) = \frac{c(m)}{c(0)}$$

This procedure can readily be written as a computer program and arrangements made to plot the correlation coefficients automatically.

An example of the process

Consider the data shown in the table, which lists the number of rounds fired on a certain range between the years 1957 and 1971.

The data were entered to the processor in the usual BASIC format, taking the 1957 value first.

Year	Rounds
1957	49,043
1958	35,529
1959	36,966
1960	39,944
1961	33,917
1962	35,804
1963	32,840
1964	34,217
1965	28,553
1966	31,611
1967	29,886
1968	36,377
1969	40,499
1970	34,042
1971	30,077

Note: Since we will specify 12 autocorrelation values (lags) and we require second differences also, then 15 data values must be input to the machine.

The output from the computer was as follows:

TOTALS

MEAN OF OBSERVATIONS IS 3.5237,+04

Lag	Autocovariance	Autocorrelation
0	2.454,+07	1.000,+00
1	4.102,+06	1.671,-01
2	1.788,+06	7.285,-02
3	3.376,+06	1.375,-01
4	-9.835,+05	-4.007,-02
5	-5.102,+05	-2.079,-02
6	-3.413,+06	-1.391,-01
7	-4.731,+06	-1.928,-01

Lag	Autocovariance	Autocorrelation
8	-1.322,+07	-5.387,-01
9	-4.460,+06	-1.817,-01
10	-1.279,+07	-5.211,-01
11	-2.524,+06	-1.039,-01
12	2.088,+07	8.508,-01

MEAN OF FIRST DIFFERENCES IS -1.3547,+03

Lag	Autocovariance	Autocorrelation
0	2.710,+07	1.000,+00
1	-5.680,+06	-2.096,-01
2	-3.038,+06	-1.121,-01
3	-3.808,+05	-1.405,-02
4	-2.076,+06	-7.659,-02
5	-5.458,+04	-2.014,-03
6	-1.308,+06	-4.828,-02
7	6.470,+06	2.388,-01
8	-9.771,+06	-3.606,-01
9	1.310,+07	4.835,-01
10	-2.250,+07	-8.305,-01
11	-3.072,+07	-1.133,+00
12	2.738,+07	1.010,+00

MEAN OF SECOND DIFFERENCES IS 7.3454,+02

Lag	Autocovariance	Autocorrelation
0	5.729,+07	1.000,+00
1	-2.584,+07	-4.510,-01
2	4.504,+06	7.862,-02
3	-6.529,+06	-1.140,-01
4	4.272,+06	7.457,-02
5	-2.923,+06	-5.102,-02
6	-4.862,+06	-8.423,-02
7	1.391,+07	2.429,-01
8	-2.664,+07	-4.651,-01
9	5.667,+07	9.891,-01
10	-2.345,+07	-4.094,-01
11	-7.971,+07	-1.391,+00
12	2.498,+07	4.361,-01

When the autocorrelation coefficients are plotted against the various lags for the original values and for the first and second differences, the results shown in Figures A14.1, A14.2 and A14.3 respectively are obtained. The relationship between these values is diagnostic of the type of model which should be fitted.

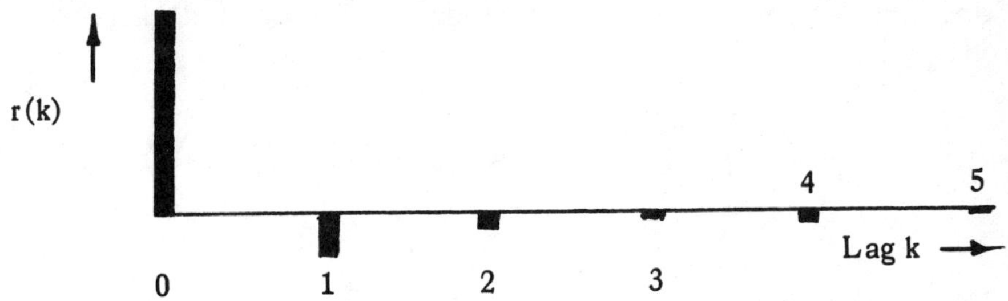

Figure A14.1 Autocorrelation coefficients for original data

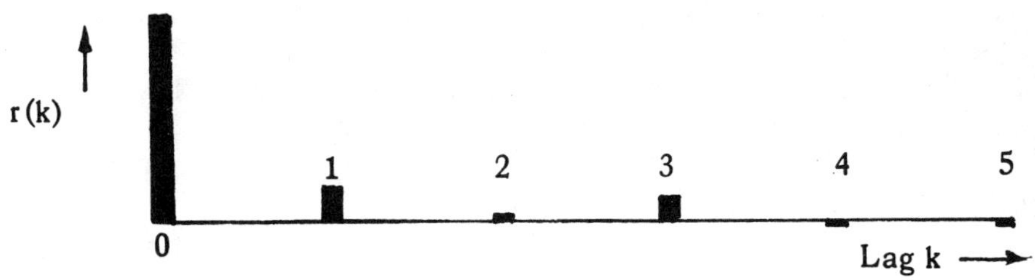

Figure A14.2 Autocorrelation coefficients for first differences

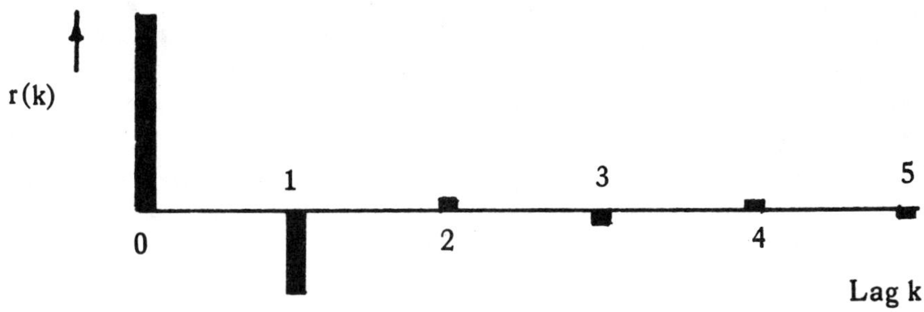

Figure A14.3 Autocorrelation coefficients for second differences

The autocorrelations for the first differences were calculated as shown earlier. The autocorrelations for the second differences were calculated using the mean given by:

$$\overline{w} = \frac{\sum_{i=1}^{n-2} w_i}{n-2}$$

and the estimated autocovariance of the second difference given by:

$$c(0) = \frac{\sum_{i=1}^{n-2} (w_i - \overline{w})(w_i - \overline{w})}{n-2}$$

For example, Figure A14.1 is vaguely diagnostic of a second order autoregressive model, AR(2), but Figure A14.3 indicates quite clearly that a first order integrated autoregressive model ARIMA (1,2,0) should be satisfactory. See various diagrams given at the end of this appendix. The method of diagnosis of the various types of models is dealt with in detail later in this appendix.

The next problem is to estimate the value of the coefficient ϕ for the chosen first order integrated autoregressive model, which for a one period forecast is of the form:

$$\hat{z}_{t+1}(1) = (2+\phi_1)z_t - (1+2\phi_1)\, z_{t-1}+\phi_1 z_{t-2}$$

(This is derived from the first equation on page 334 in the manner described on pages 129 and 130. Note z_{t-1} and z_{t-2} are the preceding values in the series.)

For this particular model ϕ_1 is equal to $r(1)$, and since the autocorrelation values of the *second* differences were used (Figure A14.3) we obtain the estimated autocorrelation value $r(1)$ for the second differences from the tabulation of values from the computer. It will be found that in this case $r(1)$ is equal to -.45, so for this particular example, where ϕ_1 equals $r(1)$, ϕ_1 is equal to -.45.

Substituting the appropriate data values in order to forecast the total rounds fired in 1972 gives:

$$\hat{z}_{72}(1) = (1.55 \times 30.077)-(.1 \times 34{,}042)-(.45 \times 40{,}499)$$

i.e.

$$z_{72}(1) = 24{,}990 \text{ rounds.}$$

In practice the total number of rounds fired in 1972 proved to be 20,722. It must be admitted that this forecast is in error by approximately 20 per cent, but forecasting from a simple least squares linear regression line (which from the scattered nature of the data was all that superficially appeared reasonable) gave an estimate of 30,652 rounds, or an error of nearly 48 per cent. After studying the data for some time, one might well have decided to fit a second-order polynomial, and this would have given a prediction of 36,825 rounds, or an error of nearly 78 per cent.

Other models

Other frequently used models are the first- and second-order autoregressive models, the first- and second-order moving average models, the first- and second-order mixed autoregressive moving average models, and the most general case, which is the autoregressive integrated moving average model.

Basically, the autoregressive models use past data values which are given differing emphasis in order to provide a forecast of the future. The moving average model takes the mean value and adds the weighted (i.e. differing emphasis) sum to all previous errors.

The autoregressive moving average is a combination of both the foregoing models, whilst the autoregressive integrated moving average is a rather similar model for a non-stationary series.

Model classification

A simple shorthand notation is used to describe the various models. The most general model is written as (p,d,q) where p, d and q are defined as below for the general equation shown on page 334.

The value of p indicates the order of the autoregressive process,

d is the power of a 'difference' operator which is the difference between successive values,

and q indicates the order of the moving average process.

Thus:

(1,0,0) is a first-order autoregressive model for a stationary series,

(1,1,0) is a first-order integrated autoregressive model for a non-stationary series,

(0,0,1) is a first-order moving average model for a stationary series,

(0,1,1) is an integrated first-order moving average model for a non-stationary series,

(1,0,1) is a first-order autoregressive combined with a first-order moving average model, and

(1,1,1) is a first-order autoregressive integrated moving average model.

The general autoregressive process

The general form of the autoregressive process is given by:

$$(1 - \phi_1 B - \phi_2 B^2 - \ldots\ldots \phi_p B^p)\,\dot{z}_t = a_t$$

The general moving average process

The general moving average process is given by:

$$\dot{z}_t = (1 - \theta_1 B - \theta_2 B^2 - \ldots\ldots \theta_q B^q) a_t$$

The general autoregressive moving average process

The general form of the autoregressive moving average process is given by:

$$(1 - \phi_1 \mathrm{B} - \phi_2 B^2 - \ldots\ldots \phi_p B^p)\,\dot{z}_t = (1 - \theta_1 B - \theta_2 B^2 - \ldots\ldots \theta_q B^q) a_t$$

Non-stationary series

Non-stationary series often represent a more practical situation, and may involve forecasting where seasonal changes occur, or where there is slow drift in the values. In order to obtain a solution the series must have 'homogeneity', although it may be non-stationary. For example, in Figure A14.4 the patterns inside the 'boxes' on the graph are repetitive but are located differently.

Models which fulfil this situation are the autoregressive integrated moving average models, known more briefly as ARIMA, the most common being the first- and second-order models. The models are obtained by summing, or integrating, the stationary process *d* times. Hence the *d* in the model description (p,d,q), where p is the order of the stationary autoregressive process and q is the order of the stationary moving average process.

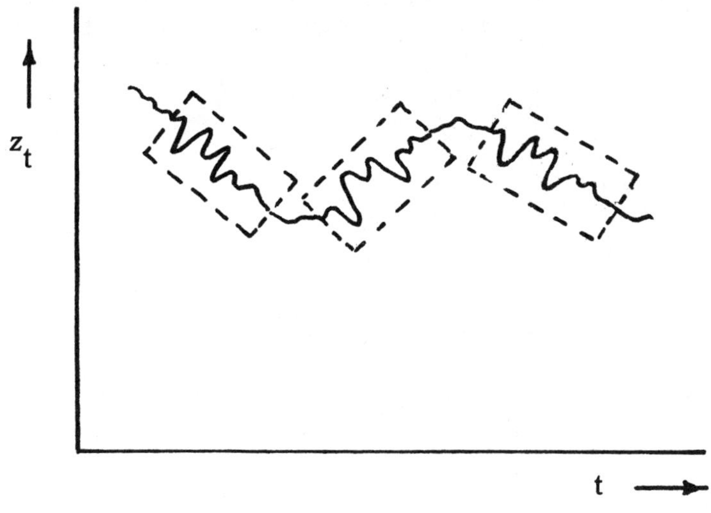

Figure A14.4 An example of a non-stationary series with a repetitive pattern

The general autoregressive integrated moving average process

The general form of the autoregressive integrated moving average process is given by:

$$(1-\phi_1 B-\phi_2 B^2 \ldots\ldots \phi_p B^p)\ \nabla^d z_t = (1-\theta_1 B-\theta_2 B^2 \ldots\ldots \theta_q B^q)a_t$$

The symbols are as previously defined.

Model identification

The first-order autoregressive model ARIMA (1,d,0)

Equation of function:

$$(1 - \phi_1 B)\nabla^d z_t = a_t$$

From Figures A14.5a and A14.5b it can be seen that ρ_k decays exponentially. If $\phi > 0$ the values are all positive (Figure A14.5a) and if $\phi_1 < 0$ the values are alternately positive or negative. If all the terms in the second case are made positive the graph appears similar in shape (smooth exponential decay) to the first one where ϕ_1 was positive.

By definition ρ_o is equal to one.

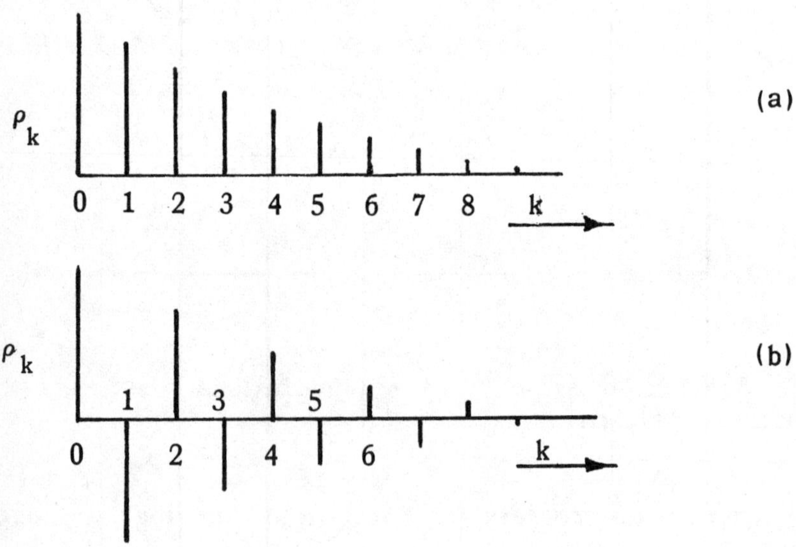

Figure A14.5a and b

Preliminary estimates from:

$$\phi_1 = \hat{\rho}_1 = r(1)$$

Admissible values:

$$-1 \quad \phi_1 \quad 1$$

Note: These diagnostics also apply in the case where d is equal to zero, i.e. The AR(1) model.

The equation in this case is given by:

$$(1 - \phi_1 B)\, \dot{z}_t = a_t$$

where $\dot{z}_t = (z_t - \bar{z})$, and ϕ_1 is estimated from $r(1)$ as before.

The second order autoregressive model ARIMA (2,d,0)

Equation of function:

$$(1 - \phi_1 B - \phi_2 B^2) \quad \nabla^d z_t = a_t$$

From Figures A14.6a and A14.6b it can be seen that ρ_k decays exponentially with a superimposed sine wave. In the first case $\phi_1 \quad 0$ and in the second $\phi_2 \quad 0$.

Again, by definition ρ_k is equal to one.

Preliminary estimates from:

$$\phi_1 = \frac{\rho_1(1-\rho_2)}{1-\rho_1^2} \quad \text{and} \quad \phi_2 = \frac{\rho_2-\rho_1^2}{1-\rho_1^2}$$

Admissible values:

$$(\phi_1+\phi_2) \quad 1 \; ; \quad (\phi_2-\phi_1) \quad 1 \; ; \quad -1 \quad \phi_2 \quad 1$$

Note: The diagnostics also apply to the case where d is equal to zero, i.e. The AR(2) model.

The equation in this case is given by:

$$(1 - \phi_1 B - \phi_2 B^2)\, \dot{z}_t = a_t$$

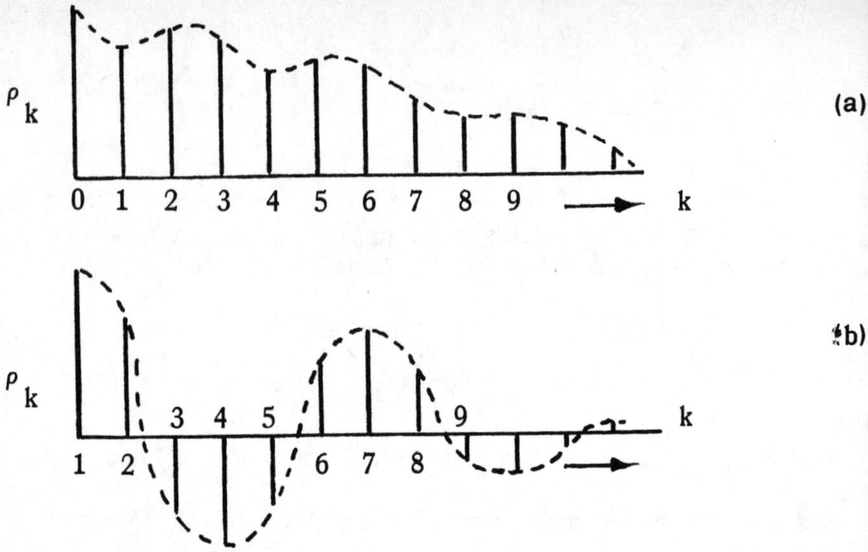

Figure A14.6a and b

The first order moving average model ARIMA (0,d,1)

Equation of function:

$$\nabla^d z_t = (1 - \theta_1 B)a_t$$

From Figures A14.7a and A14.7b it can be seen that only ρ_1 and ρ_2 are non-zero. In the first case, $\theta < 0$, i.e. θ is negative. In the second case, $\theta > 0$, i.e. θ is positive.

Preliminary estimates from:

$$\theta_1 = -\frac{1}{2\rho_1} \pm \frac{1}{(2\rho_1)^2} - 1$$

Admissible values:

$$-1 < \theta_1 < 1$$

Note: This also applies to the case where d is equal to zero, i.e. the MA(1) model.

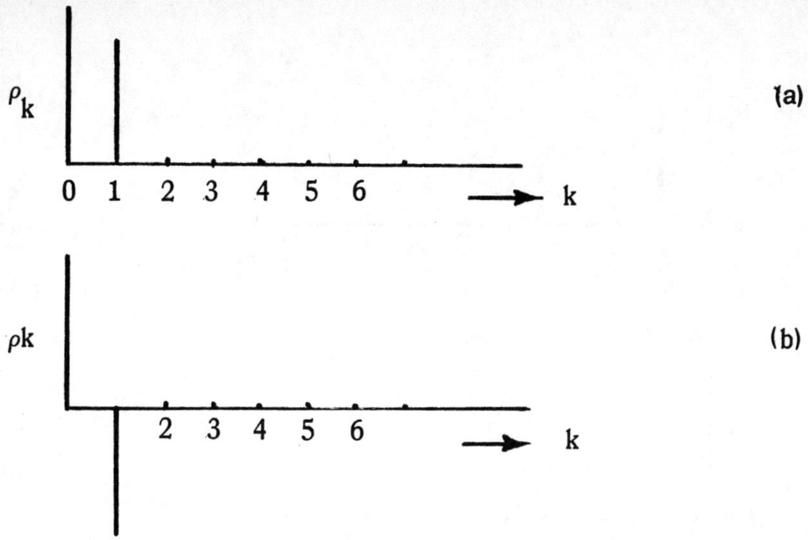

Figure A14.7a and b

The equation in this case is given by:

$$\dot{z}_t = (1 - \theta_1 B)a_t$$

where $$\dot{z}_t = (z_t - \bar{z}).$$

The second order moving average model ARIMA (0,d,2)

Equation of function:

$$\nabla^d z_t = (1 - \theta B - \theta B^2)a_t$$

Figures A14.8a and A14.8b show that only ρ_0, ρ_1 and ρ_2 are non-zero. That is to say, in practice all subsequent values of ρ are almost zero.

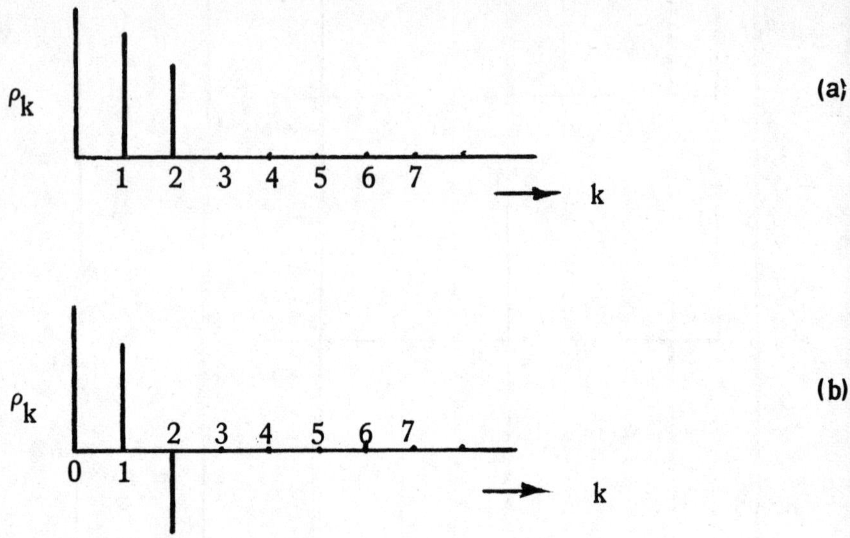

Figure A14.8a and b

Preliminary estimates from:

$$\rho_1 = \frac{-\theta_1 (1-\theta_2)}{1 + \theta_1^2 + \theta_2^2}$$

and

$$\rho_2 = \frac{-\theta_2}{1 + \theta_1^2 + \theta_2^2}$$

Admissible values: $(\theta_2+\theta_1) < 1$; $(\theta_2-\theta_1) < 1$; $-1 < \theta_2 < 1$

Note: This also applies where d is equal to zero, i.e. the MA(2) model.

The equation in this case is given by:

$$\dot{z}_t = (1 - \theta_1 B - \theta_2 B^2) a_t$$

where

$$\dot{z}_t = (z_t - \bar{z})$$

The first order mixed process ARIMA (1,d,1)

Equation of function:

$$(1 - \phi_1 B) \quad \nabla^d z_t = (1 - \theta B)a_t$$

From Figures A14.9a, A14.9b and A14.9c it can be seen that ρ_k decays exponentially after lag 1, which is relatively large and in the positive or negative direction.

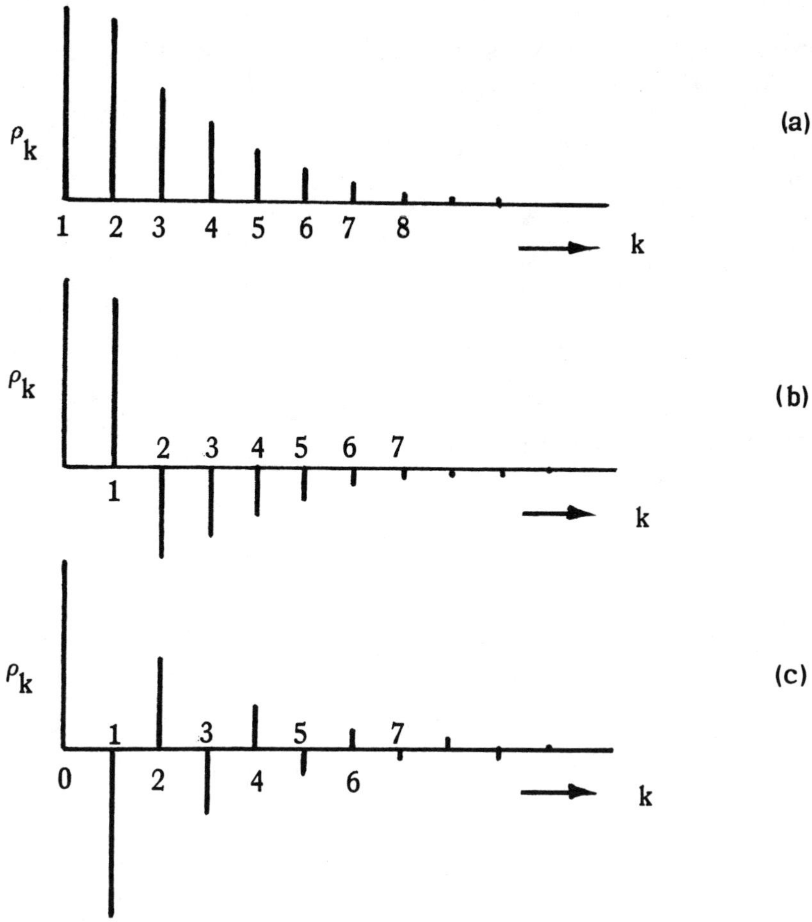

Figure A14.9 a, b and c

Preliminary estimates from:

$$\rho_1 = \frac{(1-\theta_1\phi_1)(\phi_1-\theta_1)}{1+\theta_1{}^2-\phi_1\theta_1}$$

$$\rho_2 = \rho_1\phi_1$$

Admissible values:

$$1 \quad \phi_1 \quad -1, \qquad -1 \quad \theta_1 \quad 1$$

Note: This also applies to the case where d is equal to zero, i.e. the ARMA (1,1) model.

The equation in this case is given by:

$$(1 - \phi_1 B)\,\dot{z}_t = (1 - \theta_1 B)a_t$$

where

$$\dot{z}_t = (z_t - \overline{z})$$

An ARIMA $(2,d,1)$ would have superimposed damped sine waves on the exponential.

Primary data

ARMA : family

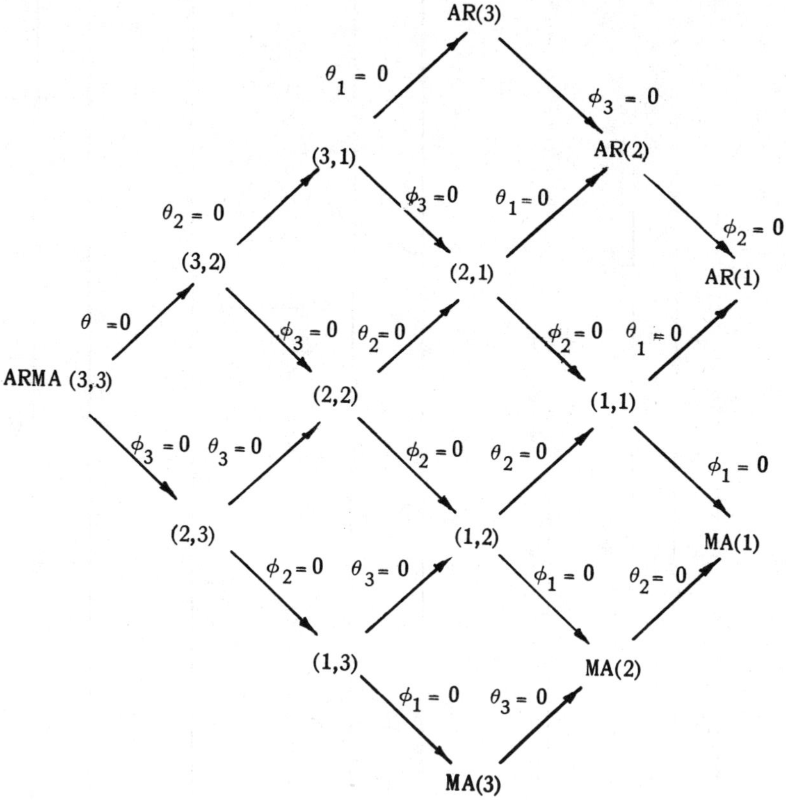

Figure A14.10 Diagram illustrating the way in which higher orders of the ARMA model degenerate into simple moving average or autoregressive processes as the various ϕ and θ values become zero.

First order differences

ARIMA : family one

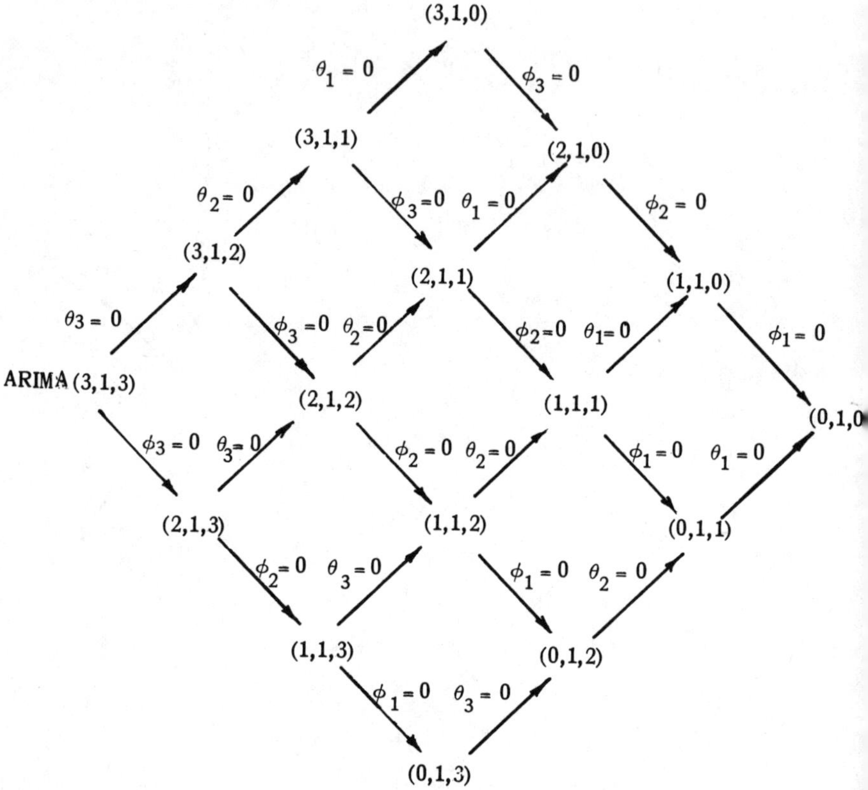

Figure A14.11 Diagram illustrating the way in which the ARIMA (p,l,q) models simplify as various ϕ and θ values become zero.

ARIMA : family two

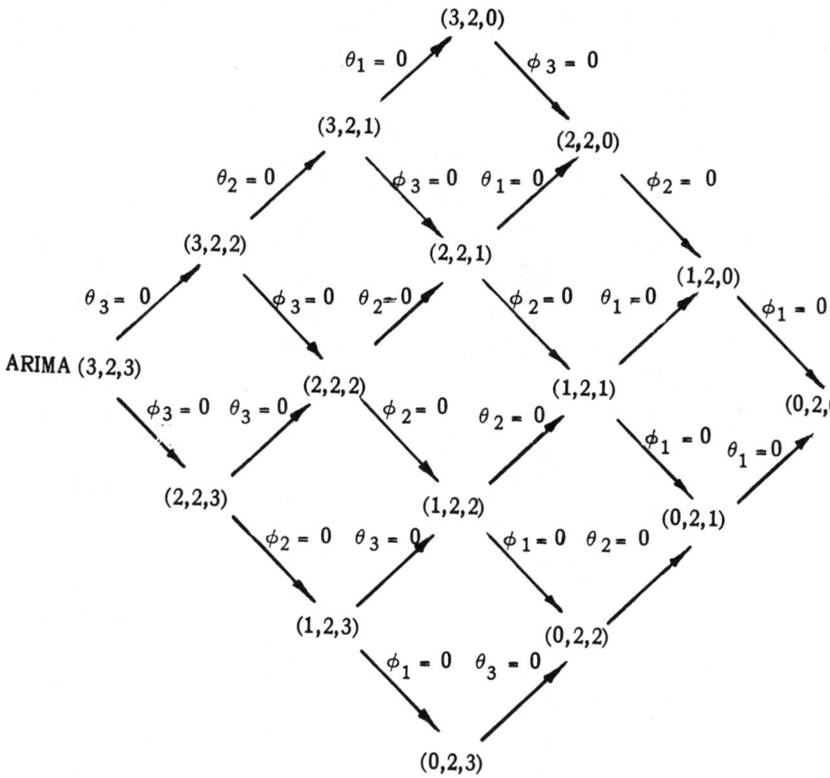

Figure A14.12 Diagram illustrating the way in which the ARIMA (p,2,q) models simplify as various ϕ and θ values become zero.

Note: These families of models virtually form a three-dimensional array when this page is viewed in conjunction with the two preceding pages, each page being held over the other.

Estimation of ϕ and θ values from the following table

Table A14.1 The relationship of ρ_1 to θ for a first-order moving average process

θ	ρ_1	θ	ρ_1
0.00	0.000	0.00	0.000
0.05	−0.050	−0.05	0.050
0.10	−0.099	−0.10	0.099
0.15	−0.147	−0.15	0.147
0.20	−0.192	−0.20	0.192
0.25	−0.235	−0.25	0.235
0.30	−0.275	−0.30	0.275
0.35	−0.315	−0.35	0.315
0.40	−0.349	−0.40	0.349
0.45	−0.374	−0.45	0.374
0.50	−0.400	−0.50	0.400
0.55	−0.422	−0.55	0.422
0.60	−0.441	−0.60	0.441
0.65	−0.457	−0.65	0.457
0.70	−0.468	−0.70	0.468
0.75	−0.480	−0.75	0.480
0.80	−0.488	−0.80	0.488
0.85	−0.493	−0.85	0.493
0.90	−0.497	−0.90	0.497
0.95	−0.499	−0.95	0.499
1.00	−0.500	−1.00	0.500

This table may be used to obtain first estimates of the parameters

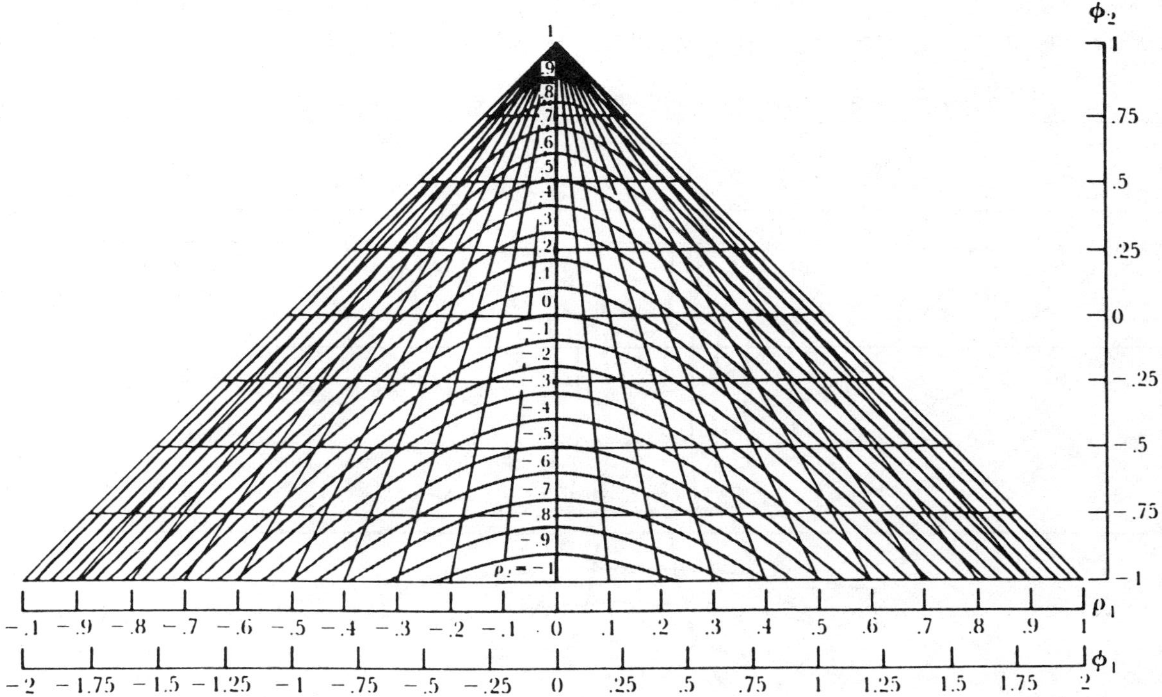

This chart may be used to obtain estimates of the parameters.

Figure A14.13 The relationship of ρ_1 and ρ_2 to ϕ_1 and ϕ_2 for a second-order autoregressive process

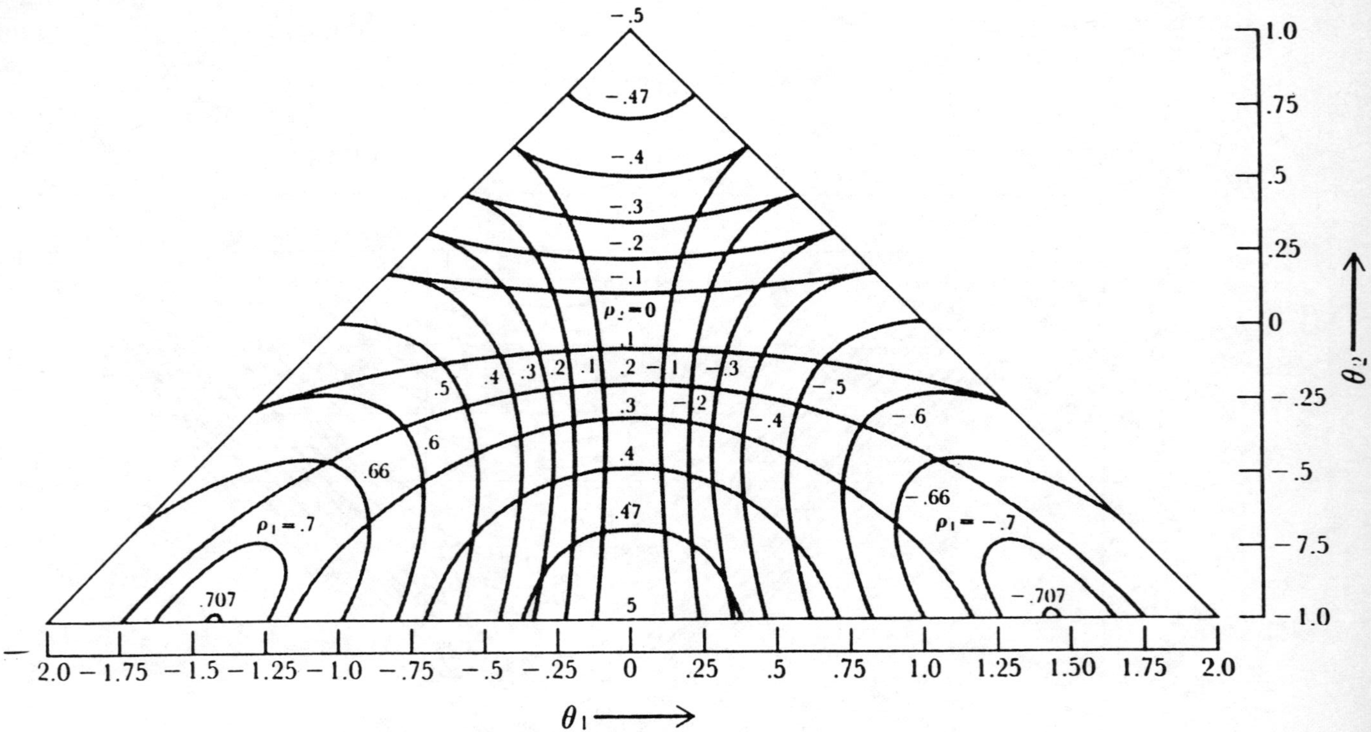

This chart may be used to obtain first estimates of the parameters

Figure A14.14 The relationship of ρ_1 and ρ_2 to θ_1 and θ_2 for a second-order moving average process

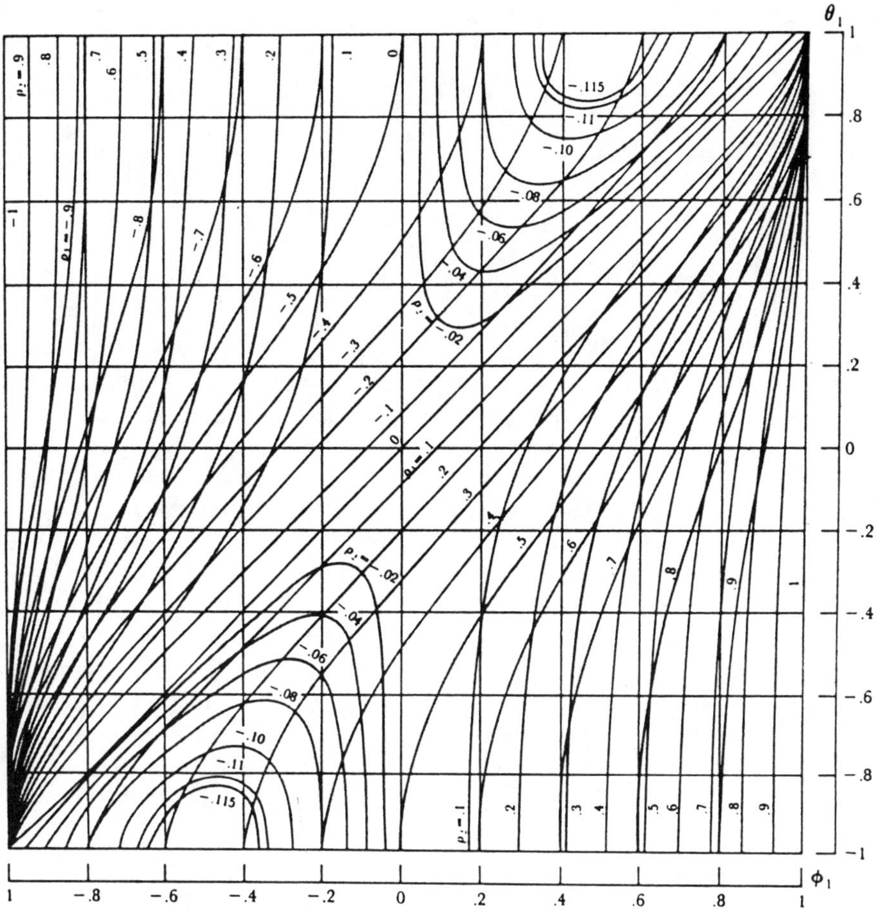

This chart may be used to obtain first estimates of the parameters

Figure A14.15 The relationship of ρ_1 and ρ_2 to ϕ and θ for a mixed first–order autoregressive–moving average process

Appendix 15 The random division of linear quantities

We often talk about random events, random variations or random arrivals in a rather general manner, without considering their precise meaning. According to the *Concise Oxford Dictionary* the word 'random' means ' ... haphazard, without aim of purpose or principle ...'.

However, the impact of truly random events on a continuum does, in fact, give rise to effects, or continua, which indeed do have a certain aim, purpose or principle.

Perhaps the simplest way to illustrate this concept is to consider dropping a point, completely at random, anywhere along a line of unit length (see Figure A15.1). If we consider that the line is divided into two equal parts, A and B, then the point will fall either in A or B with equal probability. Suppose that it falls into the A section at some point X; then, on average, it will fall halfway along A, or $\frac{1}{4}$ of the distance from one end, and $\frac{3}{4}$ the distance from the other. Similar remarks apply to the other half, B, of the unit length.

Thus, if a large number of rods of unit length are each broken into two parts at random and in every case the shorter piece is placed on a pile which we will call pile A, and the larger pieces are placed on a pile B; then it will be found that the average length in pile A will be one-quarter of the original unit length and in pile B it will be three-quarters. Of course, there may be appreciable variations from these values if any individual piece is considered, since the quarter and three-quarter values mentioned are the average of many experiments.

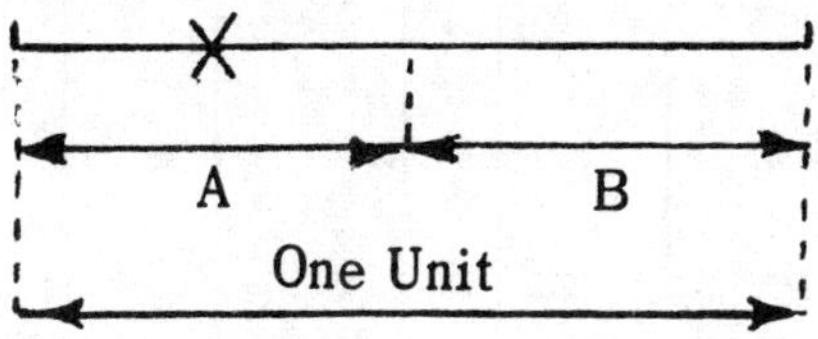

Figure A15.1 Determining the average size of the pieces in a single random break

It can be said, more generally, that if N is the number of parts into which the unit length is divided, the sizes can be calculated as follows:

The smallest expected part (or rank) is given by:

$$\frac{1}{N} \times \frac{1}{N}$$

The largest expected part (or rank) is given by:

$$\left(\frac{1}{N} \times \frac{1}{N}\right) + \left(\frac{1}{N} \times \frac{1}{N-1}\right)$$

The next largest expected part (or rank) is given by:

$$\left(\frac{1}{N} \times \frac{1}{N}\right) + \left(\frac{1}{N} \times \frac{1}{N-1}\right) + \left(\frac{1}{N} \times \frac{1}{N-2}\right)$$

and so on.

This can be expressed in rather more mathematical terms as:

$$E_i = \frac{1}{N} \sum_{x=0}^{x=N-i} \frac{1}{N-x} \qquad i=1,2,3, \dots N$$

where E_i is the expected length, N is the number of parts into which the unit length is divided and i is the rank (i.e. 1 to N).

Hence, when certain items are divided in a random fashion, such as the intervals between speeches in a play, it is likely that the length of these will be roughly in proportion to the values given by the formula.

If we rank, say, 10 values in order, so that the largest is first and the second largest next, etc., then the expected distribution of values (with *N* equal to 10) will be as shown in Figure A15.2.

A fascinating feature of the distribution is the way in which a distribution of *N* values contains all the characteristics of other distributions which have values of less than *N*.

For example, if we have the distribution of expected values for *N* equal to 10 (as shown in Figure A15.2) and we wish to find the distribution of expected values for *N'* equal to 5, then it is simply necessary to raise the *X*-axis until it rests on the top of the new *N* value plus one, as shown at *X'X'* in Figure A15.3.

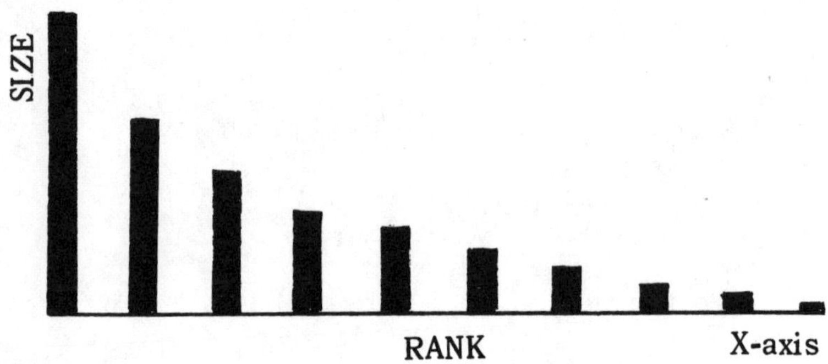

Figure A15.2

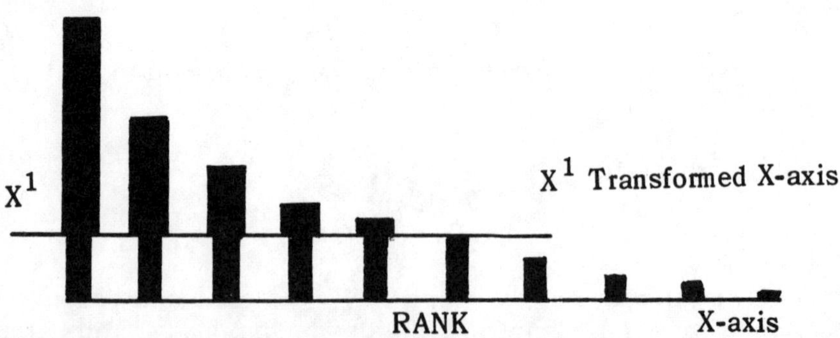

Figure A15.3

The ordinates arising from the new axis X'X' are then in the correct proportions.

It is therefore suggested that the formula provides a good basis for estimating and ranking random partitions of this kind, and that quite a wide variety of natural phenomena follow this pattern of expected values.

The rate of *X*-axis movement for the distribution, as illustrated in Figure A15.3, can sometimes be a useful prediction aid. The direction of movement of the axis will depend on the phenomena being studied. For example, if we consider the number of branches of science which may be expected to exist at any time, we find that the *Z*-axis is gradually being lowered as time passes and new ranks are revealed as the new sciences unfold.

Care must be taken when ranking only those items above a certain size, to subtract this minimum value, multiplied by the number of items to be ranked, from the total value to be partitioned, before the E_i values are calculated. This minimum, or threshold, value must then be added to the calculated value in order to produce the correct estimate.

Clearly, when N is very large, the determination of the expected values without mechanical or electrical computing assistance becomes an enormous task. However, the application of Euler's Theorem leads to an expression for the expected values (the E_i values) which involves the logarithm of a function of N and the rank number i, plus an error term.

A compact formula, which is accurate to about 1.5 per cent, is given by the expression:

$$E_i = \frac{1}{N} \log_e \frac{N+.57}{i-.43}$$

Expressing the formula in a slightly different form gives:

$$i = .43 + (N+.57)e^{-NE_i}$$

It is interesting to note that the first of these formulae can be used to generate random arrivals for computer simulation studies in the place of the Poisson distribution. It has value in certain kinds of prediction processes and can be useful for assessing the expected distribution of markings in judgemental situations.

Another practical application of this formula is the way in which one may expect a number of projects to contribute to the overall profitability of a company. A company may set out quite deliberately to have, say, five major activities which will contribute equally to the total profit of the company. Over the years the randomising action of nature will cause the expected average values of profit from the five activities to migrate into the kind of distribution just described.

It should also be noted in passing that the random division of multidimensional figures, such as areas and volumes, has some practical significance. For example, the random division of area may be important in estimating the expected sizes of states, counties or towns. The random division of volume into expected sizes may provide an approximate indication of the natural elements, and so on. Such items as have been investigated using this approach have conformed remarkably well to the expected pattern.

Appendix 16 The use of PERT and CPM

PERT, which stands for 'Programme (Project) Evaluation and Review Technique', began to assume importance in the UK in the early 1960s. Perhaps, rather unkindly, some would say that many of the early users of PERT were excessively zealous in its use and occasionally gave the impression that they regarded the means to the end as almost more important than the end itself. Some of the enthusiasm for PERT was due to the fact that many of the users of the method had little knowledge of any other form of scheduling. In the long term this ultimately may have proved to be a disadvantage as they were not aware of the problems which can be encountered with almost any form of scheduling.

CPM, which stands for 'Critical Path Methods', was evolved in the USA in the late 1950s. Originally, PERT and CPM were two distinct and quite separate techniques; but today one name often implies the use of the other. A variety of additional names are also sometimes used, such as critical path analysis, and network analysis.

The whole process, which henceforth will simply be referred to as PERT, is concerned with determining the minimum time necessary, and sometimes the required expenditure in terms of resources, for the fulfilment of various goals. In the West PERT has been used mainly for job scheduling in workshops, but Soviet books on military operational research imply that it is used for planning combat operations, since these involve both time and space.

Initially, by way of explanation of the technique, we will consider a simple workshop scheduling task. A more complex military example will be given later to demonstrate the scope of the method.

A PERT diagram gives a number of explicit statements about 'events' and 'activities'. Basically, an 'activity' is simply a job, or task, which must be performed. Activities are time-consuming and may also absorb resources. They are illustrated on the diagram by means of a line, where the tail of the arrow denotes the start of the job and the head of the arrow indicates the completion of the work. Unlike Gantt charts, the duration of the task is not usually represented by drawing the length of the line to some suitable scale. The nature of the activity is usually written in close proximity to the line.

'Events' are usually denoted on the diagram by circles and indicate points in time where one or more activities start and/or finish. The circles are numbered for reference purposes and are often referred to as nodes on the diagram.

In constructing the network diagram there are five main questions to ask about each activity:

1 What activities precede it?
2 What activities run concurrently?
3 What activities follow it?
4 What controls the start of the activity?
5 What controls the finish?

Thus in considering a workshop schedule of the form shown in Figure 122 it will be clear that Job B follows Job A.

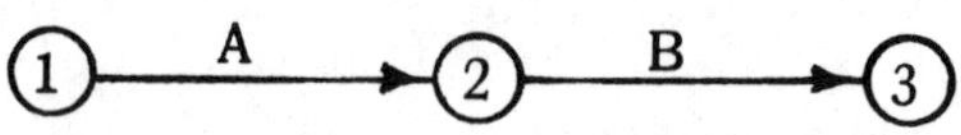

Figure A16.1 Representation of sequential tasks A and B

Similarly, concurrent tasks are shown by two lines separated from each other as shown in Figures A16.2a, A16.2b and A16.2c.

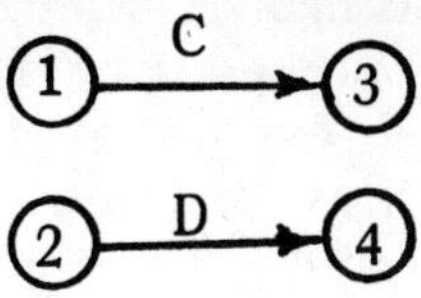

Figure A16.2a

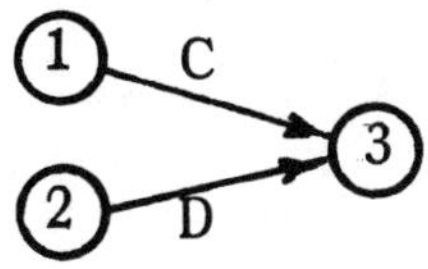

REPRESENTATION OF CONCURRENT TASKS C AND D

Figure A16.2b

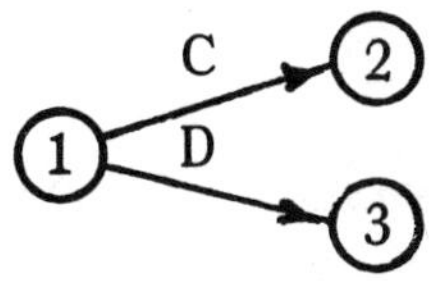

Figure A16.2c

Arranging the network in diagrammatic form usually makes it possible to see the problem more clearly. It is also helpful in constructing a tabulation of topologically ordered events and this is usually a necessary precursor to any work on a computer.

A PERT chart has the following properties:

1 No event can take place until all the operations entering it have been completed.
2 No operation leaving an event can begin until the given event has taken place.
3 No subsequent operation can begin before all the preceding operations have been completed.

A special procedure is adopted in numbering the events. The diagram of activities is generally arranged to run from left to right. The first event, or start of the plan, is numbered zero and all activities leaving event number zero are deleted. The events at the end of these deleted activities are considered and those which do not have other operations entering them are said to be events of the first rank and are arbitrarily numbered from one, in integer steps upwards (number 1 in Figure A16.3).

All activities leaving events of the first rank are then deleted. The events thus exposed are said to be of the second rank and are numbered sequentially (see events 2 and 3 in Figure A16.3). This procedure is continued.

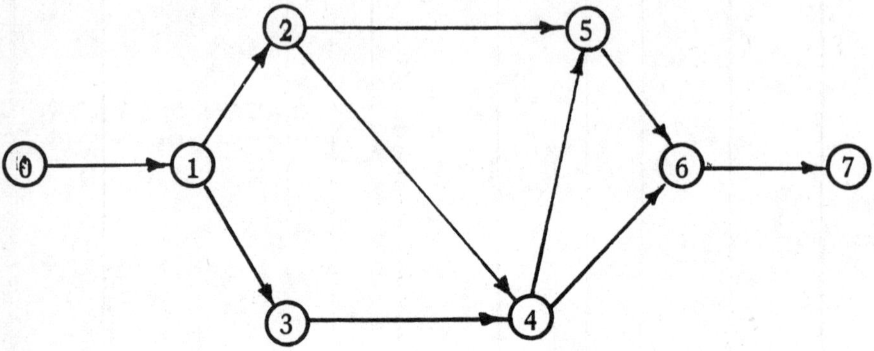

Figure A16.3 The events are numbered sequentially

Thus after using the procedure for the (k-1) step, the events of the (k-1) rank are defined. All operations leaving the events of (k-1) rank are deleted and the events where these operations end are considered. First, events are selected which do not have other operations leading into them. These events are of the kth rank and are numbered sequentially. The process is continued to the final event.

In complicated processes situations often arise where one activity or part of the process, hitherto conducted in parallel, cannot be undertaken until another part of the process is completed. In this case it is necessary to introduce a dummy, or fictitious, activity lasting for zero time. Some show the existence of a dummy procedure on the chart as a dotted line but this is not always necessary.

In constructing the diagram a number of empirical rules have developed which assist in achieving clarity. Thus, as far as possible, in constructing the diagram one should not have crossovers in the lines or bends in them, and one should try to keep the numbers reading from left to right.

Additional information is sometimes shown on the network diagram. This often includes the earliest and latest arrival times for certain activities. Earliest possible arrival times are entered in square box symbols close to the event nodes, and the latest possible arrival times which can occur without delaying the completion of the overall plan are entered in triangular symbols located against each node, see Figure A16.4. Some prefer to make the nodes slightly larger and to divide them into several sectors, rather like cutting a cake. The event number of the node and the earliest and latest completion times are then entered in the different sectors of the partitioned node. The precise diagrammatic arrangements are a matter of personal preference.

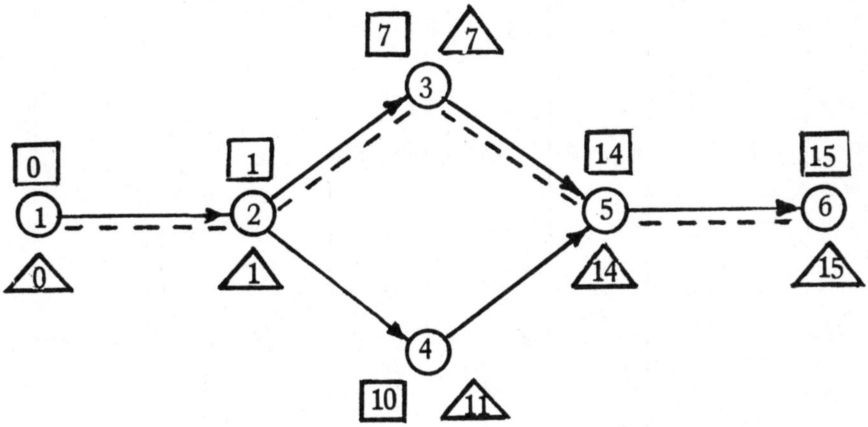

Figure A16.4 Earliest possible arrival times are shown in 'boxes' and latest possible arrival times in triangles

The critical path, which is shown by dotted lines in Figure A16.4, is the one in which any delay in the completion of a task would cause a delay in the final completion date. This path is also sometimes referred to as the 'most stressed path'.

Two other terms are often encountered in considering PERT diagrams. These are the terms 'slack' and 'float'. The difference between the expected and latest dates of an event is known as the 'slack' and represents the amount of time that event could be delayed without affecting the scheduled completion date for the whole project. If a longer time can be taken for the activity without affecting the completion date, it is known as the primary slack of the activity, or the total float of the activity. A negative float occurs where insufficient time is available for the activities.

As mentioned earlier, the PERT diagram can be set out in tabular form. Figure A16.6 is compiled from the network diagram shown at Figure A16.5 and is a complete description of that network. In Figure A16.5 it should be noted that the numbers against the lines show the times to complete an activity, whilst the numbers in the circles simply indicate the occurrence of an event.

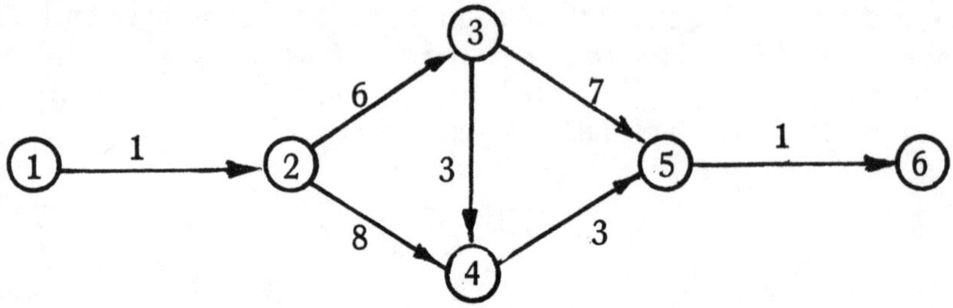

Figure A16.5 A 'PERT' network diagram

Initial Event (Entry)

		1	2	3	4	5	6	Earliest time for event
*	1		1					0
*	2			6	8			0 + 1 = 1
*	3				3	7		6 + 1 = 7
	4					3		8 + 1 3 + 7 ~ 10
*	5						1	7 + 7 3 + 10 ~ 14
*	6							1 + 14 = 15

NOTE: Asterisk denotes critical path

Figure A16.6 PERT network information displayed in tabular form

The network can, of course, be laid out to a time base, but it is not customary to do this. Figure A16.7 shows the example given in Figure A16.5 and listed in Figure A16.6 arranged in this way.

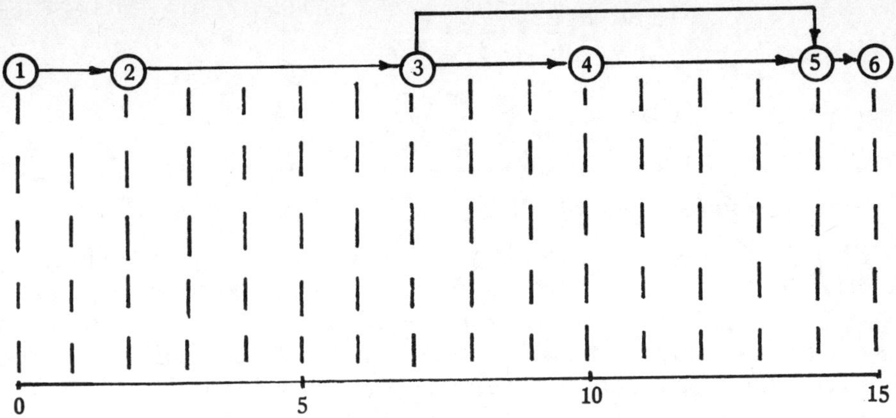

Figure A16.7 Network diagram laid out to a time base

In the early days PERT normally included estimates for the probability of completion of each task. It was assumed that the errors in estimating the times were independent. This is not necessarily the case, since the estimator may tend to be generally over-optimistic or pessimistic about many of the jobs.

The estimate of the probable completion date for the total task depends on the mean values of the individual jobs, since in a sample the sum of the means provides the best estimate. The distribution of probable completion dates depends on the sum of the individual variances.

Further estimates were sometimes made by making assumptions about the completion date distribution using the most optimistic, most pessimistic and most likely times. This rather empirical approach has tended to fade from use as a consequence of the difficulty of making accurate estimates but the procedure was briefly as follows.

Suppose that:

A is the most optimistic time for job finish
B is the most pessimistic time for job finish
M is the most likely time for job finish

Then by making some rather doubtful assumptions about the shape of the distribution the mean and variance are given by:

$$\text{Mean} = \frac{A}{6} + \frac{2M}{3} + \frac{B}{6}$$

$$\text{Variance} = \frac{(B - A)^2}{36}$$

Military applications of PERT

PERT has direct military applications in various scheduling activities. It can be used, for example, for logistic problems, for planning military build-up during mobilisation, for the deployment of troops engaged on mining and demolition work and for general supply scheduling. It can be used for loading, unloading, refuelling and servicing of ships, aircraft and large vehicles.

Some would say that PERT is rather more useful for military scheduling than it is for civilian activities since the latter activities often tend to be rather more relaxed. After being greeted with much enthusiasm, PERT frequently did not live up to expectations. Perhaps one of the principal reasons for this was that, originally, the so-called 'latest possible target dates' were usually revealed in the plan. As a consequence, these very frequently became the actual target dates in the minds of the busy departmental managers. Quite naturally, when one aims to complete a task by the latest possible target date, occasional slippage will occur due to exogenous effects over which a manager has no control. In the event that the activity proves to be on the critical path the whole plan then fails. Later, a tendency developed of not revealing to the managers the extent of the available 'slack'. Unfortunately, some of them then tended to assume the existence of much more than was actually available. Further waste of time and interdepartmental ill-feeling could then be created as a consequence of acrimonious 'post mortems'.

Nevertheless, on large jobs the project officer cannot carry all the dates and times in his head and so some scheduling system has to be employed. It is likely that for most civilian activities it is desirable for the project officer, with suitable assistance, to construct the PERT diagram himself and to reveal only the target dates, and not the slack, to his contractors and staff. Indeed it is often inadvisable to indicate that a PERT plan, as such, exists. This method has proved highly successful in completing some very major projects on schedule.

Military scheduling is somewhat different. In wartime the pressure to achieve the objectives is enormously greater so that difficulties which can be experienced by the latest possible scheduled date unconsciously becoming the actual target date no longer apply in the majority of cases.

In this case, therefore, PERT can present an effective and useful means of planning a military campaign. Tarakanov (50) claims that the fact that the whole of a combat operations plan can be displayed in this way makes the perception of the substance of the plan a much easier task and also simplifies the management of the operation itself.

Let us consider the following scenario: the 1st, 2nd, 3rd and 4th Red Group Forces are based some miles behind the frontier which roughly follows the River Fen. The Red Commander is given instructions to attack and to capture the Port Ilo with the minimum delay (see map at Figure A16.8). He decides that rather than make an attack through the area adjacent to the river town of Lin he will only pretend to attack in this direction. His main thrust will be southwards where a crossing of the River Fen may be more easily executed. He then plans to thrust north-westwards to attack Port Ilo at the same that that it is being bombarded by naval forces in a simultaneous attack from the sea.

A diagrammatic layout of this plan might look roughly as shown in Figure A16.9.

The diagram contains only a fairly simple analysis of how the plan might appear. Several sections of the plan could be split down into much greater detail if necessary. For example, the engineers might possibly have a separate PERT chart for building temporary bridges and semi-permanent bridges across the River Fen.

Purists will notice that one arm of the diagram remains unclosed but one could quite easily add another node to state 'operation completed' and thus for both remaining arms of the diagram. However, in this case there is a good reason for leaving the PERT plan as shown since the battle will have to continue in the future. A review of actual performance against the original plan can be carried out in order to obtain more realistic assessments for the next plan.

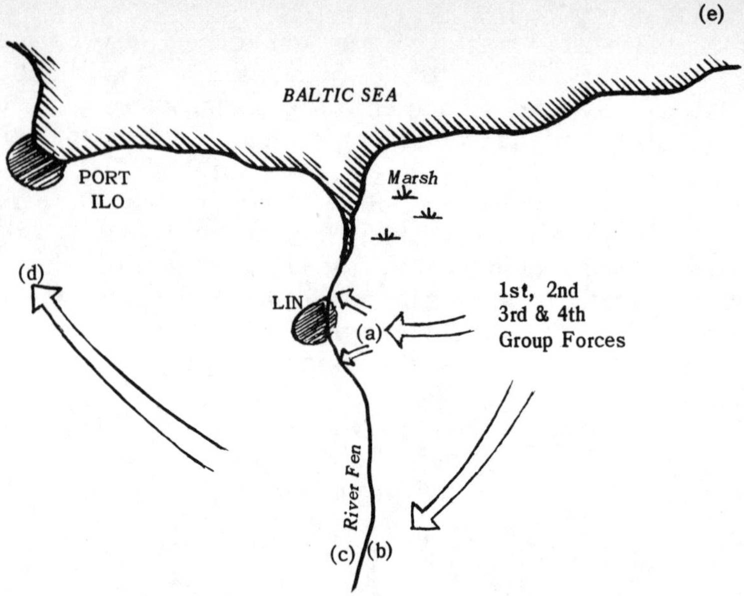

Figure A16.8 Map of the battle area

As mentioned earlier there are, of course, many other arbitrary applications of PERT which could be developed within this particular plan, e.g. loading ships, deploying mines, constructing defensive positions and planning manpower uses. Many other examples will surely come to the mind of the reader.

Perhaps the greatest value of PERT is in fairly deterministic situations. For example, if the 2nd, 3rd and 4th Groups had failed to cross the River Fen, then the existing plan would have been completely invalid and a new one would have to be drawn up. This might have involved moving some units up from the south and producing a combined naval and army attack at the estuary of the River Fen. Thus, it is necessary to have a wide range of PERT plans available in military situations to meet various contingencies.

In many ways the whole process of battle can be regarded as a stochastic dynamic program not unlike the 'taxi-driver' or the 'toymaker' problems. The only difficulty is that the number of dimensions involved is so large that only the barest essentials of any campaign can be considered in this way. PERT has been used in industry for workshop scheduling and project control. A

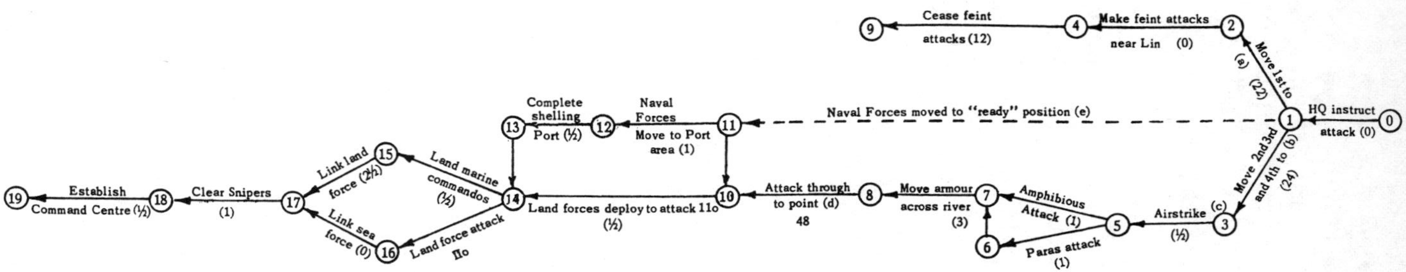

Figure A16.9 A PERT diagram for the proposed attack plan. Note that this diagram, contrary to normal convention, reads from right to left.

number of software packages are available to enable computers to assist in the scheduling task but the problems of stochastic and exogenous effects remain.

This appendix has only given a brief outline of the PERT process and many of the refinements have been omitted. The reader who finds use for the method is advised to read further. (50,84,85,86)

Appendix 17 Subjective assessments in military battle models

Military battle models require the estimation of the attrition coefficients, which describe the relative lethality and casualties to the participants. The coefficients have a strong effect on the battle outcome irrespective of whether the Lanchester equations [87], Weale's heterogeneous model [88] or simulation methods are used.

One of the more interesting aspects of war is the frequency with which surprises occur. In the Second World War consider the unexpected Russian attack on Poland, the collapse of the Allies in France, the extent of the British success in fighting off a numerically superior German air force, the Russian attack on Finland, the casualties inflicted by the Finns on the Russians and the German attack on Norway. Indeed, perhaps the only feature which proceeded in roughly the expected manner was the final invasion of Europe by Britain and America. It should be noted that by the time that this event took place weapons on both sides were of broadly similar quality and tactical procedures were also reasonably alike.

Thus, in the unfortunate event of war, one must be prepared for surprises and be ready to take full advantage of every possible opportunity. Short of war, it is necessary to be ready to accommodate scenario changes, and to ensure that both our foreign policy and armed forces have flexibility and versatility.

Fitting Weale's battle model to a number of historical battles, ranging from the 14th century to fairly recent events, in no circumstances initially yielded the correct results no matter how carefully the attrition coefficients were estimated beforehand. The error was by no means small. For example, in the Crécy battlefield it was necessary to bring about a tenfold change in the coefficients in order to get approximately correct results. In the case of a British–German tank battle at Gabr Saleh, in the Second World War, very careful estimates were made of the probable weapon lethalities by relative expert assessors, but still the wrong side won and a two-and-a-half–fold change in the coefficients was needed to achieve the correct outcome.

The difficulties lie in assessing the subjective variables and in obtaining intelligence data on enemy intrinsic weapon lethality. If no wars have occurred for some years, there is difficulty in assessing the effectiveness of enemy weapons and this, in itself, probably involves a great deal of subjective judgement. The problems with regard to human effects, such as training and tactics, are even greater.

Some would say that training and certain other similar factors cannot be brought into the model. But if one cannot bring these factors into account, can one really be surprised if an otherwise elaborate model proves to have an error of two to tenfold in practice? Of course, if one does not have to fight a battle, it does not matter over-much, except for the fact that in paying the defence insurance premium (which after all is all the defence budget is) one may be wildly under- or over-insuring.

Figure 3.1 on page 40 contains some subjective variables related to human abilities which need to be assessed in battle modelling. It will be seen that training affects many other aspects, such as deployment and tactics, accuracy, rate of fire and maintenance, whilst these in turn influence several others. Ultimately, they all affect the probable capability of our forces. Terrain is another master variable which is very important. It must be evident that if we just include such things as intrinsic lethality, expected rate of fire, expected accuracy under 'ideal' conditions, and omit some of the more practical variables simply because they are not easy to express in mathematical terms, there is an increased chance that our model will deviate from representing truth when more rugged conditions are applied.

The question arises how do we estimate, or quantify, variables such as training? It is likely that this can be done adequately, even if the way in which it is accomplished may seem rather crude. For example, we have a good idea whether national armies spend a lot of time in training and in exercises. Even if we only have a few categories such as excellent, good, normal, moderate and poor, we may be improving the realism of the model to an enormous extent. It seems likely, judging from an analysis of historical battles, that in moving from 'poor' to 'excellent', a factor of at least three is involved. This factor may, in itself, be dependent on other parameters such as the relative vulnerability of the attacker. This, in turn, is related to the mobility of warfare, where in highly mobile wars the scope for brilliance of manoeuvre and command is often quite large.

An interesting practical example can be found in relation to the Second World War. By counting the number of men and arms the Allies controlled at the outset of the war it was possible to say to ourselves 'we have slightly more of everything than they have, so it will be a hard struggle but we will win in the end'. As it happened this forecast was quite right, but for almost entirely the wrong reasons. If we had asked ourselves how well our troops were trained in comparison to the Germans, we might have said 'moderate', whilst we might well have rated the German troop training as 'excellent' or 'good'.

If this had caused us to apply a factor of two to the battle coefficients, we would have been far less optimistic than we were. Moreover, the bad subjective estimates we made at the time for the intrinsic lethality of our weapons would have been less important. In addition, had we been able to appreciate the new power of the attacker in comparison to the defender, we may have understood that a new mobility situation had arisen. The folly of declaring war until we were more closely able to match our men and machines to those of the enemy would then have been evident.

Luckily, the unexpected did move in our favour with regard to the aerial Battle of Britain. The 3,000-strong Luftwaffe firmly expected to deal with 600 British aircraft quite easily. This was especially true as the abilities of the machines on both sides were known to be very broadly similar. In practice, however, pilot supply, and not aircraft, proved to be the limiting factor. Many of the British pilots were able to parachute safely to their own territory whilst the Germans, if they escaped from their aircraft, were taken prisoner. The ability to recover men and materials from the battlefield can therefore have a significant effect on the ultimate outcome of a conflict.

Thus, against all the apparent odds (though not so having regard to recovery of men and materials) the tiny British air force brought the advance of the mighty German war machine to a halt in the West. The magnitude of this task can be judged by the fact that the German forces still came close to crushing a moderately equipped Soviet army many times its own size. Indeed the loss of the aerial Battle of Britain represented the first major reverse of the German armed forces in the Second World War and had a morale effect at the higher staff levels in the German forces that is not always appreciated. 'Morale' is not mentioned in the conceptual diagram shown in Figure 3.1, but it could be shown in the diagram between 'training' and 'our probable attrition rate on the enemy'. It will obviously be affected by the rate at which casualties are suffered and should also be included in any part of the battle models where casualty levels for defeat are considered.

As long as all the important subjective variables are included in models it does not matter if, for most of the time, the switches are set at 'normal' or 'average' conditions. The mere fact that the variables have been included will mean that somebody has had to think about them quite seriously before accepting the 'normal' switch setting. If only minor relative effects and changes are being considered, then a 'normal' setting may be quite satisfactory. Hopefully, however, if a major study is under way, someone will occasionally push the switch over in order to discover the more unpleasant side to the situation.

In business subjective variables are equally important, and some attempt should be made in developing any commercial model to assess the magnitude of the possible variations. Only in this way can reasonable precautions be taken against unexpected events which could lead to bankruptcy.

Appendix 18 Notes on statistical testing including t-tests, chi-squared and analysis of variance

Most of statistics is ultimately concerned with the determination of probabilities and the assessment of variability. Variability occurs naturally in the real world, and difficulties can arise when we attempt to treat nature and events in an entirely deterministic manner. Appendix 15 shows how expectation (or average) variability can become deterministic and in practice we often use the Gaussian distribution as a simple mathematical representation of probable deviation about some expected mean value.

Of course, if we consider a practical example when only a few samples are taken, we will not get a smooth Gaussian curve but a histogram approximating to this distribution. A typical histogram is shown in Figure A18.1. A theorem, known as the Central Limit Theorem, states that if random samples of n observations are drawn from a population with finite mean, μ, and standard deviation, σ, then, when n is large, the sample mean, $\overline{y}$, will be approximately normally distributed with mean equal to μ and standard deviation $\sigma/\sqrt{n}$. The approximation will become more and more accurate as n becomes large. The Central Limit Theorem can also be restated to apply to the sum of the sample measurements:

$$\sum_{i=1}^{n} y_i$$

The sample measurements also tend to possess a normal distribution, in repeated sampling, with mean equal to $n\mu$ and standard deviation $\sqrt{n}\sigma$ as n becomes large (see Figure A18.2).

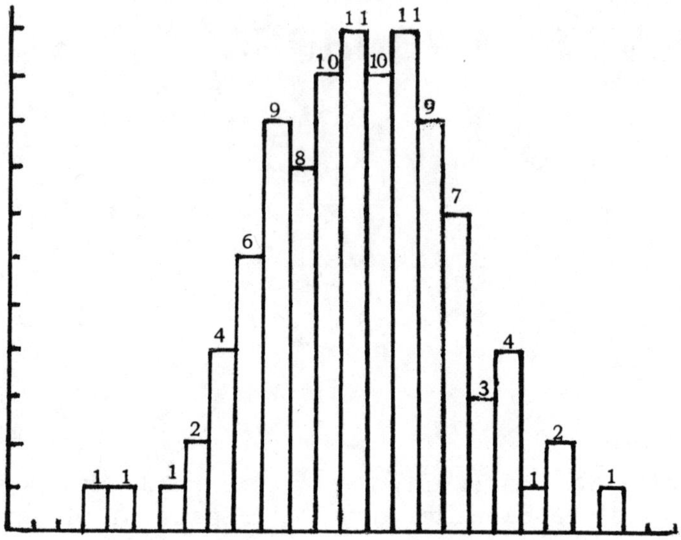

Figure A18.1 A simple histogram showing typical variability when the numbers in the sample are fairly small

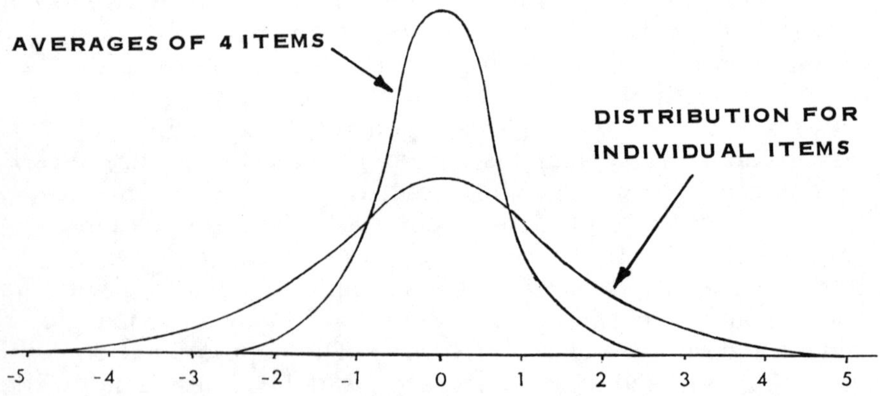

Figure A18.2 Curves showing how sample distributions become more 'peaky' as the average of more samples is taken

Many estimators which are used to make inferences about a whole population are the sums or averages of the sample measurements. When this is correct, and when the sample size n is sufficiently large to give a fair representation, we can expect the estimator to have a normal probability distribution according to the Central Limit Theorem.

The normal probability distribution is symmetrical about the mean so we can list the areas between the mean μ and a specified number, z, of standard deviations to the right of μ. The distance from the mean to a given value y is $(y-\mu)$. If we express this in terms of units of standard deviation, σ, we obtain:

$$z = \frac{y-u}{\sigma}$$

According to the Central Limit Theorem the statistic z possesses an approximately normal distribution in repeated sampling where n is large and where:

$$z = \frac{\bar{y}-\mu}{\sigma/\sqrt{n}}$$

In many cases we wish to test some hypothesis, for example, that the mean of some results is a particular specified value against the possibility that is greater than this value. If we substitute the sample deviation s for the population deviation σ in the above formula, there would be some value of z which could be deduced from many such experiments, which would provide a suitable basis for the acceptance or rejection of the hypothesis.

W.S. Gosset, writing under the pen name of 'Student', in 1908 discovered a distribution for samples drawn from a normally distributed population of the form:

$$t = \frac{\bar{y}-\mu}{s/\sqrt{n}}$$

He named the quantity he was studying 't' and so the distribution become known as 'Student t' distribution.

Unlike z, the t-distribution is much more variable, tailing out to the right and left. The variability of z in repeated sampling is due solely to $\bar{y}$; the other quantities appearing in z (n and σ) are not random. On the other hand, the variability of t is contributed by two random quantities, namely $\bar{y}$ and s, which can be shown to be independent of one another. Thus, when $\bar{y}$ is very large, s may be small and vice versa. As a result, t is more variable than z in repeated sampling. Finally, as we might surmise, the variability of t decreases as n increases because the estimate of σ,s, is based upon more and more information. When n is infinitely large, the t and z distributions are identical. Thus, Gosset discovered that the distribution of t depended upon the sample size, n.

The divisor of the sum of squares of deviations (n-1) that appears in the formula for s^2 is called the number of degrees of freedom associated with s^2. The origin of the term 'degrees of freedom' is linked to the statistical theory underlying the probability distribution of s^2. The test statistic t is based upon a sample of n measurements or that it possesses (n-1) degrees of freedom.

Applications of the t-distribution

It is worth very briefly considering some of the ways in which the t-distribution is used in practice. One use is to determine the confidence limits for the mean value of a population when we have taken only a limited number of samples from that population.

Let us suppose that we wish to be 95 per cent certain that a mean value lies within certain confidence limits. In this case the area under the normalised probability distribution curve, α, must not exceed .05 (i.e. 1 - .95).

The critical value of t will always be larger than the corresponding critical value of z for a specified α. For example, where α = .05, the critical value of t for n = 2 is t = 6.314, which is very large when compared with the corresponding z = 1.645. It will be found from t-tables that when looking down the $t_{.05}$ column, that the critical value of t decreases, reflecting the effect of a larger sample size on the estimation of σ. Finally, when n is infinitely large, the critical value of t equals 1.645.

With some algebra it can be shown (89) that the 95 per cent confidence limits for the mean are given by:

$$\bar{y} - t_{\frac{.05}{2}} \frac{s}{\sqrt{n-1}} < \mu < \bar{y} - t_{\frac{.05}{2}} \frac{s}{\sqrt{n-1}}$$

Since we have limits in both directions for the tails of the distribution (as there are two tails) and α must not exceed .05, in this case we should look up the value of t in t-tables from the column headed .025 (i.e. .05/2).

It should be noted that 95 per cent confidence limits means that if we could conduct the experiment 100 times on 95 occasions we would expect the outcomes to be within the limits specified.

If we wish to test whether the mean from two samples is the same, then we can calculate what is known as the 'pooled variance' as follows:

$$s^2 = \frac{\sum_{i=1}^{n_1} (y_i - \bar{y}_1)^2 + \sum_{i=1}^{n_2} (y_i - \bar{y}_2)^2}{n_1 + n_2 - 2}$$

The test static is then given by:

$$t = \frac{\bar{y}_1 - \bar{y}_2}{s\sqrt{\frac{1}{n_1} + \frac{1}{n_2}}}$$

We would then accept the hypothesis that $\mu_1 - \mu_0 = 0$ (that the means are equal) when the test statistic is greater than some t value.

Sometimes, having shown that the means are not equal, we may wish to know how large a difference we may expect.

In this case if we assume the difference of the population means to be $\mu_1 - \mu_2$, then the difference of our sample $\bar{y}_1$ and $\bar{y}_2$ provides the best estimate of this value. We can then write our confidence limits as follows:

$$(\bar{y}_1 - \bar{y}_2) \pm t_{\alpha/2} \; s\sqrt{\frac{1}{n_1} + \frac{1}{n_2}}$$

Another important test is called the 'paired *t*-test'. This test takes into consideration the fact that it may be possible to account for some of the variance.

It is important to take care to ensure that independent variables are used in constructing a *t*-test and, where the experimenter has designed the experiment in such a way that some of the variance can be removed from consideration, full advantage should be taken of this effect by using a paired t-test.

For example, if we consider the data contained in the table taken from Mendenhall and Reinmuth (90), reproduced below, we find a good example of this situation.

AUTOMOBILE	A	B
1	10.6	10.2
2	9.8	9.4
3	12.3	11.8
4	9.7	9.1
5	8.8	8.3
	$\bar{y}_1 = 10.24$	$\bar{y}_2 = 9.76$

The example concerns a manufacturer who wishes to test the qualities of two different types of tyre. In order to do this he runs all the tyres for a specified number of miles (let us say 10,000 miles) on five different cars. Each car is fitted with one of each of the different types of tyre on the rear wheels. The tyres are fitted randomly to the off-side or near-side. The table shows the wear for the different types of tyre.

We can carry out a straightforward test to compare the means using the formula:

$$s^2 = \frac{\sum_{i=1}^{n} (y_i - \bar{y}_1)^2 + \sum_{i=1}^{n} (y_i - \bar{y}_2)^2}{n_1 + n_2 - 2}$$

When this is done it will be found that the pooled variance is so large in the experiment that we cannot reject the hypothesis that both types of tyre are equally good.

However, in carefully designing this experiment we have taken out some of the variability and really should be comparing the difference between *pairs* of tyres fitted to the same car, which after all have been through roughly the same conditions.

This leads to the table given below:

AUTOMOBILE	d= A-B
1	.4
2	.4
3	.5
4	.6
5	.5
	$\bar{d}$= .48

It will be found that the mean and standard deviation of these measurements are as follows:

Mean	$\bar{d}$	= .48
Standard deviation	s_d	= .0837

so really we should consider:

$$t = \frac{\bar{d}}{s_d/\sqrt{n}} = \frac{.48}{.0837/\sqrt{5}} = 12.8$$

The critical value for a two-tailed test with $\alpha = .05$ and four degrees of freedom is 2.776. The t value 12.8 is large and highly significant and so we should conclude that the tyre wear for type B is less than for type A.

Some applications of the χ^2 distribution

There are some situations where we are more concerned with the variances than the comparison of means. Quite frequently we need to deduce information about the population variance. In this situation it is necessary to employ a different technique to the ones already outlined.

We know that the sample variance given by:

$$s^2 = \frac{\sum_{i=1}^{n} (y_i - \bar{y})^2}{n - 1}$$

is an unbiased estimator of the population variance σ^2.

A distribution of sample variances generated by repeated sampling will therefore have a probability distribution which starts at $s^2 = 0$ (as s^2 cannot be negative) and which also has a mean value equal to σ^2. We could repeatedly sample from a specified normal distribution to obtain the critical values for s^2 in the tail zones. Fortunately it was found that a distribution called the χ^2 distribution was useful for this purpose.

$$\chi^2 = \frac{(n - 1)s^2}{\sigma^2}$$

Let us consider some simple examples of the use of the χ^2 distribution.

An air pistol manufacturer claims that his air pistol fires slugs at 300 feet per second within a range of 40 feet per second.

A sample of ten measurements was found to give a mean of 305 feet per second and a variance of 190.

If we suppose that the manufacturer considered that the results would be within the range 95 per cent of the time then this is equivalent to 4σ and therefore $\sigma = 10$.

Thus the hypothesis is that the population variance, σ^2, is 100 and the alternative hypothesis is that σ^2 is greater than 100.

The alternative hypothesis would depend on a one-tailed test using the χ^2 distribution and with $\alpha = .05$ and $n = 10$ the χ^2 value is 16.919. If the test statistic exceeds this value we will reject the hypothesis that the variance is no greater than 100:

$$s^2 = 190, \quad n = 10, \quad \sigma^2 = 100$$

so

$$\chi^2 = \frac{(n - 1)s^2}{\sigma^2} = \frac{(10 - 1) \times 190}{100} = 17.1$$

The test statistic is greater than the 16.919 value so we should reject the manufacturer's claim.

If we wish to calculate, from the available data, the 90 per cent confidence interval for the population variance we find that:

$$\frac{(n - 1)s^2}{\chi^2_{.05}} < \sigma^2 < \frac{(n - 1)s^2}{\chi^2_{.95}}$$

or

$$\frac{(10 - 1)\ 190}{16.919} < \sigma^2 < \frac{(10 - 1)\ 190}{3.325}$$

i.e.

$$101.07 < \sigma^2 < 514.28$$

Before leaving χ^2 tests it is important to outline briefly the use of contingency tests as these can sometimes prove valuable. They can probably also be most easily demonstrated by a simple example. The following data are reputed to have been compiled from an assessment of the intelligence and smartness of the clothing of some 1,725 school children.

	Dull	Intelligent	Very Intelligent
Very well clothed	81	322	233
Well clothed	141	457	153
Poorly clothed	127	163	48

We are required to test whether intelligence is independent of the economic level at the .1 per cent level.

The first step is to draw up the following simple table inserting the row, and column, totals around the edge. The grand total of all events is shown at the corner of the table.

	Dull	Very Intel-ligent	Intel-ligent	Total
Very well clothed	81	322	233	636
Well clothed	141	457	153	751
Poorly clothed	127	163	48	338
Total	349	942	434	1725

It is next necessary to find the values we might reasonably expect in each of the cells if no effects were to be expected.

We would expect the children in each clothing category to have the same proportion of dull children and intelligent children. The expected proportion of dull children is 349/1725, or .202 of the total children. Each of the row totals in the first column should therefore be multiplied by this number to give the expected numbers of dull children in each clothing category.

When the same is done to obtain the expected proportion of children in the intelligent and very intelligent categories it becomes possible to carry out a χ^2 test. It will be found that the expected values for column 1 are 128, 152, 68; for column 2 are 347, 410, 185 and for column 3 are 160, 189 and 85.

Thus our test statistic is given from:

$$\sum_{\text{all cells}} \frac{(\text{Observed—Expected})}{\text{Expected}}$$

or

$$\frac{(81-128)^2}{129} + \frac{(141-152)^2}{152} + \frac{(127-68)^2}{68} + \frac{(322-347)^2}{347}$$

$$+ \frac{(457-410)^2}{410} + \frac{(163-185)^2}{185} + \frac{(233-160)^2}{160} + \frac{(153-189)^2}{189}$$

$$+ \frac{(48-85)^2}{85} = 135.92$$

If there are n rows and m columns in the table the number of degrees of freedom for the χ^2 test is given by $(n - 1)$ times $(m - 1)$ so that in this case we seek the χ^2 value with 4 degrees of freedom at the .1 per cent level.

We find from tables that:

$$\chi^2_{4,\ .1\%} = 18.47$$

Since our test statistic is much greater we conclude that it is very significant and that there is an association between intelligence and economic level.

Some applications of the F distribution

Quite frequently the need arises to compute the variances of two different populations. In this situation an obvious first reaction is to compare the ratio of the sample variances s_1^2, and s_2^2 and hope that these are representative of the population variances σ_1^2 and σ_2^2.

The question then arises just how large or small should

$$s_1^2/s_2^2$$

be in order to reject the hypothesis that the variances are equal. An answer can be obtained by repeated sampling from the populations in order to develop the distribution of

$$s_1^2/s_2^2$$

This distribution is, in fact, known as the F distribution and is plotted in statistical tables.

Let us consider a simple example of the use of the F distribution to compare variances.

A control laboratory is interested in whether two methods of analysis are equally reliable. Do the following data obtained by the two different methods support the assumption that equal variances will be obtained by the two methods given the following data:

Method 1:

4.27, 4.32, 4.29, 4.30, 4.31, 4.30, 4.30, 4.32, 4.28, 4.32
(10 results)

Method 2:

4.28, 4.32, 4.32, 4.29, 4.31, 4.35, 4.29, 4.32, 4.33, 4.28, 4.37, 4.38, 4.28, 4.32
(14 results)

Mean of the first set of measurements $x_1 = 4.301$

Mean of the second set of measurements $x_2 = 4.317142857$

The variance of the first set of measurements is:

$$s_1^2 = \frac{\sum_{i=1}^{n} (x_i - \overline{x}_1)^2}{n - 1} = .0002988889$$

$$s_2^2 = \frac{\sum_{i=1}^{n} (x_i - \overline{x}_2)^2}{n - 1} = .0010527473$$

$$\frac{s_2^2}{s_1^2} = 3.522202732$$

We can assume χ^2 distributions for each set of measurements with n-1 degrees of freedom. Thus there are 9 degrees of freedom for the first set and 13 degrees of freedom for the second set.

In this case we need to do a two-tailed test and the F values for the various degrees of freedom at $\alpha = .05$ and $\alpha = .025$ (corresponding to the 10 per cent and 5 per cent levels respectively as there are two tails) are as follows:

$$F_{13,9,.05} = 3.05$$

$$F_{13,9,.025} = 3.85$$

Thus we cannot reject the hypothesis that the variances are equal at the 10 per cent level of significance.

Analysis of variance

Finally let us briefly consider some of the procedures known as analysis of variance, A of V, or ANOVA. As the name suggests, analysis of variance procedures seeks to analyse the variations between independent sets of variables. It is a useful technique and if an experiment is carefully designed, it can minimise the number of experiments which need to be conducted and also assess the effects of interference which tend to mask the inter-relationships between the variables.

The basic process involves partitioning the sum of the squared values of the deviations into parts, each part being attributed to one of the independent variables, or to a remainder, which is said to be random error. The test therefore basically depends on the ratio of the variance attributable to variation in one particular variable (often called variation between treatments) with the variance due to the variation which is known to be due to the random error. This ratio can then be used as the basis for an F test. Thus we have:

$$F = \frac{\text{Mean square between treatments}}{\text{Mean square due to error}}$$

We can then conduct two one-tailed F tests with degrees of freedom appropriate to the variables.

Let us now consider a simple example of the analysis of variance process.

Four examiners at a military academy are suspected of having rather different standards. They share in the task of final assessment of officer cadets individually rather than collectively, in order to spread their work load. This has resulted in some cadets complaining of unequal treatment. The Commanding Officer at the Academy has conducted an inquiry into these allegations by having the four assessors mark photocopied papers of the original scripts of six individual cadets. The results are given in the table.

	Cadet					
	1	2	3	4	5	6
Examiner 1	31	27	68	69	51	61
Examiner 2	30	29	71	72	56	65
Examiner 3	34	33	64	64	49	60
Examiner 4	24	28	66	65	52	58

Begin by summing the values across the rows and down the columns to give the totals R_i and C_j respectively. We also obtain G, the grand total, by summing either the R_i or C_j values. This then gives the table shown overleaf.

	Cadet						
	1	2	3	4	5	6	$\sum R_i$ values
Examiner 1	31	27	68	69	51	61	305
Examiner 2	30	29	71	72	56	65	323
Examiner 3	34	33	64	64	49	60	305
Examiner 4	24	28	66	65	52	58	293
$\sum C_j$ values	119	117	267	270	208	244	G=1225

The sum of the squares due to the rows (where I and J are number of columns and number of rows) is given by:

$$\frac{\sum_{i=1}^{n} R_i}{J} - \frac{G^2}{IJ}$$

$$= \frac{376{,}228}{6} - \frac{1{,}500{,}625}{24}$$

$$= \underline{178.63}$$

The sum of the square due to the columns is given by:

$$\frac{\sum_{j=1}^{m} C_j}{I} - \frac{G^2}{IJ}$$

$$= \frac{274{,}839}{4} - \frac{1{,}500{,}625}{24}$$

$$= \underline{6{,}183.71}$$

The total sum of the squares is given by:

$$\sum_{i=1}^{n} \sum_{j=1}^{m} y^2_{ij} - \frac{G^2}{IJ}$$

$$= 69{,}171 - 62{,}526.04$$

$$= \underline{6{,}644.96}$$

The residual value for the sum of the squares can then be obtained by subtracting the sum of the squares for the rows and columns from the total sum of squares, i.e.

$$6{,}644.96 - 6{,}183.71 - 178.63$$

$$= \underline{282.62}$$

It is now possible to construct the analysis of variance table as follows.

	Sum of squares	Degrees of freedom	Mean squares	Mean square ratios
Difference between rows (examiners)	178.63	3	59.54	3.16
Difference between columns (cadets)	6,183.71	5	1,236.74	65.64
Residual	282.62	15	18.84	
Total	6,644.96	23		

The F values for the 5 per cent and .1 per cent levels of significance with the appropriate degrees of freedom are as follows:

$$F_{3,15,.05} = 3.29$$
$$F_{5,15,.001} = 7.57$$

The mean square ratios are 3.16 and 65.64 so we see that whilst there is a very significant difference indeed between cadets there is also some difference between examiners at near to the 5 per cent level. In practice we would probably carry out a further test in order to increase the sample size. In the interim period it might be wise to make some allowance for the differences between examiners.

There are many other aspects to the analysis of variance technique which have not been described. For example, the method can be expanded to three variables, but unless a computer package is available in situations where there are many cell values, the method can be long and tedious and involve many hours of work. The model for three-way analysis of variance is given in Kendall and Stuart. (91) This work contains one of the best treatments on the more advanced aspects of the technique.

In conclusion it is important to remember the need for care in conducting the statistical tests. Tests yielding results which run contrary to common sense should be treated with extreme caution and further work should be undertaken. The following is a true example which occurred during military proof testing. It illustrates how incorrect conclusions can be drawn through failure to use statistics, and also how statistics can be applied incorrectly.

The first situation was fairly straightforward in that it simply involved the failure to apply any statistical tests when a straightforward t-test would have enabled the correct conclusions to be drawn. Having explained that t-tests should have been used in this particular instance, the mathematician concerned (a first-class honours graduate with some statistical training) used a t-test to compare the results from two different shell velocity measuring instruments when both were tested on the same projectile. A number of rounds were fired and the conclusion was drawn that there was no significant difference between the measurements obtained by different instruments. However, further investigation suggested that round-to-round variations in projectile velocities might be masking any difference in mean values between the instruments. The application of a paired t-test rather than a normal t-test eliminated effects due to round-to-round variations in the projectile velocity and made it quite clear that there was indeed a very real difference between the instruments.

Finally, with regard to computer software packages it is worth remembering that they need careful application and that not all are 'user friendly' (i.e. easy to use). Advice should be sought from a good statistician when in doubt. Not all the packages have been written by statisticians, and hence it is sometimes only too easy for the user to apply them incorrectly.

Appendix 19 Hypothesis test example

Hypothesis testing has been briefly outlined in the main text on pages 187 to 191. A simple example is given here of the application of hypothesis testing to a problem involving the use of the binomial probability distribution.

Example: An operational research man devised a simple model for forecasting commodity price movements. He decided from analysis of past data that turning points in price occur after a period of sideways movement in the market, during which the price movement corresponds to a 'random walk' and can move upwards or downwards with equal probability. If he can forecast the market movement, his company can make a profit by buying 'long' or selling 'short'. He uses as a decision rule the fact that the market moves the same way in four successive periods, and this indicates to him that the random period is ended and a genuine movement is taking place which will force prices higher or lower. How frequently is he likely to be wrong?

In this case the hypothesis H_0 is that the situation is random at the market turn, and so the probability p of a rise, or the probability q of a fall is equally likely, i.e. $p = q = \frac{1}{2}$ (note: the movement is hardly ever likely to be precisely sideways and so this possibility is not considered).

H_o : movement is random

and the alternative hypothesis is:

H_1 : (the movement is up)
(the movement is down)

In this case H_1 contains two possibilities so the hypothesis is said to be *composite*. If it contained only one it would have been called *simple*.

From the binomial and probability theorems the chance of r successes of an event which occurs with a probability p in a total of n events is:

$$\frac{n!}{r!\,(n-r)\,!} \quad (p)^r(q)^{n-r}$$

where $q = 1 - p$

Hence if r is the number of times a rise occurs, we have the probability of $1r$ (rises) in the first period with a random situation (i.e. $p = q = \frac{1}{2}$) is therefore:

$$\frac{4!}{1!\,(4-1)\,!} \quad (\tfrac{1}{2})^1(\tfrac{1}{2})^3 = 1/4$$

also the probability of $2r$ is:

$$\frac{4!}{2!\,(4-2)\,!} \quad (\tfrac{1}{2})^2(\tfrac{1}{2})^2 = 3/8$$

and so on for $3r$, $4r$ and $0r$.

Tabulating these possibilities we have:

Probability of 4 rises	1/16
Probability of 3 rises	1/4
Probability of 2 rises	3/8
Probability of 1 rise	1/4
Probability of 0 rises	1/16

In this case using the normal statistical jargon, and remembering our test was for four successive rises or four successive falls, we would say that the 'critical' region was $r = 4$ (all rises) or $r = 0$ (all falls). Hence the probability of a Type 1 error (α) would be $1/16 + 1/16 = 1/8$. Hence, if the model was otherwise satisfactory in every respect, the operational research man's forecast of a developing price trend would be successful seven out of eight times.

Appendix 20 Fuzzy goal programming using LP techniques

There are many aspects of real life where intentions are not precise, or clearly defined. On a great many occasions this is primarily due to lack of clear thought, or even not thinking sufficiently about the problem. As analysts many of us have helped others by using the technique of simply splitting the problem down into a series of small component parts about which, individually, a decision can be made more easily.

In practice, however, there remains a number of problems in which precision of aim is lacking. There are good practical reasons why it is often sensible to bring about some sort of precision by simply quoting values to our client until we reach some figure, or set of figures, which are acceptable to him, but from which movement in one direction is generally unacceptable. Fuzzy goal programming deals with this kind of problem.

For example, most of us would not be indifferent to being given a million pounds. Yet it all depends on our utility for money. If we were suddenly given 50 million pounds and a few hours later were told that it was regretted that a mistake had been made, and that it was only 49 million pounds, then most of us would be indifferent to the apparent loss of that particular one million pounds.

There is, therefore, a kind of so-called 'fuzzy' region in which it is not easy to detect any difference in our utility valuation of certain effects. In actual fact it is not really fuzzy at all, but simply means that our utility curve is flat in that region. In a similar way we might have insufficient data on the magnitude, or required magnitude, of some variable, but merely have a rather vague idea of its size. 'Fuzziness' is, therefore, in itself a rather vague concept, with our uncertainty taking the form of a distribution, so that we tend to regard some particular value as the most likely and others as less likely. However, let us look at the way some people deal with these matters.

Defence problem example*

The Army has two types of equipment, A and B. The battle effectiveness of 100 pieces of A is 80 units, and that of 100 pieces of B is 40 units. Each of the types of equipment is virtually identical in cost, whilst from experience and field trials it is believed that the most efficient force structure is obtained with *about* three pieces of Type A equipment to two pieces of Type B. It is also believed that *around* 630 effectiveness units will be required to provide a reasonable counter to the threat. This leads the Army to suggest that it needs *about* 600 Type A equipments and *about* 400 of Type B. Thus we have a problem with three fuzzy goals.

The procedure in solving this problem is to place bounds outside which the goals will be regarded as not being achieved, and within which a valuation can be given in comparison to the other goals. The aim is to maximise this valuation within the constraints set.

As you may expect, in this case the way to solve this problem is to split it up into a series of small problems with variations on the constraints, and to compare the optimum values obtained for each of the sub-problems.

* This example is adapted from one given in Reference 92, which in turn is adapted from one given in Reference 93.

Let X_1 be the number (in hundreds) of equipment Type A
Let X_2 be the number (in hundreds) of equipment Type B

Δ_1 is the amount of deviation which can be tolerated, positively or negatively, from the effectiveness goal.

Δ_2 is the amount of deviation (in hundreds) which can be tolerated, positively or negatively, from the average goal for equipment Type A.

Δ_3 is the amount of deviation (in hundreds) which can be tolerated, positively or negatively, from the average goal for equipment Type B.

If we set Δ_1 equal to $\pm$ 10 the effectiveness can range between 620 and 640.

Let us say that the quantity of equipments for Types A and B (Δ_2 and Δ_3) can range by a factor $\pm$ 2 (i.e. 200).

Now if we wish to rate our attainment of the effectiveness goal to some value between 1 and 0 when the value of our force effectiveness ranges between 630 and 620, then our proximity to this goal may be described by some function:

$$\frac{80X_1 + 40X_2 - 620}{\Delta_1} = \frac{80X_1 + 40X_2 - 620}{10}$$

Thus we can write that the effectiveness E in achieving the military effectiveness goal u_1 is:

$$0 \text{ if } 80X_1 + 40X_2 \leqslant 620$$

and is

$$\frac{80X_1 + 40X_2 - 620}{10} \quad \text{if } 620 \leqslant 80X_1 + 40X_2 \leqslant 630 \tag{1}$$

so

$$E \leqslant 8X_1 + 4X_2 - 62$$

or

$$62 \leqslant 8X_1 + 4X_2 - E \tag{2}$$

Our goal for the equipment Type A is 6 units $\pm$ 2. We wish the goal for type A (u_2) to be given by some value E, such that it is 0 if $X_1 \leqslant 4$.

So $\dfrac{X_1 - 4}{2}$ fulfils this if $4 \leqslant X_1 \leqslant 6$ (3)

Similarly, the E value which shows the efficiency of the equipment Type B goal is 0 when $X_2 \leqslant 2$.

So that

$$\frac{X_2 - 2}{2} \quad \text{if } 2 \leqslant X_2 \leqslant 4 \tag{4}$$

Rewriting Equation 3 in terms of E gives:

$$E \leqslant .5X_1 - 2 \quad \text{or} \quad 2 \leqslant .5X_1 - E \tag{5}$$

And similarly for Equation 4:

$$E \leqslant .5X_2 - 1 \quad \text{or} \quad 1 \leqslant .5X_2 - E \tag{6}$$

Thus we have our first linear program formulation, which is as follows:

$$- E + 8X_1 + 4X_2 \geqslant 62$$

$$- E + .5X_1 \geqslant 2$$

$$- E + .5X_2 \geqslant 1$$

$$620 < 80X_1 \qquad .40X_2 \leqslant 630$$

$$4 < X_1 < 6$$

$$2 < X_2 < 4$$

We need to maximise E and the solution is $E = .958$ for $X_1 = 5.92$ and $X_2 = 3.92$.

But there are quite a lot of other linear programs which need to be solved in order to be sure that we have the right solution.

Thus, as a consequence of the discontinuities created by the fuzzy goals, it is necessary to try all combinations to cover the ranges. In this particular example there are eight combinations of the goal equations, and so there are eight linear programs.

First set of linear programs

$$620 \leqslant \text{effectiveness} \leqslant 630$$

	Case 1	Case 2	Case 3	Case 4
X_1	4 - 6	4 - 6	6 - 8	6 - 8
X_2	2 - 4	4 - 6	2 - 4	4 - 6

Second set of linear programs

$$630 \leqslant \text{effectiveness} \leqslant 640$$

	Case 5	Case 6	Case 7	Case 8
X_1	4 - 6	4 - 6	6 - 8	6 - 8
X_2	2 - 4	4 - 6	2 - 4	4 - 6

Let us write out the goals and solutions fully for this problem, listing all the eight cases.

Case 1

The formulation for this case has already been given. The solution was:

$$E = .958, \quad X_1 = 5.92, \quad X_2 = 3.92$$

Case 2

$$- E + 8X_1 + 4X_2 \geqslant 62$$

$$- E + .5X_1 \geqslant 2$$

$$E + .5X_2 \leqslant 3$$

$$620 \leqslant 80X_1 + 40X_2 \leqslant 630$$

$$4 \leqslant X_1 \leqslant 6$$

$$4 \leqslant X_2 \leqslant 6$$

Solution

$$E = .937, \quad X_1 = 5.87, \quad X_2 = 4$$

Case 3

$$-E + 8X_1 + 4X_2 \geqslant 62$$

$$E + .5X_1 \leqslant 4$$

$$-E + .5X_2 \geqslant 1$$

$$620 \leqslant 80X_1 + 40X_2 \leqslant 630$$

$$6 \leqslant X_1 \leqslant 8$$

$$2 \leqslant X_2 \leqslant 4$$

Solution

$$E = .875, \quad X_1 = 6, \quad X_2 = 3.75$$

Case 4

$$-E + 8X_1 + 4X_2 \geqslant 62$$

$$E + .5X_1 \leqslant 4$$

$$E + .5X_2 \leqslant 3$$

$$620 \leqslant 80X_1 + 40X_2 \leqslant 630$$

$$6 \leqslant X_1 \leqslant 8$$

$$4 \leqslant X_2 \leqslant 6$$

Infeasible

Case 5

$$
\begin{aligned}
E + 8X_1 + 4X_2 &\leq 64 \\
-E + .5X_1 &\geq 2 \\
-E + .5X_2 &\geq 1 \\
630 \leq 80X_1 + 40X_2 &\leq 640 \\
4 \leq X_1 &\leq 6 \\
2 \leq X_2 &\leq 4
\end{aligned}
$$

Solution

$$E = .96, \quad X_1 = 5.92, \quad X_2 = 3.92$$

Case 6

$$
\begin{aligned}
E + 8X_1 + 4X_2 &\leq 64 \\
-E + .5X_1 &\geq 2 \\
E \quad .5X_2 &\leq 3 \\
630 \leq 80X_1 + 40X_2 &\leq 640 \\
4 \leq X_1 &\leq 6 \\
4 \leq X_2 &\leq 6
\end{aligned}
$$

Solution

$$E = .94, \quad X_1 = 5.88, \quad X_2 = 4$$

Case 7

$$
\begin{aligned}
E + 8X_1 + 4X_2 &\leq 64 \\
E + .5X_1 &\leq 4 \\
E + .5X_2 &\leq 1
\end{aligned}
$$

$$630 \leqslant 80X_1 + 40X_2 \leqslant 640$$

$$6 \leqslant X_1 \leqslant 8$$

$$2 \leqslant X_2 \leqslant 4$$

Solution

$$E = .889, \quad X_1 = 6, \quad X_2 = 3.78$$

Case 8

$$E + 8X_1 + 4X_2 \leqslant 64$$

$$E = .5X_1 \leqslant 4$$

$$E + .5X_2 \leqslant 3$$

$$630 \leqslant 80X_1 + 40X_2 \leqslant 640$$

$$6 \leqslant X_1 \leqslant 8$$

$$4 \leqslant X_2 \leqslant 6$$

Solution

$$E = 0, \quad X_1 = 6, \quad X_2 = 4$$

Thus, Case 5 yields the best solution, with $E = .96$, for 592 pieces of equipment Type A and 392 pieces of equipment Type B.

General comments

This solution is achieved by means of weighting factors which put greater emphasis on the very important goals than on the moderately important goals.

It is necessary to be quite precise about the nature of the fuzziness of the problem or impractical answers will be obtained. For example, it must be stated if the fuzziness is distributed in a Gaussian fashion, or a step function, or triangular shape, etc.

The need for common sense in setting the constraints and shaping these in a practical manner is emphasised. Unfortunately, in very large goal programming problems, the number of linear programs to be solved may be considerable. To some extent the machine time may be shortened by adopting special methods but the run time can be large when many items and constraints are involved.

For some situations, the dynamic programming approach (described earlier) may well have some advantages over the LP method, especially if it is linked to branch and bound techniques. However, in the case of triangular constraints the approach outlined above may well be worth using, especially if it can be combined with some form of sensitivity analysis. (94)

This kind of multi-criteria decision problem is now receiving increasing attention and two international conferences on the subject were held in Belgium at Brussels and Namur (95) during 1985. There appears to be two rather different schools of thought on the solution methodology. One school, largely based in the US, seems to place emphasis on utility theory and the belief that all fuzziness and criteria can be precisely defined at the outset; whereas the other school, largely based in Europe, believes that if the problem is explored properly changes in utility values and fuzziness of attitude are likely to occur during the process.

Bernard Roy (96) has suggested the advisability of using multiple criteria decision aids (MCDA) as a means of progressing to a good robust solution rather than attempting to optimise purely on initial thoughts and data. Accordingly, at the Universite de Paris Dauphine he has developed a series of computer methods called Electre I, II, III and IV for exploring the better solutions.

Many other methods are being developed which progress towards a decision solution rather than provide a single so-called 'optimum' solution. It is suggested that in this way the decision maker learns much more about the true nature of his problem. J.P. Brans (97) uses a series of outranking models called Promethee to move to a good decision.

There are now very many software packages used for multi-criteria decision problems. They range from small packages quietly operated in universities to the more commercial packages which are the concern of E. Jacquet-Lagreze (also at the Universite de Paris Dauphine).

Undoubtedly these methods are helping the decision maker to better understand the true nature of his problems and to achieve better decisions. Quite often in the case of relatively simple decisions, this kind of evolutionary decision technique provides solutions which more closely match the intuitive thoughts of the decision maker than the older, more formal approaches. In the case of very complex situations it is usually necessary to split the problem into parts so that it can be handled more conveniently during the computer analysis process.

We will leave the last word with the great scientist Descartes, who perhaps never had a truer thought than when he once wrote in a slightly different context, 'there is a need of a method of finding the truth'.

References

1 Braithwaite, R.B. 'Models in the Empirical Sciences', *Proceedings of the International Congress on Logic Methodology and Philosophy of Science*, edited by E.Nagel et al., Stanford University Press, 1962.

2 Ackoff, R.L. and Sasieni, M.W. *Fundamentals of Operational Research*, John Wiley and Sons, New York, 1967.

3 Rivett, B.H.P. and Ackoff, R.L. *A Manager's Guide to OR*, John Wiley and Sons, 1963.

4 Laing, G.J. 'The Classification of Models', *Interdisciplinary Science Reviews*, December 1981, Heydon Press.

5 Popper, Sir K.R. *The Logic of Scientific Discovery*, Hutchinson, 1960.

6 Kuhn, T.S. *The Structure of Scientific Revolutions*, University of Chicago Press, 1962.

7 Lakatos, I. *Proofs and Refutations : Logic of Mathematical Discovery*, Cambridge University Press, 1976.

8 Cottrell, Sir A.H. *Dislocations and Plastic Flow in Crystals*, Oxford University Press, 1953.

9 Bowen, K.C. 'A Conflict Approach to the Modelling of Problems of and in Organisations', *IFORS Proceedings 1981*, North Holland.

10 Rivett, B.H.P. and Ackoff, R.L. *Policy Decision by Structural Mapping*, Proceedings of the Royal Society of London, A354, pp.407-23, 1977.

11 Eden C. *Thinking in Organisations*, Macmillan, 1979.
12 De Bono, E. *Lateral Thinking for Management*, McGraw-Hill, 1971.
13 Huntley, H.E. *Dimensional Analysis*, Macdonald and Co., 1952.
14 Bridgman, P.E. *Dimensional Analysis*, Yale University Press, 1963.
15 Staicu, C.I. *Dimensional Analysis : treatment of experimental data*, Abacus Press, 1982.
16 Finney, D.J. 'Dimension of Statistics', *Applied Statistics*, vol.26, no.3, 1977, pp.285-289.
17 Vajda, S. *Linear Programming : algorithms and applications*, Chapman and Hall, 1981.
18 Beale, E.M.L. (ed.) *Applications of Mathematical Programming Techniques*, EUP, 1970.
19 Williams, H.P. *Model Building in Mathematical Programming*, Wiley, 1978.
20 Garfinkel, R.S. and Nemhauser, G.L. *Integer Programming*, Wiley, 1972.
21 Williams, H.P. 'Logical Programs and Integer Programming', *Bullock Institute of Mathematical Applications*, vol.13, no.18, 1977.
22 Laing, G.J. 'Resource Allocation on Ranges', M.Sc. Dissertation, Sussex University, 1970.
23 Dantzig, G.B. and Wolfe, P. 'The Decomposition Algorithm for Linear Program', *Econometrica*, no.29, 1961.
24 Charnes, A. and Cooper, W.W. 'Programming with Linear Fractional Functionals', *Naval Res. Log.*, Quest 9, 1962.
25 Isbell, J.R. and Marlow, W.H. 'Attrition Games', *Naval Res. Log.*, Quest 3, 1956.
26 Duffin, R.J, Peterson, E.L. and Zener, C. *Geometric Programming : Theory and Application*, Wiley, 1968.
27 Tintner, G. 'A Note on Stochastic Linear Programming', *Econometrica*, vol.28, no.2, 1960.
28 Dantzig, G.B. 'Linear Programming under Uncertainty', *Management Science*, no.1, 1955, pp.197-206.
29 Charnes, A. and Cooper, W.W. 'Chance-constrained programming', *Management Science*, vol.6, 1960, pp.73-79.
30 Land, A.H. and Doig, A.G. 'An Automated Method of Solving Discrete Programming Problems', *Econometrica*, no.28, 1960.
31 Gomory, R.E. 'An Algorithm for Integer Solutions to Linear Programs' in Greaves, S.L. and Wolfe, P. (eds),

Recent Advances in Mathematical Programming, McGraw-Hill, 1963.
32 Laing, G.J. 'Limits to Growth in Business Organisations', Ph.D., University of London.
33 Cooper, D.F. 'The Superior Commander : A Methodology for the Control of Crisis Games', *J. Opl. Res. Soc.*, vol.30, no.6, 1979, pp.529-537.
34 Dando, M.R. and Bee, A.J. 'Operational Research for Complex Conflicts : a gaming methodology for the development of a decision-making monitor', *Opl. Res. Q*, vol.28, no.4(i), 1977, pp.853-64.
35 Huxham, C.S., Bennett, P.G., Lozowski, M.V. and Dando, M.R. 'The Study of Human Decision-Making : a cautionary tale from the world of experimental gaming', *J. Opl. Res. Soc.*, vol.32, no.3, pp.173-85, 1981.
36 Lansus, N. A business game devised by staff in the Operational Research Departments at Sussex University and Lancaster University.
37 White, D.J. *Decision Methodology*, Wiley, 1975.
38 Green, P.E. and Carroll, J.D. *Mathematical Tools for Applied Multivariate Analysis*, Academic Press, 1976.
39 Kendall, M.G. 'A Course in Multivariate Analysis', Griffin Monograph, 1968.
40 Jolliffe, I.T. 'Discarding Variables in a Principal Component Analysis', *J. Royal Statistical Society* (Series C), vol.21, no.2, 1972 and vol.22, no.1, 1973.
41 Holzinger, K.J. and Harman, H.H. *Factor Analysis*, Chicago UP, 1941.
42 Gosnell, H.F. and Schmidt, M. 'Factorial and Correlation Analysis of the 1934 Vote in Chicago', *J. Am. Statistical Assoc.*, no.31, 507.
43 Rummell, R.J. *Applied Factor Analysis*, Northwestern University Press, 1970.
44 Kendall, M.G. and Beale, M. developed a software package for assessing the proximity of grouped variables for the consulting group CEIR (now called SCICON). Professor Scott at Sussex University has also an effective software package.
45 Laing, G.J. 'How to Use a Decision Tree', *Financial Times*, Oct.2, 1970.
46 Axelrod, R. 'Effective Choice in the Prisoner's Dilemma', *Journal of Conflict Resolution*, vol.24, no.1, March 1980, pp.3-25.

47 Axelrod, R. 'More Effective Choice in the Prisoner's Dilemma', *Journal of Conflict Resolution*, vol.24, no.3, September 1980, pp.379-403.
48 Rapoport, A. 'Escape from Paradox', *Scientific American*, vol.217, no.1, July 1967, pp.50-56.
49 Howard, N. 'The Theory of Meta-Games' published in *General Systems, the Year Book of the Society for General Systems Research*, vol.X1, 1966, pp.167-86. Also same volume 'The Mathematics of Meta-Games', pp.187-200.
50 Tarakanov, K.V. *Mathematics and Armed Combat*, Izd-vo Voyenizdat, Moscow, 1974.
51 Sharpe, R. 'The Analysis of Options - an Examination', B.Sc. dissertation, Sussex University, 1975.
52 Coutie, G.A. et al. *Short Term Forecasting*, Oliver and Boyd, 1964.
53 Brown, R.G. *Smoothing, Forecasting and Prediction of Discrete Time Series*, Prentice-Hall, 1962.
54 Box, G.E.P. and Jenkins, G.M. *Time Series Analysis, Forecasting and Control*, Holden-Day, 1970.
55 Ellinger, A.G. *The Art of Investment*, Bowes and Bowes, 1955 and 1971.
56 Laing, G.J. 'An ABC of Forecasting', *Financial Times*, 6 June 1972.
57 *Cattell, R.B. A Scientific Analysis of Personality*, Penguin, 1965.
58 Fishbein, M. 'A Behavioural Theory Approach to the Relations between Beliefs about an Object and Attitude Toward the Object', *Readings in Attitude Theory and Measurement*, John Wiley, 1967.
59 Newcomb, T.M. 'An Approach to the Study of Communicative Acts', *The Journal of Psychology*, vol.21, 1946.
60 Cartwright, D. and Zander, A. (eds) *Group Dynamics, Research and Theory*, Tavistock, 1968.
61 Laing, G.J. 'The C-Link System', *Oper. Res. Soc. Annual Conference*, 1972.
62 Laing, G.J. 'The C-Link System for Studying Co-operation in Organisations', *Management Decision*, 1970.
63 Bowen, K.C. and Smith, D.G. 'The Development of Models for Application to Conflict Problems', Seminar on Use of Models in the Social Sciences, Edinburgh University, 1972.
64 Ackoff, R.L. and Emery, F.E. *On Purposeful Systems*, Tavistock, 1973.

65 Shapiro, M.J. and Bonham, M. 'Cognitive Process and Foreign Policy Decision Making', *International Studies Quarterly*, vol.17, no.2, 1973.
66 Asch, S.E. *Social Psychology*, Prentice-Hall, 1952.
67 Campbell, D.T. *Social Attitudes and Other Acquired Behavioural Dispositions*, vol.6 of *Psychology : A Study of Science*, McGraw-Hill, 1963.
68 Wiener, N. *Cybernetics or Control and Communication in the Animal and the Machine*, MIT Press, 1961.
69 Rose, J. (ed.) *Survey of Cybernetics*, Iliffe, 1969.
70 Lerner, A.Ya. *Fundamentals of Cybernetics*, Chapman and Hall, 1972.
71 Harvey, A.C. *Time Series Models* (see Chapter IV), Phillip Allen, 1981.
72 Felsen, J. 'Artificial Intelligence Techniques Applied to Reduction of Uncertainty in Decision Analysis Through Learning', *Opt. Res. Q.*, vol.26, no.3ii, 1975.
73 Felsen, J. *Cybernetic Approach to Stock Market Analysis versus Efficient Market Theory*, Exposition Press, New York, 1975.
74 Laing, G.J. *Reliability and Human Communication*, 5th Symposium on Reliability Technology, 1978.
75 Laing, G.J. 'Communication and its Constraints on the Structure of Organisations', *Omega*, vol.8, no.3, 1980.
76 Rivett, B.H.P. 'Policy Selection by Structural Mapping', *Proc. Royal Society*, A354, 1977, pp.407-423.
77 Kaplan, M. and Schwarz, S. (eds) *Human Judgement and Decision Processes in Applied Settings*, Academic Press, 1977.
78 Saaty, T.L. *The Analytic Hierarchy Process*, McGraw-Hill, 1980.
79 Linstone, H.A. and Turoff, M. (eds) *The Delphi Method : Techniques and Applications*, Addison-Wesley Publishing Co., 1975.
80 Roberts, P.C. 'Measurements of Performance', Informal Papers for the Civil Service College Heads of OR Meeting, 1978.
81 Bellman, R.E. and Dreyfus, S.E. *Applied Dynamic Programming*, Princeton University Press, 1962.
82 Balas, E. 'An Additive Algorithm for Solving Linear Programs with Zero-One Variables', *Operations Research*, vol.13, 1965, pp.517-546,.
83 Forrester, J.W. *Industrial Dynamics*, MIT Press, 1961.
84 Battersby, A. *Network Analysis*, Macmillan, 1964.

85 Kaufman, A. and Desbazeille, G. *The Critical Path Method*, Gordon and Breach, 1969.
86 Moder, J.J. and Phillips, C.R. *Project Management with CPM and PERT*, Von Nostrand-Rheinhold, 1970.
87 Lanchester, F.W. *Aircraft in Warfare*, Constable and Co., 1916.
88 Weale, T.G. 'Stochastic Lanchester Theory Using Numerical Methods', Operational Research Society Conference, Stirling University, September 1979.
89 Hoel, P.G. *Introduction to Mathematical Statistics*, Wiley, 1962.
90 Mendenhall, W. and Reinmuth, J.E. *Statistics for Management and Economics*, Wadsworth, 1971.
91 Kendall, M.G. and Stuart, A. *The Advanced Theory of Statistics*, vol.3, Griffin, 1968.
92 Narasimhan, R. 'Goal Programming in a Fuzzy Environment', *Decision Sciences*, vol.11, no.2, April 1980 (note minor error in Eqn 16 of this paper).
93 Lee, S.M. *Goal Programming for Decision Analysis*, Philadelphia, Penn., Averback, 1972.
94 Bellman, R.E. and Zadeh, L.A. 'Decision Making in a Fuzzy Environment', *Management Science*, vol.17, 1970, pp.B141-B146.
95 The Second International Summer School on Multiple Criteria Decision Making Methods Application and Software. 15th to 26th July 1985. Facultes Universitaires Notre Dame de la Paix, Namur, Belgium.
96 Roy, B. *'The Optimisation Problem Formulation: Criticism and Overstepping'*, J. Operational Research Society, vol.32, 1981, pp.427-436.
97 Brans, J.P., Mareschal, B. and Vincke, Ph. *'Promethee: A New Family of Outranking Methods in Multicriteria Analysis'*, Operational Research, IFORS, 1984.

Index

Acceleration forces 29
Ackoff, R.L. 5,151
Activity based approach 71,267
Adaptive models 6,10
Adjacency matrix 156,157
Advanced mathematical programming 55-58
Affective concepts 155
Aircraft loading using DP 234
Algol based simulation languages 268
Allocation methods 203
Allocation of aircraft to targets 203
Alternative hypothesis 189
Alternative hypothesis example 386
Analogous flow 25
Analogue models 9,10,15-18,76
Analysis of variance 42,18
Analytic models 3
ARIMA 128,334,342,343
ARMA 332,340,341,347
Armour and Co 135
Arrival time simulation for ships 68
Artificial variables 216
Asch, S.E. 155
Attitudes 149
Autocorrelation in time series 324
Autocorrelation coefficients 124
Autocovariance in time series 324
Autoregressive models 330,332
Autoregressive moving average 332
Axelrod, R. 401
Axioms in utility theory 297

B activities in simulation 267

Backward formulation in DP 62
Balas, E. 258
Ballistic pendulum 23,24
Ballistic trajectories 125
Bartlett's test 91,192
Battersby, A. 403
Battle courses 77
Bayes strategy 100
Beale, E.M.L. 93,400
Bearish triangles 133,138
Beauty contests 164-167
Bee, A.J. 401
Behaviour in games 147,148
Behavioural models 6,10,147
Beliefs 154
Bellman, R.E. 224,404
Bernoulli 37
Best test 190
Bimodal stability 109
Binomial probability 387
Blast impulse 23,24
Blending problem LP example 212
Blending problem in oil refinery 53
Bonham, M. 152
Boolean facilities in languages 268
Bound based activities in simulation 267
Boundary conditions 41
Bounding 259
Bowen, K.C. 39,151
Box-Jenkins forecasting concept 224
Box-Jenkins forecasting example 327
Box-Jenkins forecasting method 322-328
Box, G.E.P. 402
Braithwaite, R.B. 399
Branch and bound methods 246-258
Branch and bound models 63
Branching 259
Brittle lacquer technique 25
Brown, R.G. 123
Business games 75

C activities in simulation 267
CA models 9,10,17,65,71
Campbell, D.T. 155
Cannonical correlation analysis 90,92
Cardinal utility theory 292
Cattell, R.B. 148
Cartwright, D. 150
Case study number one 22
Case study number three 31
Case study number two 27
Causal models 123
Centurion tank model 29
Chance constrained programming 58
Charnes, A. 400
Chartists 132
Check list for CA modelling 71
Check list for LP formulation 55
Chemical plant optimisation 217
Chi-squared 91
Chi-squared tests 42,186
Choice 80,164-167,291
Civil engineering models 27

Classification of models, a new method 7,10
Classification of time series models 331,332
Cluster analysis 90,93
CM models 9,10,19,59,63,67
Co-operation 149,150
Cognisant problems 38
Cognitive concepts 155
Cognitive mapping 151,152
Cognitive models 151-155
Collaborative modelling 7
Comment cards in programs 263-265
Commodity exchange 132
Communication 150,151
Company model using simulation 259-265
Complex decision trees 301-314
Complex decisions 67
Complex models 7
Complexity 46
Complexity and size problems with DP 63
Complexity in simulation 18
Complexity of interactions 74
Component analysis 90,91,192
Composition of forces for set cost 234
Computational analogue models 8,9,66,70,71,73
Computational mathematical models 8,9,17,59,63,67
Computer models 17,19
Computer programs for statistics 94
Computer solutions for LP 216
Conative concepts 155,156
Conceptual flow diagrams 39,40
Conceptual framework 4
Conceptual model 4
Concrete for model construction 26
Concurrent tasks in PERT 354
Conditional activities in simulation 267
Confidence intervals 185
Constraints and restrictions in LP 53,213,215,219
Continuous flow problems 52
Continuous random variable 5
Controllable variables 5
Controlling concrete strength 27
Cooper, D.F. 401
Cooper, W.W. 57,400
Correlated variables 93
Corroborative modelling 7
Cost-effectiveness 168,169
Costs of implementation 194,195,196
Coutie, G.A. 402
Crecy, Battle of 175
Critical region 190
Cybernetic model building 158-163
Cybernetic, meaning of 158
Czechoslovakia 118

Dando, M.R. 401
Dangerous situation modelling 67
Danzig, G.B. 57

Data 44
De Bono, E. 46
Decision analysis strategies 311
Decision theory 95,311-321
Decision theory in investment 95-97
Decision trees 301-310
Decision trees in Decision Theory 95
Decisions of numbers required in battle 311-316
Decomposition algorithm 57
Degrees of freedom 185
Delphi technique 174
Dendogram 93,94
Density of material and penetration 35
Density scaling 29
Desbazeille, G. 403
Descartes 398
Descriptive models 3
Design of experiments 42
Detente and cold war 115
Deterministic models 2,3,10
Deterministic simulation in games 107
Determino-stochastic models 3
Diet programs in LP 53,58
DIF files 181
Difference operators 129
Differencing in time series 324
Differencing time series 125
Di-graphs 156
Dimensional analysis 44,47-50
Dimensional homogeneity 48
Dimensions in statistics 50
Disclosure of information to model builders 196
Discrete random variable 5
Discriminant analysis 90,93
Dispersion analysis 93
Distribution fitting 186
Dividend policies in companies 261
Dock simulation 68
Dominant strategy 80
Dow-Jones index forecast 161
Dreyfus, S.E. 403
Dual problem 57
Duffin, R.J. 400
Dummies of humans 30
Dynamic program algorithm 233
Dynamic programming example appendix 221
Dynamic program model class 20
Dynamic program method 60
Dyamic programs for instrument replacement 240-242
Dynamo systems language 272

Economic models 6,10
Economy 159
ECSL simulation language 268
Eden, C. 45,151
Effectiveness assessment 72

Einstein, Albert 20
Ellinger, A.G. 132
Emery, F.E. 151
End effects in simulations 71
Endogenous variables 4
Error terms in forecasting 130
ESP simulation language 267
Estimation of time series forecasting coefficients 344-347
Event based approach 71
Exogenous variables 4,199
Expected values of strategies 113
Explanatory models 3
Extrapolation 123,125

Factor analysis 90,91
Feedback loops 158
Felsen, J. 161
Fictitious choice 81
Find loops in languages 268
Finney, D.J. 50
Fishbein, M. 149
Fitting curves 88
Flags for program control 262-265
Flight simulators 77,122
Float in PERT 358
Force mix for maximum effectiveness 247-250
Force structure fuzzy example 390
Forecasting 122
Forecasting using panels of people 168
Foreword viii
Formula for diagrams 140
Forrester, J. 270
Forward formulation in DP 62
Fourier analyses 124
Fractional programming 57
Fragment penetration 32
Frangible panels 30
Frequency of mixed strategies 84,85
Frequency scaling 30
Fuzzy goal programming 389-398

Gabr Saleh, Battle of 175
Game strategies 79,82,84,100
Game theoretic models 78
Game theory 78
Games against nature 98,105
Games with nature involved 317-321
Garfinkel, R.S. 258
Geometric programming 57
Geometrically scaled models 13
Gomory algorithm 63
Gomory, R.E. 63
Goodness 198
Gosnell, H.F. 91
Gosset, W.S. 371
GPSS simulation language 268
Graphical explanation of LP 215
Gross national produce and defence 118
GSP simulation language 268

Harman, H.H. 91

Harvey, A.C. 403
Head and shoulder figures 133,137
Heterogeneous models 4
Heuristics 64
High order polynomials in extrapolation 125
Hill-climbing 64
Histogram 188
Hoel, P.G. 404
Holzinger, K.J. 91
Homogeneous models 4
Hotelling's T-Test 93,192
Hovercraft 118
Howard, N. 105
Human behaviour models 147
Human interaction models 147
Hungary 118
Hurwicz strategy 100
Hyperbolic programming 57
Hypothesis 1
Hypothesis test example 386-388
Hypothesis testing 187,188
Hypothetico-deductive reasoning 31

IA models 8,9,15,25,76
Iconic models 3
Idealised targets 31
Identification of time series models 334-340
IG models 8,9,14,29,30
Impact decelerations 43
Impartial attitudes 201
Implementation of models 194-195
Importance of communication in games 115
Improbability in 'Gaming' 76
Indifference in choice 165
Industrial dynamics 271,272
Inflation 159
Ingenuity in looking for correlation 88
Input-output analysis 41
Instability in systems 158,159
Instrument replacement using DP 240-242
Insurance 175
Integer linear programming 56
Integer program algorithms 63
Integrated autoregressive moving average models 333,339
Interactions 67
Interactive models 67
Interest rate effects on companies 262
Intermediate substitute algorithm 57
Interpretive analogue models 8,9,15,25,76
Interpretive approach to PG models 15
Interpretive geometric models 8,9,15,29,30
Inventory 141
Inverted head and shoulders 133
Isbell, J.R. 57

Jenkins, G.M. 124
Jolliffe, I.T. 91

Kaplan, M. 168

Kaufman, A. 403
Kendall, M.G. 90,93,384
Kinetic energy of fragments 33
Kuhn, Thomas 20

'Lag' in time series 124
Lakatos, I. 20
Lanchester, F.W. 365
Land-doig algorithm 63,258
Lansus business game 75
Laplacian strategy 100,313
Latent non-cognisant problems 38
Latent roots 192
Least cost route 223
Lee, S.M. 404
Lehner multiplication congruence 268
Lerner, A.Ya 403
Levels in dynamic programming 221
Likelihood ratio tests 190
Linear program formulation 54,55
Linear programming 52,53,212
Linear programming as an aid to fuzzy problems 389-398
Linear regression 185
Linkages between concepts 155
Linstone, H.A. 403
Long term prediction from random breaks 349
Lotteries 81,293-296
Lower value of game 83
Lunar module mock-up 11,12
Mach number 50
Machine based simulations 266
Macro-economic models 6
Marlow, W.H. 57
Material based simulations 266
Mathematical meaning of diagrams 143
Mathematical models 19,20,31,32
Mathematical programming 51-58
Matrices 51,52
Maximin strategy 83
Maximum strategy 100
Mean in PERT diagrams 360
Megaton weapon effects 31
Mendenhall, W. 374
Meta-games 105,115-117
Methodology of building models 36
Micro-economic models 6
Microdyn systems language 272
Military advance as a dynamic program 224,234
Military applications of PERT 360-363
Military attack scheduling 361
Military attack with weather hazards 317-318
Military planning using PERT 353-363
Minimax regret strategy 101,313-315
Minimax strategy 84
Misperception 121
Mistakes in simulation models 70,71

Mixed modelling techniques 29,30
Mixed strategy 84
Model 1
Models of the world 147,148
Moder, J.J. 403
Monetary value 292
Morgenstern 292
Motivation 151
Movement in conflict games 106
Moving average models 331,332
Multi-attribute judgement 168
Multi-variate analysis 90
Multiple plate armours 36
Multiple regression 89

Narasinham, R. 404
Negative feedback 159
Negotiation game 287-290
Nemhauser, G.L. 258
Newcomb, T.M. 149
Newton, Isaac 20
Neyman-Pearson Lemma 190
NIMS 54
Noise in forecasting 136,138
Non-cognisant problems 38
Non-linear programming 57
Non-reactive models 6
Non-stationarity on time series 333
Normative models 3
Nuclear safety 174
Null hypothesis 189
Null hypothesis 386,387

Objective forecasting 123
Objective function example 143,214
Objective function explanation 4
Odds for playing strategies 85
Optimal route 60
Optimality principle 61
Optimality principle in DP 222
Option analysis 118-120
Overturning moments 44

Paradigms and theories 21
Parametric programming 57
Path probability estimation in decision trees 305,306
Paths in dynamic programming 222
Penalty and reward comparison 109,110
Pennant configurations 133
Penney, Lord 21,45
Perception 151,152
Performance 198
Personality factor scales 148
PERT diagrams 353
Peterson, E.L. 400
PG models 8,9,13,28,29
Phillips, C.R. 403
Photoelastic technique 36
Physical geometric models 8,9,13,26,27
Physical models 9-10
Planning using PERT 353-364
Plaster of Paris models 26
Plastic flow 37

Plotting facilities in simulation 268
PM models 8,9,20,32-35
Point and figure charts 135
Popper, Sir Karl 20
Positive feedback 158
Positive feedback in markets 133
PP models 9-12,76
Prandtl number 50
Preferred order 167
Preferred strategies 119
Pressure coefficients 50
Pressure measurement 26
Prestige value 81
Price wars 113
Principle of optimality 61
Prisoners dilemma 78,106
Prisoners to circumstance 289
Probabilistic models 2
Processes in simulation languages 267
Projectile tracking with radar 125
Prospects in utility theory 293
Purchases of weapons with deployment restrictions 251-258
Pure mathematical models 9,10,20,31
Pure physical models 9,10,11,76

Quadratic programming 57

Raiffa and decision trees 104
Random breaks 165,348-352
Random correlations 86-88
Random division of linear quantities 348-352
Random number generation 70,268
Random variable 5
Rapoport, A. 115
Reachability 156
Reactive models 6
Really rugged solutions 199
Reflexive control 115
Regressions 89
Reinmuth, J.E. 374
Relationships between time series models 341-343
Reorder costs and quantities 141
Replacement of equipment using DP 223
Research games in simulation 74,153,154,287-290
Restricted choice games 74
Revised simplex method 216
Reynolds number 44,50
Risk aversion 291
Risk decisions 297-300
Rivett, B.H.P. 6,41,165
Roberts 198
Robust solutions 64
Robustness 198
Roles 198
Rugged solutions 64
Rupture velocity of material 35

Saddle point 84
Safety and rare events 174,175

Sasieni, M.W. 5
Satisfaction matrices 102
Satisficing 198
Saucer configuration 133
Savages minimax regret strategy 101,313-315
Scaling effect 13,14,43
Scaling laws for shock waves 24
Scan loops in languages 268
Scheduling systems 360
Schwarz, S. 168
Secant modulus of elasticity 35
Semantic differential 152
Sensitivity analyses 197
Separable programming 57
Sequential tasks in PERT 354
Shadow costs 208
Shapiro, M.J. 152
Sharpe, R. 118
Shift operators 129
Shock waves 24
Simplex method 216
Simplicity 44,45
Simscript simulation language 266
Simulation languages 178,266-269
Simulation models 18,66
Simulation using a computer (example) 259-265
Simulation using research games 74
Slack in PERT diagrams 357
Slack variables 216
Sliding arc extrapolation 125
Southern Rhodesian bonds 98
Special software 177
Spectrum of models 7
Stability in systems 120,158,159
Stages in dynamic programming 221
State in dynamic programming 221
Stationarity in time series 124
Statistical forecasting models 123
Statistical models 86,123
Statistics in testing models 183-191
Stochastic approach 72
Stochastic dynamic programming 238
Stochastic models 2,10
Stochastic variable 5
Stochastic programming 58
Stock control 141
Stock Exchange 132
Strain energy 34
Strength scaling 26,27
Stress interviews 77
Stresses in materials under fragment attack 33
Structural mapping 165
Structure problems 6
Stuart, A. 384
Studying dangerous situations 18
Subjective forecasting 123
Subjective judgements 164
Subjective probability estimation 170-172
Systems dynamics 73,270-275

t-Tests 42
Tarakanov, K.V. 116

Taxation effects on companies 261,262
Taxi-driver problem 238
Taylor, Sir Geoffrey 21,45
Teleological approach 168
Terminology 2
Testing models 182-193
Texas utilities 134
Theories and paradigms 21
Theory 1
Theory validation 42
Tied polynomials 125
Time effects in games 105
Time series 122
Time, The problem of 168
Tit for tat 115
Tintner, C. 58
Top hat simulation 68
Torsional oscillations 124
Transport algorithm 203
Triple bottoms 133
Triple tops 133
Trying to build a model 38
Turoff, M. 403
Two player game, Solution by simulated play 108,109
Two player games 74
Type I error example 388
Type I errors 189,190
Type II errors 190

Uncontrollable variables 4
Underwriters experience 175
Unexpected changes in variables 104
Upper limit of game 84
Utility in games 111-113
Utility theory 81,291
Utils 294

Value judgements 317-321
Value scales 164
Variability 67
Variability due Research and Development 72
Variables of primary importance 41
Variables of secondary importance 41
Variance in PERT diagrams 360
Verification of models 197
Vertical bar charts 134
Vertical thinking 46
Viscosity 44
Visicalc software 180
Von Kalmen filters 161
Von Neuman, J. 292

Wages, Prices, Inflation 159
Wall Street 132
War games 15
War gaming 75
WASP 54,65
Weale, T.G. 365
Weapon effectiveness 175,176
Wedge diagrams in chartism 138
Weighting coefficients 128
White, D.J. 81
Wiener, N. 158
Williams, H.P. 400
Wilson Lot Size formula 141,145
Wolfe, P. 57
Word processing 181
World War I simulation 76

Wound ballistics 32

Youngs modulus in
penetration effects 35

Zadeh, L.A. 404
Zander, A. 150
Zener, C. 348
Zero Sum Games 81
Zero-One Branch and
Bound methods 63,251-
258
Zero-One integer linear
programming with
interference 56
Zero-One integer linear
programming 56